Lecture Notes in Economics and Mathematical Systems

Managing Editors: M. Beckmann and H. P. Künzi

166

Raymond Cuninghame-Green

Minimax Algebra

Springer-Verlag
Berlin Heidelberg New York 1979

Author
Prof. R. Cuninghame-Green
Dept. of Mathematical Statistics
University of Birmingham
P.O. Box 363
Birmingham B15 2TT/England

AMS Subject Classifications (1970): 90-02, 90B99

ISBN 3-540-09113-0 Springer-Verlag Berlin Heidelberg New York
ISBN 0-387-09113-0 Springer-Verlag New York Heidelberg Berlin

Printed in Germany

Printing and binding: Beltz Offsetdruck, Hemsbach/Bergstr.
2142/3140-543210

FOREWORD

A number of different problems of interest to the operational researcher and the mathematical economist - for example, certain problems of optimization on graphs and networks, of machine-scheduling, of convex analysis and of approximation theory - can be formulated in a convenient way using the algebraic structure $(R,\oplus,\otimes)$ where we may think of R as the (extended) real-number system with the binary combining operations $x\oplus y$, $x\otimes y$ defined to be $\max(x,y)$, $(x+y)$ respectively. The use of this algebraic structure gives these problems the character of problems of linear algebra, or linear operator theory.

This fact has been independently discovered by a number of people working in various fields and in different notations, and the starting-point for the present Lecture Notes was the writer's persuasion that the time had arrived to present a unified account of the algebra of linear transformations of spaces of n-tuples over $(R,\oplus,\otimes)$, to demonstrate its relevance to operational research and to give solutions to the standard linear-algebraic problems which arise - e.g. the solution of linear equations exactly or approximately, the eigenvector-eigenvalue problem and so on. Some of this material contains results of hitherto unpublished research carried out by the writer during the years 1970-1977.

The previous lack of such an account has made the publication of further research in the field problematical: for example, the results in [46] make use of the spectral inequalities in Chapter 26 below, which have not to the writer's knowledge appeared previously in print. Once embarked upon the project, therefore, the best aim appeared to be that of developing a systematic theory, analogous to conventional linear algebra, rather than merely making a collection of particular analyses.

This desire to be comprehensive and fundamental does not, of course, always make for simplicity or obvious relevance, so we have provided several substantial sections relating the theory to the motivating problems of the first chapter. In particular, the reader may care to glance at Section 25-5 on page 207, where several different parts of the theory are brought together in the solution of a problem of machine-scheduling.

Although certain passages may be slightly abstract for some tastes, the text does not pre-suppose the possession of more than the basic vocabulary of modern algebra, as available for example from [35] (with whose authors the present writer shares the conviction that modern algebra is an important tool of the applied mathematician). The non-elementary concepts and structures which are used are defined in context.

The subject matter is organised as follows. We begin with an extended motivational essay in Chapter 1, where it is shown that a diversity of problems, mainly from the field of operational research, receive a natural formulation in terms of linear algebra over $(R, \oplus, \otimes)$ or its dual $(R, \oplus', \otimes')$. Chapters 2 and 3 present the axiomatics and basic manipulative properties of the class of structures to which $(R, \oplus, \otimes, \oplus', \otimes')$ belongs, and establish the principle of isotonicity, that algebraic expressions are isotone functions of their arguments.

Chapter 4 defines a class of structures called blogs, which model the system $(R, \oplus, \otimes, \oplus', \otimes')$ more closely, but also include a three-element system called ③ which allows much of the theory of Boolean matrices to be included in our general theory.

Spaces and matrices are introduced in Chapter 5, and their duals in Chapter 6, together with principles of isotonicity of matrix algebra.

In Chapter 7 the foundations are laid for a theory of conjugacy: for each matrix $\underline{A}$ we define its conjugate $\underline{A}^*$, and it turns out in Chapter 8 that $\underline{A}$ and $\underline{A}^*$ induce transformations which are each other's inverse, or more accurately, each other's residual. Chapter 9 applies these ideas to scheduling theory, and Chapter 10 exhibits the mappings induced by $\underline{A}$ and $\underline{A}^*$ as residuated linear transformations of spaces of n-tuples, or as we shall say: residuomorphisms.

Chapters 11 and 12 are largely technical in nature, being concerned with identifying those transformations which avoid forming the product $-\infty \otimes +\infty$, since obviously this product has no meaning in terms of an original operational research problem. The matrices which induce such transformations have formal properties similar to those of stochastic matrices and we call them G-astic matrices. These G-astic matrices play a role again in Chapters 13 and 14 where we address the problem of solving systems of linear equations. Our residuation theory leads to a certain principal solution, and we also develop all solutions for an important class of cases. This analysis depends upon use of the system ③ developed in Chapter 4, and enables us in Chapters 16 and 17 to develop a theory of rank, dimension and linear independence. In Chapters 18 and 19 we view the principal solution as in effect the solution to a minimum-norm problem, and apply this in Chapter 20 to the theory of networks.

Chapters 21 to 26 present a spectral theory for our matrix algebra, beginning with a study of projections, which have formal properties closely similar to those operating in more classical spaces. Chapter 22 makes an excursion into the shortest path problem for directed weighted graphs, and Chapter 23 shows that the solution to this problem yields a partial solution to the problem of finding the eigenvectors of a special class of matrices - the definite matrices. In Chapter 24 we introduce a class of direct similarity transformations through which the eigenvector - eigenvalue problem for general matrices can be reduced to that for the definite matrices. This enables us in Chapter 25 to present a computational regime for the general eigenvector - eigenvalue problem for matrices over $(R, \oplus, \otimes, \oplus', \otimes')$. Then Chapter 26 presents certain spectral inequalities which play something of the role of the classical spectral theorem for matrices.

Chapter 27 discusses the orbit of a square matrix A - that is the sequence: $A, A^2, \ldots$, and gives sufficient conditions that it shall converge finitely. Some material is also presented on permanents, which to a certain extent play the role of determinants. Finally Chapter 28 examines a number of canonical forms to which matrices may be reduced by similarity and equivalence transformations.

The twenty-eight chapters are each organised into titled sections, numbered within chapters. Thus, e.g.: 13-2. Compatible Trisections, being the second section of Chapter 13. Algebraic expressions within the text are indexed e.g.: (2-7), denoting the seventh indexed expression in Chapter 2.

Formal mathematical statements are called propositions, lemmas, theorems or corollaries. A proposition is a formal statement of something which is assumed to be known or to follow immediately from the general argument, or which is proved elsewhere, and no formal proof is offered here. A lemma is a formal statement which is intended as a stepping-stone to one or more theorems, which present the principal results of the theory, and from which one or more corollaries may be deduced in respect of more limited special cases. Formal proofs are given for all lemmas, theorems and corollaries.

Several dualities run through the material simultaneously, so that a given result may have as many as eight dual forms. It is a common practice in mathematics to present formal statements of only one of a set of dual forms of a given result. However, because of the proliferation of dualities and the fact that in a later argument it may be necessary to use a result in a dual form other than that in which it was stated and proved, we present a few of our theorems in all relevant dual forms simultaneously, by use of tables. Of course we give a proof of only one form.

Formal mathematical statements are numbered sequentially within chapters, regardless of status. Thus Proposition 17-3 is followed by Theorem 17-4, which is followed by Lemma 17-5. The notation ● marks the end of a proof or the end of a proposition.

A list of symbols and notations is given at the end, after the references. References are cited within the text by use of square brackets - thus e.g.: [27] .

The preparation of this typescript has been the despair of more than one secretary, but I should like to thank Ms Vivienne Newbigging and Ms Elaine Haworth for their repeated triumphs over my daunting untidyness.

The assistance given by Ir. F. Huisman, a graduate student of the author, in sketching an early draft is also gratefully acknowledged.

R.A.C-G.

CONTENTS

1. MOTIVATION

1-1. Introduction

In the past 20 years a number of different authors, often apparently unaware of one another's work, have discovered that a very attractive formulation language is provided for a surprisingly wide class of problems by setting up an algebra of real numbers (perhaps extended by symbols such as $-\infty$,etc.) in which, however, the usual operations of multiplication and addition of two numbers are replaced by the operations: (i) arithmetical addition, and (ii) selection of the greater (or dually, the less) of the two numbers, respectively.
Thus, if R is the set of (suitably extended) real numbers, we define for each $x,y \in R$:

$$\left.\begin{array}{l} x \otimes y \text{ to be the arithmetical sum } x + y \\ x \oplus y \text{ to be the quantity } \max(x,y) \text{ (or perhaps } \min(x,y)). \end{array}\right\} \quad (1\text{-}1)$$

Shimbel [1] applied these ideas to communications networks.

Cuninghame-Green [4], [7] and Giffler [5] applied them to problems of machine-scheduling. Several authors [11], [12], [17], [23] have pointed out their usefulness in relation to shortest-path problems in graphs, and recent papers discuss analogous applications in the fields of languages and automata theory. In the following section, we have collected together a number of such example problems, primarily from the field of operational research. We propose to illustrate the fact that the notation (1-1) gives these problems the character of problems of linear algebra. The discussion is intended to be intuitive rather than precise.

1-2. Miscellaneous Examples

1-2.1 Schedule Algebra

One common feature of industrial processes is that machines do not act independently, and a typical machine cannot begin a fresh cycle of activity until certain other machines have all completed their current cycles. For example, in a steelworks the activity of tapping a cast cannot begin until the furnace has finished refining and the teeming crane is able to bring a prepared ladle. Or on an assembly line a man cannot begin a new assembly until, say, two interlocking sub-assemblies have arrived from different sources with independent production rates.

A natural way of describing such a system is to label the machines, e.g., 1,...,n, and to describe the interferences by recurrence relations such as:

$$x_3(r+1) = \max(x_1(r)+t_1(r),\ x_2(r)+t_2(r))$$

This expresses the fact that machine 3 must wait to begin its $(r+1)^{st}$ cycle until machines 1 and 2 have both finished their r^{th} cycle, the symbol $x_i(r)$ denoting the starting time of the rth cycle of machine i, and the symbol $t_i(r)$ denoting the corresponding activity duration. This mode of analysis gives rise to formidable-looking systems of recurrence relations:

$$x_i(r+1) = \max(x_1(r) + a_{i1}(r), \ldots, x_n(r) + a_{in}(r)), \quad (i=1, \ldots, n) \tag{1-2}$$

where, for notational uniformity, terms $a_{ij}(r)$ and $x_j(r)$ are made to occur for all $i = 1, \ldots, n$ and all $j = 1, \ldots, n$ by introducing where necessary quantities $a_{ij}(r)$ equal to "minus infinity" for each combination (i,j) which has no physical significance; the operator max will then "ignore" these terms.

Let us now make a change of notation. Write:

$$\left.\begin{array}{lll} x \oplus y & \text{instead of} & \max(x,y) \\ x \otimes y & \text{instead of} & x + y \end{array}\right\} \tag{1-3}$$

We may refer to this notation as "max algebra".
Expression (1-2) becomes:

$$x_i(r+1) = (a_{i1}(r) \otimes x_1(r)) \oplus \ldots \oplus (a_{in}(r) \otimes x_n(r)), \quad (i=1, \ldots, n) \tag{1-4}$$

which is a kind of inner product. Introduce the obvious vector-matrix notation:

$$\underline{A}(r) = [a_{ij}(r)]$$
$$\underline{x}(r) = [x_i(r)]$$

and (1-4) becomes:

$$\underline{x}(r+1) = \underline{A}(r) \otimes \underline{x}(r) \tag{1-5}$$

Formalisms of this kind were developed by Giffler [5] and Cuninghame-Green [4].

Expression (1-5) is a very intuitive "change-of-state" equation. By iteration we have:

$$\underline{x}(r+1) = \underline{A}(r) \otimes \underline{A}(r-1) \otimes \ldots \otimes \underline{A}(1) \otimes \underline{x}(1)$$

showing how the state $\underline{x}(r)$ of the system evolves with time, from a given initial state $\underline{x}(1)$, under the action of the operator represented by the matrix $\underline{A}$.

For simplicity of exposition assume for the moment that the quantities $a_{ij}(r)$ are independent of r. Define $\underline{A}^r = \underline{A} \otimes \underline{A} \ldots \otimes \underline{A}$, r times (associativity holds!). The "orbit" of the system is then:

$$\underline{x}(1),\ \underline{A} \otimes \underline{x}(1),\ \underline{A}^2 \otimes \underline{x}(1), \ldots$$

and it is clear that the sequence of matrices:

$$\underline{A},\ \underline{A}^2,\ \underline{A}^3, \ldots$$

will determine the long-term behaviour of the system: does it oscillate? Does it, in some suitable sense, achieve a stable state?

Several authors have discussed such sequences for a particular class of matrices. Dubbed "definite" by e.g. Carré in [17], these are (in terms of the present example) matrices $[a_{ij}]$ in which no "circuit-sum":

$$a_{ii};\ a_{ij} + a_{ji};\ a_{ij} + a_{jk} + a_{ki};\ \ldots$$

is positive. This work has extended to these definite matrices some results obtained previously by Lunc for Boolean matrices (see e.g. [6]).

An interesting operational question, which occurs in relation to the problem of controlling systems of this type, is this: "how must the system be set in motion to ensure that it moves forward in regular steps; i.e. so that for some constant λ, the interval between the beginnings of consecutive cycles on every machine is λ? And what are the possible values of λ?" Reference to the notation (1-3) shows that we are concerned with the problem:

$$\underline{x}(r + 1) = \lambda \otimes \underline{x}(r)$$

But $\underline{x}(r + 1) = \underline{A} \otimes \underline{x}(r)$, so we must solve:

$$\underline{A} \otimes \underline{x}(r) = \lambda \otimes \underline{x}(r) \tag{1-6}$$

Clearly we have arrived at an eigenvector-value problem, as discussed in [7], where once again, an operational problem assumes the form of a linear algebra problem. In [7] we prove that the only possible value for the eigenvalue of the matrix $[a_{ij}]$ is the greatest of the "circuit-averages":

$$a_{ii};\ \frac{a_{ij} + a_{ji}}{2};\ \frac{a_{ij} + a_{jk} + a_{ki}}{3};\ \ldots$$

Hence the definite matrices discussed by Carré may be characterised as those having negative or zero eigenvalue.

A full discussion of the eigenvector-value problem is given in Chapters 23 to 25 below; the set of eigenvectors of a given matrix $\underline{A}$ form a space (in max-algebra) generated by a finite number of fundamental eigenvectors which may be derived by rational operations on the matrix $\underline{A}$.

In classical algebra, we know that under suitable circumstances a sequence of vectors $\underline{x}$, $\underline{A}\ \underline{x}$, $\underline{A}^2\underline{x}$, $\underline{A}^3\underline{x}$, ...
will converge to an eigenvector of the matrix $\underline{A}$. The corresponding property in max algebra is obviously that under suitable circumstances the orbit of the system shall tend in the long run to go forward by constant steps. This problem is also analysed below, in Chapter 27.

1-2.2 Shortest Path Problems

Suppose a transportation network has n cities. Certain pairs (i,j) of cities are connected by a direct road link such that the distance from i to j using this link is d_{ij}. We do not necessarily assume that $d_{ij} = d_{ji}$. For a pair of (i,j) of cities without a direct link we define formally $d_{ij} = \infty$ and we now let $\underline{D}$ be the matrix $[d_{ij}]$.

It is possible, for a given pair (i,j) of cities, that the direct road link (if any) may not represent the shortest path from i to j. It may be shorter to go via another city, thus using two links of the network:

$$d_{ik} + d_{kj} \leq d_{ij}$$

What then is the shortest distance from i to j using exactly two links?

Clearly it is $\min_{k=1,\ldots,n} (d_{ik} + d_{kj})$ (1-7)

Now we make the following change of notation:

$$\left.\begin{array}{l} x \oplus' y \text{ instead of } \min(x,y) \\ x \otimes y \text{ instead of } x + y \end{array}\right\} \qquad (1\text{-}8)$$

Notation (1-8) is thus a sort of dual to notation (1-3), and we may call it "min algebra". We shall call the operation $\oplus'$ dual addition.

Expression (1-7) now becomes:

$$(d_{i1} \otimes d_{1j}) \oplus' \ldots \oplus' (d_{in} \otimes d_{nj}) \qquad (1\text{-}9)$$

which is exactly the inner product of row i and column j of the matrix $\underline{D}$ in this notation. Formalisms of this kind occur in several of the references, including for example [1], [11] and [17].

Clearly then the matrix $\underline{D}^2$ gives the shortest intercity distances using exactly two links of the network per path and by an immediate induction the matrix $\underline{D}^p$, for integer p, gives the shortest intercity distances using exactly p links of the network per path. Suppose that the system has no circuits with negative total distance (i.e. that $\underline{D}$ is a "definite" matrix in min algebra). A moment's reflection shows that no path with more than n links can then be shorter than a path with n or less links since a path with more than n links must contain a circuit, which can then be deleted to give a path between the same endpoints as before but with fewer links and no greater length. Hence for each pair (i,j), the shortest path from i to j is the $(i,j)^{th}$ element of one of the matrices $\underline{D}$, $\underline{D}^2$, ..., $\underline{D}^n$.

Define the dual addition of matrices using the operation $\oplus'$ in the obvious componentwise fashion: $[x_{ij}] \oplus' [y_{ij}] = [x_{ij} \oplus' y_{ij}]$, and consider the matrix $\underline{\Gamma}'(\underline{D})$ defined by:

$$\underline{\Gamma}'(\underline{D}) = \underline{D} \oplus' \underline{D}^2 \oplus' \ldots \oplus \underline{D}^n \qquad (1\text{-}10)$$

Since the shortest path from i to j is the shortest of the paths from i to j using one link, or two links, ..., or n links, and bearing in mind the meaning of the symbol $\oplus'$, we see that $\underline{\Gamma}'(\underline{D})$ is the shortest path matrix for the network for which $\underline{D}$ gives the direct distances.

Actually, expression (1-10) certainly does not suggest the best way of calculating $\underline{\Gamma}'(\underline{D})$, for which one would use one of the established shortest-path algorithms given for example in [39]. However, expression (1-10) is important in the algebraic theory, and is discussed (for definite matrices) in a number of the references. In Chapter 23 below we show that all the fundamental eigenvectors of $\underline{D}$ occur as columns of $\underline{\Gamma}'(\underline{D})$. In [17], Carré shows how the elements of $\underline{\Gamma}'(\underline{D})$ may be computed using an analogue of a Gauss-Seidel iteration.

1-2.3 The Conjugate

The relation: $\underline{x}(r + 1) = \underline{A} \otimes \underline{x}(r)$ developed in Section 1-2.1 enables other operational questions to be given a linear-algebra format. For example, to the operational question, "How must the system be set in motion to ensure that, for some fixed n, the nth cycles are undertaken at preassigned times?" corresponds the "simultaneous linear equations" problem ([4]):

$$\text{Solve } \underline{B} \otimes \underline{x} = \underline{b} \quad (\text{where } \underline{B} = \underline{A}^{n-1}) \tag{1-11}$$

This problem does not in general have a solution, and in practice it must often be replaced by some problem with a criterion of best-approximation - for example by the problem of minimising the maximum earliness subject to zero lateness ([3] and [43]):

Find $\underline{x}$ such that $\underline{B} \otimes \underline{x} \leq \underline{b}$ and $\max\limits_{i=1,\dots,n} (b_i - (\underline{B} \otimes \underline{x})_i)$ is minimised.

A reformulation hereof, having rather more clearly the character of a linear algebra problem is:

Find a Chebychev-best approximate solution to (1-11), which satisfies $\underline{B} \otimes \underline{x} \leq \underline{b}$. (1-12)

In [25], we show that the solution of problems (1-11) and (1-12) requires the development of a duality. Alongside the addition operation $\oplus$ we must introduce the dual addition $\oplus'$ with the interpretation $x \oplus' y = \min(x,y)$. (Hence the name "minimax algebra" for the theory which then develops, in which the operations of min algebra and max algebra are intermingled.) For a given matrix $\underline{A} = [a_{ij}]$, we now define the conjugate of $\underline{A}$ to be $\underline{A}^* = [-a_{ji}]$, i.e. $\underline{A}^*$ is derived from $\underline{A}$ by transposing and negating. This definition of course extends to vectors (n-tuples), if we regard them as $(n \times 1)$ matrices.

It is easy to see that $(\underline{A}^*)^* = \underline{A}$ and $(\underline{A} \oplus \underline{B})^* = \underline{A}^* \oplus' \underline{B}^*$, and in general this conjugate matrix $\underline{A}^*$ has many of the formal properties of a Hermitian conjugate in conventional linear algebra. Furthermore, we may define a dual multiplication $\otimes'$ for scalars and for matrices (details are given in Chapter 6 below), such that $(\underline{A} \otimes \underline{B})^* = \underline{B}^* \otimes' \underline{A}^*$, and derive the following results, which were presented in [25]:

Proposition 1-1. Problem (1-11) has a solution if and only if $\underline{x} = \underline{A}^* \otimes' \underline{b}$ is a solution; and then $\underline{x} = \underline{A}^* \otimes' \underline{b}$ is actually the greatest solution. ●

Proposition 1-2. Problem (1-12) has the solution $\underline{x} = \underline{A}^* \otimes' \underline{b}$; moreover this is the greatest solution. ●

The proofs given in [25] depend essentially upon two facts:

(i) that matrix multiplication is an isotone operation,

(ii) that the following relations hold for arbitrary matrices, provided only that the relevant products exist:

$$\underline{A} \otimes (\underline{A}^* \otimes' \underline{B}) \leq \underline{B} \leq \underline{A} \otimes' (\underline{A}^* \otimes \underline{B})$$

$$\underline{A} \otimes (\underline{A}^* \otimes' (\underline{A} \otimes \underline{X})) = \underline{A} \otimes \underline{X}$$

No proofs of these facts are given in [25] and it is in fact one aim of the present memorandum to furnish these proofs. Our immediate purpose, however, is to illustrate how the investigation of the operational problems formulated in (1-11) and (1-12) gives rise to a fully-fledged theory of conjugacy analogous to that of conventional linear algebra.

1-2.4 Activity Networks

A complex project - let us say the construction of a hospital - presents a large number n of activities. Some activities can proceed quite independently of one another, whilst others are in the nature of things so related that one must precede the other (foundations must be built before walls are built). We wish to know the latest time at which each activity can be started if we are to meet a given completion date.

Define a conventional activity (n + 1) called "finished", with the property that all other activities must precede it. Introduce the quantities a_{ij}, where a_{ij} is the minimum amount of time by which the start of activity i must precede the start of activity j, if the activities are so related; otherwise take $a_{ij} = -\infty$. Clearly, for a meaningful physical situation we shall have:

$$\left.\begin{array}{l} a_{ii} = 0 \ (i = 1, \ldots, (n+1)) \\ \text{and for each circuit } (i_1, i_2), (i_2, i_3), \ldots, (i_p, i_1), \text{ at least one of} \\ a_{i_1 i_2}, a_{i_2 i_3}, \ldots, a_{i_p i_1} \text{ is } -\infty. \end{array}\right\} \quad (1\text{-}13)$$

Let τ be defined as the planned completion time of the project, assumed given, and for $i = 1, \ldots, (n + 1)$ define t_i to be the latest allowable starting time for any activity i. Then we seek $t_1, \ldots, t_n$ consistent with $t_{n+1} = \tau$.

A moment's reflection shows that:

$$t_i = \min_{j = 1, \ldots, (n+1)} (-a_{ij} + t_j) \quad \text{for } i = 1, \ldots, (n+1) \qquad (1\text{-}14)$$

(actually, this is an appeal to Bellman's principle of optimality).

But if we define the matrix $\underline{A} = [-a_{ij}]$ and the vector $\underline{t} = [t_i]$ we see that (1-14) is just $\underline{t} = \underline{A} \otimes' \underline{t}$ in the notation of (1-8) i.e. we are led again to the eigenvector-value problem, this time in min algebra. Furthermore, consider Bellman's iterative procedure for solving network problems, as set out for example in [29]: we guess a first approximate solution to (1-14), which we substitute for the t_j in the right-hand side of (1-14), obtaining new approximations t_i which we again substitute ... .

Clearly, if our first approximate solution vector is $\underline{t}(1)$, we generate thus the sequence:

$$\underline{t}(1), \ \underline{A} \otimes' \underline{t}(1), \ \underline{A}^2 \otimes' \underline{t}(1), \ \ldots$$

and we are led again to the question: under what circumstances will this sequence converge to an eigenvector of $\underline{A}$? And when will the convergence be bound to occur in a finite number of steps? Answers to these questions are given in Chapter 27 below, including some results presented in [6].

1-2.5 The Assignment Problem

Let there be given n jobs, each of which has to be assigned to a different one of n men. For each (i,j), $1 \leq i \leq n$, $1 \leq j \leq n$, let a_{ij} represent some numerical measure of the utility of assigning job j to man i. We seek the assignment which produces greatest total utility.

This classical assignment problem may be formulated as follows. Let $\mathcal{G}_n$ represent the symmetric group on n elements. Then we seek the permutation $\sigma \in \mathcal{G}_n$ such that:

$$\sum_{i=1}^{n} a_{i\sigma(i)} \quad \text{is maximised} \tag{1-15}$$

We may reformulate this problem in max algebra using a natural extension of the notation (1-3). Specifically, an iterated product involving the operation $\otimes$ can be written using a capital pi ($\Pi_\otimes$) and an iterated sum involving the operation $\oplus$ can be written using a capital sigma ($\Sigma_\oplus$), the suffixes $\otimes$ and $\oplus$ sufficing to distinguish these symbols from their normal use for iterated arithmetical operations.

Then expression (1-15) becomes:

$$\prod_{i=1}^{n}{}_{\otimes} \; a_{i\sigma(i)}$$

and so the value of the optimal assignment, namely:

$$\max_{\sigma \in \mathcal{G}_n} \sum_{i=1}^{n} a_{i\sigma(i)}$$

can be written:

$$\sum_{\sigma \in \mathcal{G}_n}{}_{\oplus} \left\{ \prod_{i=1}^{n}{}_{\otimes} \; a_{i\sigma(i)} \right\} \tag{1-16}$$

The formal resemblance of (1-16) to a determinant is immediately apparent. In fact (1-16) is precisely the permanent of the matrix $[a_{ij}]$, in max algebra. Such permanents play a useful role in the theory. Yoeli [6] has shown, for example, that the matrix $\underline{\Gamma}'(\underline{D})$ of (1-10) is actually the adjugate of matrix $\underline{D}$ - i.e. element (i,j) of $\underline{\Gamma}'(\underline{D})$ is equal to the permanent of the matrix obtained by deleting row j and column i from matrix $\underline{D}$. These ideas are discussed in Chapter 27 below.

1-2.6 The Dual Transportation Problem

Let there be given m producers and n consumers of some commodity. Producer i ($1 \leq i \leq m$) has production capacity p_i and consumer j ($1 \leq j \leq n$) has total demand d_j, and the cost of transporting one unit of the commodity from producer i to consumer j is c_{ij}.

The problem of determining what quantity of goods should be sent from each producer to each consumer, so as to satisfy all consumer demands, stay within all production capacities, and minimise total transportation costs, can be formulated as a linear program - see for example [40].

The dual of this linear program is well-known to be:

Minimise $\sum_{j=1}^{n} d_j y_j - \sum_{i=1}^{m} p_i x_i$

Subject to $x_i - y_j \leq c_{ij}, \quad i = 1, \ldots, m; \quad j = 1, \ldots, n$ (1-17)

Several algorithms for solving transportation problems work with this dual formulation. Now the theorem of complementary slacks (see e.g. [33]) assures us of the following:
if none of the producers is redundant (in the sense that none of that producer's product is transported in the optimal solution) then for each i = 1, ..., m the constraint (1-17) is satisfied as an equation for at least one j ($1 \leq j \leq n$) by the optimal $\underline{x}$ and $\underline{y}$.

Then the constraints (1-17) may clearly be replaced by:

$$x_i = \min_{j=1,\ldots,n} (c_{ij} + y_j) \quad (i = 1, \ldots, m) \tag{1-18}$$

But (1-18) is just a set of linear relations when rewritten in the notation (1-8)

$$x_i = \sum_{j=1}^{n}{}_{\oplus'} (c_{ij} \otimes y_j) \quad (i = 1, \ldots, m)$$

which we can write in vector-matrix notation:

$$\underline{x} = \underline{C} \otimes' \underline{y} \tag{1-19}$$

The dual transportation problem may thus be reformulated as a problem in min algebra, with linear constraints. It can be shown that the set of feasible solutions for this dual problem form a space.

1-2.7 Boolean Matrices

Instead of considering the entire system of real numbers, we could take some additive subsemigroup of the real numbers such as the rational numbers, or the integers, since the operations $\otimes$ and $\oplus$ then retain their properties and meanings. In particular it is natural in many situations to restrict ourselves to the non-negative real numbers when we are formulating problems of distance, time or cost.

A more fundamental step is to consider the subsemigroup consisting of the two elements 0 and $-\infty$ for which the operations $\oplus$ and $\otimes$, interpreted as usual as max and +, have the following operation tables:

x	y	$x \oplus y$	$x \otimes y$
$-\infty$	$-\infty$	$-\infty$	$-\infty$
$-\infty$	0	0	$-\infty$
0	$-\infty$	0	$-\infty$
0	0	0	0

In the sequel, we shall refer to this system by the symbol ②.
We see at once that ② is a two-element Boolean algebra with the

operations ⊕ and ⊗ as Boolean sum and product respectively. With a little care, therefore, we can develop a formalism which covers useful portions of the theory of Boolean matrices also.

For example, let there be given an abstract directed graph with n nodes. Let the $(n \times n)$ matrix $\underline{D} = [d_{ij}]$ have $d_{ij} = 0$ if there is a directed arc from node i to node j, otherwise $d_{ij} = -\infty$. Clearly $\underline{D}$ plays the role of the familiar adjacency matrix of the graph - see e.g. [30] or [34]. Form the successive powers $\underline{D}^2$, $\underline{D}^3$, ..., in max algebra (notation (1-3)) and define:

$$\underline{\Gamma}(\underline{D}) = \underline{D} \oplus \underline{D}^2 \oplus \ldots \oplus \underline{D}^n \tag{1-20}$$

(Obviously, expressions (1-10) and (1-20) are formal duals of one another. We are using min algebra in (1-10) and max algebra in (1-20).) It is not difficult to see that element (i,j) of the matrix $\underline{D}^p$ is now the greatest arithmetical sum of the form:

$$d_{ii_1} + d_{i_1 i_2} + \ldots + d_{i_{p-1} j},$$

i.e. that element (i,j) of the matrix $\underline{D}^p$ is zero if and only if there exists a directed path having exactly p arcs from node i to node j. Hence the matrix $\underline{\Gamma}(\underline{D})$ in expression (1-20) has element (i,j) zero if and only if the graph contains a directed path from node i to node j.

These ideas are extensively discussed in e.g. [30] and [34], using the conventional notation of Boolean algebra; our present aim is merely to illustrate that certain aspects of Boolean matrix theory may be subsumed under the formalism of max algebra.

1-2.8 The Stochastic Case

In many problems of machine-scheduling which arise in practice, it is not realistic to assume that the time taken by a given machine to perform one cycle of work is constant. In general, statistical variation will be observed and variances may be very significant. Matrices $\underline{A}(1)$, $\underline{A}(2)$, ... in equation (1-5) are sequential realisations of some stochastic process. The problem of the "convergence" of $\underline{x}(r)$ as $r \to \infty$ becomes a more complicated problem of convergence in probability in some suitable sense.

A related problem is the following. A stock of a certain commodity is increased by a constant amount c at the end of each unit of time, as a result of a steady production process. At each such time, a demand arises, following some statistical law, and the stock is then depleted so as to meet the demand as far as possible, consistently with the nett stock never falling below a given amount d. Clearly, if the stock at time r is x(r), we have:

$$x(r + 1) = \max\,(x(r) + c - u(r),\ d) \tag{1-21}$$

where u(r) is the demand arising at time r. Writing a(r) for c - u(r), we may reformulate (1-21) in max algebra as:

$$x(r + 1) = (a(r) \otimes x(r)) \oplus d$$

We obtain thus a linear recurrence relation in max algebra, in which the coefficient a(r) is stochastic.

Such stochastic problems lead us to the consideration of matrix products in max algebra:

$$\underline{A}(r) \otimes \underline{A}(r - 1) \otimes \ldots \otimes \underline{A}(1)$$

Such a product is a stochastically determined operator defining how the $(r + 1)^{st}$ state of the system depends upon the first state. Algebraically, we are here concerned with the structure of a semigroup of matrices. This problem is discussed in .

1-3. Conclusion: Our Present Aim

Through the above examples we have attempted to illustrate how a variety of different applications all give rise to formal problems of linear algebra when formulated using min algebra or max algebra, where by "linear algebra" we mean the theory of matrix transformations of spaces of n-tuples.

Now, we may also define "infinite dimensional" spaces over $(R, \oplus, \otimes)$, in particular the space of real-valued functions. It turns out that a number of interesting problems in optimisation, approximation and convex analysis can be formulated in terms of linear transformations of such function spaces over $(R, \oplus, \otimes)$. These topics are briefly touched on in [25], but we shall not develop them further here since the present study is almost exclusively devoted to providing an adequate foundation for a matrix-vector calculus over $(R, \oplus, \otimes)$. We hope, however, that enough has been said to show that the motivation for a theory of linear operators for general spaces over $(R, \oplus, \otimes)$ is quite broadly-based.

As far as the "finite-dimensional case" is concerned, elements of a theory of matrix minimax algebra have certainly existed since about 1957, but our aim in the present memorandum is to present a comprehensive theory, with new material related to the eigenvector-value problem, projections, the spectral theorem, subspaces, approximation, norms, canonical forms, and an extensive duality theory. The individual topics will be illustrated by reference to the operational research problems discussed above.

2. THE INITIAL AXIOMS

2-1. Some Logical Geography

Fig.2-1 presents thirteen axioms for an algebraic structure S having four binary combining operations $\oplus$, $\otimes$, $\oplus'$ and $\otimes'$. The mathematical systems with which we shall be concerned in this memorandum will generally be realisations of these axioms.

In the remainder of this chapter we shall discuss these axioms in some detail, but in this first section it may be helpful to look at Fig.2-1 as a whole. There are, in fact, two rather different ways of seeing these axioms.

If we divide the table by the horizontal line marked α, then we recognise the axioms above the line as being those of a lattice, with lattice operations $\oplus$ and $\oplus'$. If we now add the axioms X_4, X_5, X_6, we have a lattice -ordered semigroup with semigroup-operation $\otimes$. Finally, the axioms X_4', X_5', X_6' introduce another semigroup operation $\otimes'$ with respect to which the structure is a dual lattice-ordered semigroup. A suitably diligent search of a text on lattice-ordered semigroups (e.g. [27] or [32]) will therefore unearth a good many of the propositions which we discuss in the rest of the chapter. The aim of the chapter, however, is to bring these propositions together in a way which gives the axioms of Fig.2-1 a rather different character from that of a double lattice-ordered semigroup.

Let us, in fact, divide the table of Fig.2-1 by the vertical line marked β, ignoring axiom L_3. The axioms to the left of this line now resemble those of a ring, or semiring, with a multiplication operation $\otimes$ and an addition operation $\oplus$. The addition operation has some strange properties (X_3) but the axioms are enough to set up (for example) a matrix calculus, as repeatedly illustrated in Chapter 1 with $\oplus$ interpreted as max, and $\otimes$ interpreted as +. Axiom systems such as X_1 to X_6 have been studied by a number of authors, and were introduced by Yoeli [6]. With these axioms alone, some headway can be made, particularly in branches of the theory which are essentially generalisations of Boolean matrix theory - for example in studying the sequences:

$$\underline{A}, \underline{A}^2, \ldots$$

and:

$$\underline{A} \oplus \underline{A}^2 \oplus \ldots$$

for "definite" matrices ([6], [17]). Some analysis of the equation $\underline{A} \otimes \underline{x} = \underline{b}$ and of the eigenproblem $\underline{A} \otimes \underline{x} = \lambda \otimes \underline{x}$ is also possible ([4], [7]).

L_3:	$x \oplus (y \oplus' x) = x \oplus' (y \oplus x) = x$		
X_1:	$x \oplus (y \oplus z) = (x \oplus y) \oplus z$	X_1':	$x \oplus' (y \oplus' z) = (x \oplus' y) \oplus' z$
X_2:	$x \oplus y = y \oplus x$	X_2':	$x \oplus' y = y \oplus' x$
X_3:	$x \oplus x = x$	X_3':	$x \oplus' x = x$
X_4:	$x \otimes (y \otimes z) = (x \otimes y) \otimes z$	X_4'	$x \otimes' (y \otimes' z) = (x \otimes' y) \otimes' z$
X_5:	$x \otimes (y \oplus z) = (x \otimes y) \oplus (x \otimes z)$	X_5':	$x \otimes' (y \oplus' z) = (x \otimes' y) \oplus' (x \otimes' z)$
X_6:	$(y \oplus z) \otimes x = (y \otimes x) \oplus (z \otimes x)$	X_6':	$(y \oplus' z) \otimes' x = (y \otimes' x) \oplus' (z \otimes' x)$

(Lines labelled β and α are marked on the figure.)

Fig. 2-1. The Initial Axioms

For a comprehensive theory of linear operators, however, it is necessary as in other branches of mathematics, to introduce the notion of a conjugate space, for which purpose we require a duality which runs through the theory from the outset. We contemplate therefore, systems which have four algebraic operations consisting of the dual pairs $(\oplus, \otimes)$ and $(\oplus', \otimes')$, for which the axioms X_1 to X_6 and X_1' to X_6' all hold simultaneously. Such a system is a kind of double semiring.

The axioms to the right of line β in Fig.2-1 are just a carbon copy of those to the left, except that the operations $\oplus$ and $\otimes$ are replaced by the operations $\oplus'$ and $\otimes'$, to which we may give the interpretations min and + respectively, for example.

We see, however, that the axioms to the left of line β and those to its right do not interact with one another, since they have no operation in common. In order to give cohesion, the lattice absorption law L_3 is added, to give the axiom system of Fig.2-1. The force of axiom L_3 is that the partial ordering induced by the operation $\oplus$, and that induced by the operation $\oplus'$, are consistent, as we shall discuss below.

Although the language of lattice theory will therefore from time to time occur in the present text, we are much more concerned to view our algebraic structures as double semirings than as lattices with two multiplication operations. For this reason the symbols $\oplus$ and $\oplus'$ are used instead of the more usual lattice-theoretic symbols $\vee$ and $\wedge$, and the whole notation is designed to bring out later the analogy to linear operator theory. In later chapters lattice concepts as such will hardly be mentioned.

In some of the structures we consider, the operations $\otimes$ and $\otimes'$ may coincide. We shall speak then of self-dual multiplication.

We begin, therefore, with the axioms of Fig.2-1 in order to cover the basic properties of scalars and operators with the same set of propositions. For most of the theory we do not need to assume that the multiplication of scalars is commutative.

There are, of course, many different ways of deducing the elementary properties of systems satisfying axioms such as those of Fig.2-1. In order to give cohesion to the material, and lay the foundations for a theory of operators, we shall make extensive use of the idea of homomorphism.

2-2. Commutative Bands

Let S be a set which is linearly ordered under a binary relation $\geqslant$ i.e. for which the following axioms Y_0, Y_1, Y_2 hold:

Y_0 : $\forall\, x, y \in S$, either $x \geqslant y$ or $y \geqslant x$

Y_1 : $\forall\, x, y \in S$, if $x \geqslant y$ and $y \geqslant x$ then $x = y$

Y_2 : $\forall\, x, y, z \in S$, if $x \geqslant y$ and $y \geqslant z$ then $x \geqslant z$

From Y_0 we infer:

Y_3 : $\forall\, x \in S$, $x \geqslant x$.

The notation $x > y$ will mean $x \geqslant y$ but $x \neq y$.

Let us now change the notation by introducing a binary combining operation $\oplus$ on S by means of the definition:

$$\forall\, x, y \in S,\ x \oplus y = x \quad \text{if } x \geqslant y$$
$$= y \quad \text{otherwise.}$$

Trivially, we have:

X_0 : $\forall\, x, y \in S$, $x \oplus y = x$ or $x \oplus y = y$

and it is well known (and very easy to prove) that the system $(S, \oplus)$ is an abelian semigroup in which every element is idempotent, i.e.:

X_1 : $\forall\, x, y, z \in S$, $x \oplus (y \oplus z) = (x \oplus y) \oplus z$

X_2 : $\forall\, x, y \in S$, $x \oplus y = y \oplus x$

and X_3 : $\forall\, x \in S$, $x \oplus x = x$

A system $(S, \oplus)$ satisfying X_1, X_2, X_3 is called a commutative band, or semilattice. We shall adhere to the former term.

Equally well-known, and equally easy to prove, is the following.

Proposition 2-1. If $(S, \oplus)$ is a commutative band, then the binary related $\geqslant$ defined on S by means of the definition:

$\forall$ x, y ε S, $x \geq y$ <u>if and only if</u> $x = y \oplus z$ <u>for some</u> z ε S (2-1)
<u>also satisfies:</u>

$\forall$ x, y ε S, $x \geq y$ if and only if $x = y \oplus x$

<u>and is a partial ordering of S in the sense that axioms</u> Y_1, Y_2 <u>and</u> Y_3 <u>are satisfied. If, moreover, axiom</u> X_0 <u>holds, then axiom</u> Y_0 <u>is satisfied so that S becomes a linearly ordered set under the relation</u> $\geq$.

We shall call a commutative band <u>linear</u> if axiom X_0 holds.
Clearly any subset of a linear commutative band is a linear commutative band under the same operation $\oplus$.
The terms <u>linearly ordered set</u> and <u>linear commutative band</u> are thus equivalent in the sense made clear above. Given the relation $\geq$ we shall speak of the <u>corresponding operation</u> $\oplus$, and vice versa. For any one given structure, we shall employ both notations indifferently and without further comment.

Let F consist of the finite real numbers, under the usual ordering $\geq$. The corresponding operation is max, where:

$$\begin{aligned} \max(x,y) &= x \quad \text{if } x \geq y \\ &= y \quad \text{otherwise.} \end{aligned}$$

The system (F,max) is a commutative band, and so is the system (F,min), where

$$\begin{aligned} \min(x,y) &= x \quad \text{if } y \geq x. \\ &= y \quad \text{otherwise.} \end{aligned}$$

Both the systems (F,max) and (F,min) give examples of linear commutative bands, as also do the systems (R,max) and (R,min) where $R = F \cup \{-\infty\} \cup \{+\infty\}$ denotes the extended real numbers. These latter two systems will be called the <u>principal interpretation</u> and the <u>dual principal interpretation</u> respectively (of axioms X_1, X_2, X_3).

The set of all subsets of a given set Ω provides an example of a commutative band which is not linear. The operation $\oplus$ is set-theoretic union, and the relation $\geq$ is set-inclusion.

If a subset S of a commutative band T is itself a commutative band under the same operation $\oplus$, we shall say that $(S,\oplus)$ is a <u>commutative sub-band</u> of $(T,\oplus)$.

2-3. Isotone Functions

If S, U, are given sets then the notation S^U will denote the set of all functions from U to S. If U = {1, 2, ..., n} is the set of the first n natural numbers, then we shall write S^n as an alternative to S^U. If U is an arbitrary set and S is a commutative band, then the set S^U also becomes a commutative band when, for each pair f, g of such functions we define:

$$\forall x \in U, \quad (f \oplus g)(x) = f(x) \oplus g(x) \tag{2-2}$$

provided as usual that two functions which take identical values in S for every xεU are regarded as equal. Corresponding to the operation $\oplus$ introduced by (2-2), the set S^U acquires by (2-1) an order relation $\geqslant$ which satisfies:

$$f \geqslant g \text{ if and only if } f(x) \geqslant g(x) \qquad (\forall\, x \in U)$$

Even when S is linear, S^U is in general not linear. Thus if U is the set $\{1,2,\ldots,n\}$ of the first n natural numbers, $S^U = S^n$ is the set of all n-tuples of elements of S, with $f \oplus g$ defined componentwise; for the principal interpretation for example it is obvious that S^n is partially but not linearly ordered, if $n > 1$.

Now suppose that $(U, \oplus)$, $(S, \oplus)$ are both commutative bands. (The use of the same symbol $\oplus$ will cause no confusion) A function $f \in S^U$ is a <u>(band-) homomorphism</u> if and only if:

$$\forall\, x, y \in U,\ f(x \oplus y) = f(x) \oplus f(y) \qquad (2\text{-}3)$$

Similarly the words <u>isomorphism</u>, <u>endomorphism</u>, <u>automorphism</u>, will be given their usual meanings.
It is immediate that the homomorphisms form a commutative sub-band of the commutative band S^U.
Introducing the partial order relation corresponding to $\oplus$, we easily infer that (2-3) implies:

$$f(x) \geqslant f(y) \qquad \text{if } x \geqslant y \qquad (2\text{-}4)$$

so that f is an <u>isotone</u> function. Conversely, if U and S are both <u>linearly</u> ordered sets, every isotone function $f \in S^U$ is a homomorphism relative to the corresponding operations $\oplus$.

In particular, the endomorphisms of a linear commutative band S are just the isotone functions belonging to S^S. For the principal interpretation: the endomorphisms of (R, max) are just the monotone non-decreasing (extended) real-valued functions of one real variable.

The endomorphisms of S include in particular i_S, the <u>identity mapping</u>, which satisfies:

$$\forall\, x \in S,\ i_S(x) = x$$

For each given element a of a commutative band S, we may define the <u>translation of S induced by a</u> as the function $f_a \in S^S$ such that:

$$\forall\, x \in S, \quad f_a(x) = a \oplus x \qquad (2\text{-}5)$$

It is a straightforward matter to verify that each such translation is an endomorphism of S, and that $f_{a\oplus b} = f_a \oplus f_b$ for all $a, b \in S$.
The set of translations forms a commutative sub-band of the commutative band of endomorphisms of S, and the mapping $\tau: a \mapsto f_a$ is a homomorphism of S onto the set of translations of S. As is usual in algebra, we call such a mapping τ a <u>representation</u>. It is indeed a <u>faithful</u> representation, i.e. τ is an isomorphism, for suppose $f_a = f_b$. Then:

$$a = a \oplus a = f_a(a) = f_b(a) = b \oplus a = a \oplus b = f_a(b) = f_b(b) = b \oplus b = b$$

(making free use of X_2, X_3 and (2-5)). Summarising:

Proposition 2-2. A commutative band has a faithful representation as a commutative band of translations of itself. A linear commutative band has a faithful representation as a commutative band of isotone functions from a linearly ordered set to itself. ●

We may also regard the operation $\oplus$ as defining a function from S^2 to S, namely:

$$\oplus: (x,y) \mapsto x \oplus y$$

and it is easily verified that $\oplus$ is a homomorphism. We call the function $\oplus$ an addition (not to be confused with the function + in a real-number interpretation, which will always be called arithmetical addition).

From the fact that, for a commutative band S, addition and translations are homomorphisms and hence isotone functions, we infer further:

For all w, x, y, z in a commutative band S there hold:

$$\left.\begin{array}{llll} \text{If } y \geq z & \text{then} & w \oplus y \geq w \oplus z & \\ \text{If } w \geq x & \text{and} & y \geq z & \text{then } w \oplus y \geq x \oplus z \end{array}\right\} \qquad (2\text{-}6)$$

Now let S be a given set, and let K be a collection of functions, each of which is from S^m to S, for some integer $m \geq 1$ (not assumed to be the same for each function). By $\hat{K}$, the composition algebra generated by K, we understand the smallest class of functions having the following properties:

(i) $K \subset \hat{K}$

(ii) If $f \in K$ and $f_1,\ldots,f_r \in \hat{K}$ are such that $f \in S^{(S^r)}$ and $f_i \in S^{(S^{m_i})}$ $(i=1,\ldots,r)$ then $f(f_1,\ldots,f_r) \in \hat{K}$, where $f(f_1,\ldots,f_r)$ is by definition a function from $S^{\sum_{i=1}^{r} m_i}$ to S such that for all $x_i \in S^{m_i}$ $(i=1,\ldots,r)$ we have

$$f(f_1,\ldots,f_r): (x_1,\ldots,x_r) \mapsto f(f_1(x_1),\ldots,f_r(x_r)).$$

In other words, $\hat{K}$ is precisely the collection of functions each of which can be defined as a fixed program of applications of the given elements of K.

Each element of $\hat{K}$ is a function from S^m to S for some integer $m \geq 1$. A straightforward induction on the number of function applications necessary to create a given element of $\hat{K}$, yields the following result:

Proposition 2-3. Let S be a commutative band and let K be a collection of functions, each of which is from S^m to S for some integer $m \geq 1$. If all elements of K are homomorphisms, then so are all elements of $\hat{K}$. If all elements of K are isotone, then so are all elements of $\hat{K}$.

In particular, the composition algebra generated by the addition and translation functions together with the identity mapping consists of homomorphisms, and hence of isotone functions. ●

2-4. Belts We consider now an algebraic structure $(V,\oplus,\otimes)$ consisting of a set V together with two binary combining operations $\oplus$ and $\otimes$.
We assume that the system $(V, \oplus, \otimes)$ satisfies the following axioms:

X_1, X_2, X_3: $(V, \oplus)$ is a commutative band

X_4: $\forall\, x, y, z \in V, \quad x \otimes (y \otimes z) = (x \otimes y) \otimes z$

X_5: $\forall\, x, y, z \in V, \quad x \otimes (y \oplus z) = (x \otimes y) \oplus (x \otimes z)$

X_6: $\forall\, x, y, z \in V, \quad (y \oplus z) \otimes x = (y \otimes x) \oplus (z \otimes x)$

Yoeli [8] uses the term Q-semirings for certain algebraic structures of this type. Such structures are undoubtedly semirings as defined in [6], although the idempotency of all elements under the operation $\oplus$ gives them a very special character. Backhouse and Carré speak of regular algebras, with reference to the relevance of these structures to the theory of regular expressions.

Since X_4 defines $(V, \otimes)$ to be a semigroup, with distributive laws X_5 and X_6, it would also be appropriate to call $(V, \oplus, \otimes)$ a associative m-semilattice, or a semilattice - ordered semigroup (see [27], [32]).

The terms semiring, regular and algebra are, however, already so overloaded with disparate connotations in mathematics, and terms such as semilattice-ordered semigroup are so unhandy, that it seems appropriate to coin a new term for structures satisfying X_1 to X_6.

We shall call them belts, evoking thereby a little of the flavours of the words ring and band. The term sub-belt will be used in the obvious way.

If, moreover, $(V, \oplus)$ is a linear band, we shall say that $(V, \oplus, \otimes)$ is a linear belt. Obviously, if $(V, \oplus, \otimes)$ is a linear belt and $(U, \otimes)$ is a subsemigroup of $(V, \otimes)$, then $(U, \oplus, \otimes)$ is a linear sub-belt of $(V, \oplus, \otimes)$.

It is easy to verify that every commutative band is already a belt, if we take the operation $\otimes$ to be identical with the operation $\oplus$. We may then speak of a degenerate belt.

As an example of a non-degenerate non-linear belt, consider the set of all subsets of some multiplicative semigroup $(S, \otimes)$, with $\oplus$ interpreted as set-theoretic union, and $\otimes$ as a set-product:

$$\forall\, P, Q \in S,\ P \otimes Q = \{p \otimes q \mid p \in P,\ q \in Q\}$$

The principal interpretation of axioms X_1 to X_6 will be the system $(\mathbf{R}, \max, +)$, i.e. the extended real numbers, under the operations max and arithmetical addition. The dual principal interpretation of axioms X_1 to X_6 will be the system $(\mathbf{R}, \min, +)$.

These interpretations give examples of a linear belt. Other examples are given in [17] and [21].

2-5. Belt Homomorphisms

If S is a commutative band, let W be the set of all endomorphisms of S. Then

W is a belt if for all f, g ε W we define f $\oplus$ g ε W as in (2-2) and the product f $\otimes$ g ε W as the composition of f and g, i.e.:

$$\forall\ x \ \varepsilon\ S,\ (f \otimes g)(x) = f(g(x)) \tag{2-7}$$

In particular, for each given element a of a belt (V, $\oplus$, $\otimes$), we may define the left multiplication induced by a as the function $g_a \ \varepsilon\ V^V$ such that:

$$\forall\ x \ \varepsilon\ V,\ \ g_a(x) = a \otimes x \tag{2-8}$$

By X_5, g_a is an endomorphism of commutative band S = (V, $\oplus$), and $g_{a \otimes b} = g_a \otimes g_b$, where $g_a \otimes g_b$ is defined as in (2-7).

Now, when we use the terms homomorphism, isomorphism, endomorphism, etc, in relation to a belt we have in mind of course a function f which not only satisfies (2-3) but for which $f(x \otimes y) = f(x) \otimes f(y)$ also. For emphasis we shall occasionally say belt - homomorphism, etc.

So let V = (V, $\oplus$, $\otimes$) be a given belt. If W, the set of all endomorphisms of the commutative band S = (V, $\oplus$), is made into a belt as in (2-2) and (2-7), then it is easy to see that the mapping:

$$\tau\ :\ a \rightarrow g_a$$

where g_a is the left multiplication defined in (2-8), is a homomorphism of the belt V into the belt W, which we shall as usual call a representation. Summarising, using Proposition 2-2:

Proposition 2-4: A belt has a representation as a belt of endomorphisms of a commutative band. A linear belt has a representation as a system of isotone functions of a linearly ordered set to itself, the system of functions being closed under composition. ●

We may define right multiplications by obvious analogy to left multiplications. Both left multiplications and right multiplications are band-endomorphisms and hence isotone. Now define the multiplication $\bar{\otimes} \ \varepsilon\ V^{(V^2)}$ by $\otimes$: $(x,y) \mapsto x \otimes y$. In general this is not a homomorphism, but it is an isotone function, since if $w \geqslant x$ and $y \geqslant z$ then using the isotone mapping τ above, we have $\tau(w) \geqslant \tau(x)$ i.e. $g_w \geqslant g_x$ so

$$g_w(y) \geqslant g_x(y) \geqslant g_x(z) \quad .$$

But this is $\otimes(w,y) \geqslant \otimes(x,z)$. We now apply Proposition 2-3 to derive the following result.

Proposition 2-5: Let (V, $\oplus$, $\otimes$) be a belt, and let K consist of $\oplus$, together with the identity mapping and all left multiplications, right multiplications and translations. Then $\hat{K}$, the composition algebra generated by K, consists entirely of band-homomorphisms, and hence of isotone functions. If $\otimes$ is now adjoined to K, then $\hat{K}$ still consists of isotone functions. ●

For easy reference in future arguments, we record in particular:

For all w, x, y, z in a belt V there hold:

$$\left.\begin{array}{lll} \text{If } w \geq x & \text{then} & y \otimes w \geq y \otimes x \\ & \text{and} & w \otimes y \geq x \otimes y \\ \text{If } w \geq x & \text{and} & y \geq z \text{ then } w \otimes y \geq x \otimes z \end{array}\right\} \qquad (2\text{-}9)$$

2-6. Types of Belt

A given belt $(V, \oplus, \otimes)$ may or may not satisfy one or more of the additional axioms X_7, X_8, X_9, X_{10} below.

X_7: $\forall\, x, y \in V, \quad x \otimes y = y \otimes x$

A belt which satisfies axiom X_7 will be called a commutative belt. The principal interpretation (R,max,+) defines a commutative belt. On the other hand, the system of monotone non-decreasing real-valued functions of one real variable, closed under composition, is clearly a non-commutative belt since in general $f(g(x)) \neq g(f(x))$.

X_8: $\exists \phi \in V$ such that the products $\phi \otimes x$ and $x \otimes \phi$ both equal x, for all $x \in V$.

A belt satisfying axiom X_8 will be called a belt with identity. The symbol ϕ is used for the identity element because it is suggestive of the arithmetical zero which is the identity element in the principal interpretation (R,max,+).

X_9: $\forall\, x \in V, \exists x' \in V$ such that $x \otimes x'$ equals ϕ.

A belt satisfying axioms X_8 and X_9 will be called a division belt (by analogy with division ring). We infer as usual that the multiplicative inverse x' of x is two-sided and unique, that $(x')' = x$, that $\phi' = \phi$, and that $(x \otimes y)' = y' \otimes x'$, etc.

The system (F, max, +) supplies an example of a commutative division belt. The monotone strictly increasing real-valued functions of one real variable form a non-commutative division belt under max and composition.

From now on we use the standard notation x^{-1} to denote the unique two-sided multiplicative inverse of x.

For a division belt V, we note:

$$\forall\, x, y \in V,\ x \geq y \text{ if and only if } y^{-1} \geq x^{-1} \qquad (2\text{-}10)$$

(From $(x \oplus y) = x$ on left-multiplying by x^{-1} and right-multiplying by y^{-1}, we obtain $y^{-1} \oplus x^{-1} = y^{-1}$, and similarly for the converse).

X_{10}: $\exists\ \theta \in V$ such that: for all $x \in V$, $x \oplus \theta = x$, and $x \otimes \theta$, $\theta \otimes x$ both equal θ.

An element (necessarily unique) which satisfies axiom X_{10} is called a zero or a null. It is easily verified that if $S = F \cup \{-\infty\}$ consists of the real numbers extended by $-\infty$, then the system (S,max,+) has the element $-\infty$ as a null.

2-7. Dual Addition

Suppose two binary operations, $\oplus$ and $\oplus'$, are defined on a set S, such that:

L_1: $(S, \oplus)$ is a commutative band

L_2 : $(S, \oplus')$ is a commutative band

L_3 : $\forall x, y \in S, x \oplus(y \oplus' x) = x \oplus'(y \oplus x) = x$

Of course, L_3 presents the lattice absorption laws, which taken in conjunction with the fact that S is a semilattice under $\oplus'$ and a semilattice under $\oplus$, give necessary and sufficient conditions that $(S, \oplus, \oplus')$ be a lattice [32]. We call the operation $\oplus'$ a dual addition. By analogy with (2-1), we can define a binary relation $\leqslant$ corresponding to $\oplus'$.

As is well-known, the force of the lattice absorption laws is that the partial order under the relation $\geqslant$, and the partial order under the relation $\leqslant$ are consistent:

$$\forall x, y \in S, \quad x \geqslant y \quad \text{if and only if } y \leqslant x \tag{2-11}$$

Hence a mapping which is isotone with respect to either of $\geqslant$ and $\leqslant$ is isotone with respect to the other, and we shall merely say that it is isotone.

We shall use whichever of the symbols $\geqslant$ or $\leqslant$ is typographically convenient in a given context. We note en passant that from L_3 and (2-11) there holds for all $x, y \in S$:

$$x \oplus y \geqslant x \geqslant x \oplus' y \tag{2-12}$$

Finally, we shall write $x < y$, to mean $x \leqslant y$ but $x \neq y$.

Because of the symmetry between L_1 and L_2, we may repeat all the reasoning of Sections 2-2 and 2-3, with the symbol $\oplus$ replaced by $\oplus'$. As usual, we shall say that an axiom or expression or line of reasoning is the dual of another if the one can be formally derived from the other by systematically interchanging the symbols $\oplus$ and $\oplus'$, and the symbols $\geqslant$ and $\leqslant$. Because L_3 is self-dual, it is standard that the dual of any statements deducible from L_1, L_2 and L_3 jointly, is also so deducible.

As in Fig.2-1, we shall use e.g. the notation X_1' to denote the dual of axiom X_1, etc.

Now, from the dual of Proposition 2-3, it is clear that a dual translation:

$$f'_a : x \mapsto a \oplus' x$$

and dual addition:

$$\oplus' : (x,y) \mapsto x \oplus' y$$

are isotone with respect to $\leqslant$, and so are merely isotone. Combining these facts with the last statement of Proposition 2-3, we infer:

Proposition 2-6. Let $(S, \oplus, \oplus')$ satisfy axioms L_1, L_2, L_3, and let K consist of $\oplus$, $\oplus'$, together with the identity mapping and all translations and dual translations. Then $\hat{K}$, the composition algebra generated by K, consists entirely of isotone functions.

In particular, for all $w, x, y, z \in (S, \oplus, \oplus')$ there hold:

$$\left.\begin{array}{llll} \text{If} & w \geqslant x & \text{then} & y \oplus' w \geqslant y \oplus' x \\ \text{If} & w \geqslant x & \text{and} & y \geqslant z \quad \text{then } w \oplus' y \geqslant x \oplus' z \end{array}\right\} \tag{2-13}$$

It does not, of course, necessarily follow that a dual translation is an endomorphism of $(S, \oplus)$ or that a translation is an endomorphism of $(S, \oplus')$, **unless**:

$$\left.\begin{array}{l} a \oplus (x \oplus' y) = (a \oplus x) \oplus' (a \oplus y) \\ a \oplus' (x \oplus y) = (a \oplus' x) \oplus (a \oplus' y) \end{array}\right\} \quad (\forall a, x, y \in S) \qquad (2\text{-}14)$$

These are, of course, the lattice distributive laws. It is known that either of (2-14) in the presence of L_1, L_2, L_3, implies the other, and $(S, \oplus, \oplus')$ is then a distributive lattice. We shall not assume in general that our systems $(S, \oplus, \oplus')$ form distributive lattices. However, the following distributive inequalities always hold for all $a, x, y \in (S, \oplus, \oplus')$:

$$\left.\begin{array}{l} a \oplus (x \oplus' y) \leq (a \oplus x) \oplus' (a \oplus y) \\ a \oplus' (x \oplus y) \geq (a \oplus' x) \oplus (a \oplus' y) \end{array}\right\} \qquad (2\text{-}15)$$

These inequalities are standard lattice-theoretic results, but it is useful from our point of view to see them as consequences of the first two clauses of the following **lemma**:

Lemma 2-7. Let $(S, \oplus, \oplus')$ and $(T, \oplus, \oplus')$ both satisfy L_1, L_2, L_3. If $f:(S,\oplus) \to (T,\oplus)$ is a homomorphism then for all $x, y \in S$ there holds:

$$f(x \oplus' y) \leq f(x) \oplus' f(y)$$

and if $g: (S, \oplus') \to (T, \oplus')$ is a homomorphism then for all $x, y \in S$ there holds:

$$g(x \oplus y) \geq g(x) \oplus g(y)$$

If either f or g is a bijection, then the corresponding inequality becomes an equality.

Proof: Since f is isotone, (2-12) gives

$$f(x \oplus' y) \leq f(x) \text{ and similarly } f(x \oplus' y) \leq f(y)$$

Hence using (2-13) we have

$$f(x \oplus' y) = f(x \oplus' y) \oplus' f(x \oplus' y) \leq f(x) \oplus' f(y)$$

as asserted. Now, if f is bijective, then f possesses an inverse

$f^{-1}: f(x) \mapsto x$ for which we have:

$$\begin{aligned} f^{-1}(f(x) \oplus f(y)) &= f^{-1}(f(x \oplus y)) \qquad \text{(since } f \text{ is a homomorphism)} \\ &= x \oplus y \\ &= f^{-1}(f(x)) \oplus f^{-1}(f(y)) \end{aligned}$$

Hence f^{-1} is a homomorphism of $(T, \oplus)$ onto $(S, \oplus)$. Applying to f^{-1} the inequality proved above for homomorphisms, we have:

$$\begin{aligned} f^{-1}(f(x) \oplus' f(y)) &\leq f^{-1}(f(x)) \oplus' f^{-1}(f(y)) \\ &= x \oplus' y \end{aligned}$$

Applying the isotone function f to this inequality, we obtain

$$\begin{aligned} f(x) \oplus' f(y) &\leq f(x \oplus' y) \\ &\leq f(x) \oplus' f(y) \qquad \text{(as already proved).} \end{aligned}$$

Hence $f(x \oplus' y) = f(x) \oplus' f(y)$ in this case. The argument for g is dual. ●

If, for a given commutative band $(S, \oplus)$ it is possible to find a binary combining operation $\oplus'$ such that L_1, L_2, L_3 are satisfied, we shall say that S has a duality, or that a dual addition is defined. It is not difficult to show that this can be done in at most one way. If $(S, \oplus)$ is linear, then it is easy to verify that a dual addition is defined in the obvious way by:

$$\left.\begin{aligned} x \oplus' y &= y \text{ if } x \oplus y = x \\ &= x \text{ otherwise} \end{aligned}\right\} \qquad (2\text{-}16)$$

The resulting system is a distributive lattice.

2-8. Duality for Belts

So far, our discussion of duality is just elementary lattice theory. We extend the ideas to belts as follows. Suppose four binary operations, $\oplus$, $\otimes$, $\oplus'$, $\otimes'$, are defined on a set V such that:

M_1: $(V, \oplus, \otimes)$ is a non-degenerate belt

M_2: $(V, \oplus', \otimes')$ is a non-degenerate belt

L_3: $\forall\, x, y \in V, \quad x \oplus (y \oplus' x) = x \oplus' (y \oplus x) = x.$

We shall call the operations $\oplus'$ and $\otimes'$ a dual addition and dual multiplication respectively. As before, any axiom or expression or line of reasoning will be called the dual of another if the one can be formally derived from the other by systematically interchanging the symbols $\oplus$ and $\oplus'$, the symbols $\otimes$ and $\otimes'$ and the symbols $\geq$ and $\leq$. Obviously the dual of any statement deducible from M_1, M_2 and L_3 is also so deducible.

Since left multiplications are endomorphisms of $(V, \oplus)$, Lemma 2-7 gives for a system satisfying M_1, M_2 and L_3:

$$\left.\begin{array}{ll} \forall\, x, y\ z \in V: & x \otimes (y \oplus' z) \leq (x \otimes y) \oplus' (x \otimes z) \\ \text{and dually} & x \otimes' (y \oplus z) \geq (x \otimes' y) \oplus (x \otimes' z) \\ \text{Similarly for right multiplications:} & (y \oplus' z) \otimes x \leq (y \otimes x) \oplus' (z \otimes x) \\ \text{and dually} & (y \oplus z) \otimes' x \geq (y \otimes' x) \oplus (z \otimes' x) \end{array}\right\} \qquad (2\text{-}17)$$

Hence a system which satisfies M_1, M_2 and L_3 satisfies many distributive laws: X_5 and X_6 and their duals, and (2-15) and (2-17). Furthermore, combining the last statement of Proposition 2-5 with its dual, we have:

Proposition 2-8. Let $(V, \oplus, \otimes, \oplus'\ \otimes')$ satisfy axioms M_1, M_2 and L_3, and let K consist of $\oplus$, $\oplus'$, $\otimes$, $\otimes'$ together with the identity mapping and all translations, dual translations, left and right multiplications and dual left and right multiplications. Then $\hat{K}$, the composition algebra generated by K, consists entirely of isotone functions. ●

A number of trivial interpretations of M_1, M_2, L_3 are possible. If $(V, \oplus, \oplus')$

is a lattice, the trivial interpretation in which $\otimes$ and $\otimes'$ are defined to be identical with $\oplus$ and $\oplus'$ respectively, is excluded by the assumption of non-degeneracy in M_1, M_2, but the trivial interpretation in which $\otimes$ and $\otimes'$ are defined to be identical with $\oplus'$ and $\oplus$ respectively, would satisfy M_1, M_2, L_3 if $(V, \oplus, \oplus')$ were a distributive lattice. Accordingly, we exclude this possibility by saying that a non-degenerate belt $(V, \oplus, \otimes)$ <u>has a non-trivial duality</u> if we can find two binary combining operations $\oplus'$ and $\otimes'$ and elements a, b, c, d ε V such that M_1, M_2 and L_3 hold, together with:

$$\left.\begin{array}{l} a \oplus b \neq a \otimes' b \\ c \oplus' d \neq c \otimes d \end{array}\right\} \qquad (2\text{-}18)$$

Relations (2-18) also exclude the possibility of $(V, \oplus, \oplus')$ being a trivial lattice, i.e. a lattice in which $\oplus$ and $\oplus'$ coincide, since for such a lattice the relations L_3 immediately yield:

$$x \oplus y = x \qquad \text{for all } x, y, \text{ whence}$$
$$x = x \oplus y = y \oplus x = y$$

showing that such a lattice has at most one element, whilst (2-18) clearly implies the existence of at least two elements.

The possibility also arises that $\otimes$ and $\otimes'$ coincide. This is by no means to be seen as a trivial or degenerate case, and when it occurs, we shall speak of a <u>self-dual multiplication</u>.

For example, suppose $(V, \oplus, \otimes)$ is a non-degenerate division belt. In the language of lattice theory, such a structure is a semilattice-ordered group, and from the theory of such groups ([27], [32]), it is known that the definition:

$$\forall\, x, y \in V: x \oplus' y = (x^{-1} \oplus y^{-1})^{-1} \qquad (2\text{-}19)$$

defines a semilattice operation, such that $(V, \oplus, \oplus')$ becomes a (distributive) lattice. The relations (2-17) are all satisfied as equalities, when $\otimes'$ is taken as $\otimes$. In our terms, V acquires a duality with a self-dual multiplication.

Conversely, if $(V, \oplus, \otimes)$ is a non-degenerate division belt having a duality with self-dual multiplication, then V is a lattice-ordered group and from the theory of such groups, it is known that the relation (2-19) holds between the two semilattice operations. Summarising:

<u>Proposition 2-9.</u> <u>If $(V, \oplus, \otimes)$ is a non-degenerate division belt then $(V, \oplus, \otimes)$ has one, and only one, duality with a self-dual multiplication; and the distributive laws (2-17) all hold with equality.</u> ●

The system (F, max, +) gives an example of a commutative division belt having a duality with self-dual multiplication. A non-commutative example is provided by the belt of monotone strictly increasing real-valued functions of a real variable, under max and composition.

2-9. Some Specific Cases

Let S be a commutative band and let a, b ε S be arbitrary with $a \leq b$. It is a straightforward matter to confirm that the following subsets of S are commutative sub-bands:

$$T = \{x \mid x \in S,\ x \leqslant b\}$$
$$U = \{x \mid x \in S,\ a \leqslant x\}$$
$$V = \{x \mid x \in S,\ a \leqslant x \leqslant b\} = T \cap U$$

Further, if S is a linear commutative band, so are T, U, V. And if $a < b$ then W, X, Y are also linear commutative bands, where:

$$W = \{x \mid x \in S,\ x < b\}$$
$$X = \{x \mid x \in S,\ a < x\}$$
$$Y = \{x \mid x \in S,\ a < x < b\} = W \cap X.$$

In general, if S is a belt, we cannot immediately infer that T, U etc are also belts. However, if S is a belt with an identity element $\emptyset$, let us say that $a \in S$ is non-negative when $a \geqslant \emptyset$ and that $b \in S$ is non-positive when $b \leqslant \emptyset$. Then it is easy to verify (using (2-9)) that T is a sub-belt when b is non-positive and that U is a sub-belt when a is non-negative. In particular, if $a = b = \emptyset$ then we call T the non-positive cone of S and U the non-negative cone of S.

Again, if $b \leqslant \emptyset \leqslant a$ and S is a linear belt then T, U, W and X are linear sub-belts and in the particular case that $a = b = \emptyset$ we call W the negative cone of S and X the positive cone of S. (Of course, in particular cases, some of the above-mentioned systems may turn out to be empty).

Furthermore, if S is a commutative band with a duality then the same duality operates in T, U and V, whilst if S is a belt with a duality, having self-dual multiplication with identity element $\emptyset$ then the same duality operates in T and U under the condition $b \leqslant \emptyset \leqslant a$; if S is linear then these statements extend to W, X and Y as appropriate.

In Chapter 1, we mentioned that it was natural in certain applications to consider systems such as the non-negative, or non-positive, real numbers or integers. Such systems coincide with their own non-negative or non-positive cone - that is to say in addition to satisfying axiom X_8, they satisfy the following axiom X_{11} or its dual:

$$X_{11} : \ \forall x \in V, \quad x \leqslant \emptyset$$

3. OPENING AND CLOSING

3-1. The Operators $\sum_{\oplus}$, $\sum_{\oplus'}$, $\Pi_{\otimes}$ $\Pi_{\otimes'}$

In this memorandum we shall use the symbols $\sum$ and Π, precisely as in elementary algebra, to denote iterated sums and products using the operations $\oplus$, $\oplus'$, $\otimes$ and $\otimes'$. For example, if $u_1, u_2, \ldots, u_n$ are given (not necessarily distinct) elements of a commutative band, we shall write $\sum_{i=1}^{n}{}_{\oplus} u_i$ to denote the sum:

$$u_1 \oplus u_2 \oplus \ldots \oplus u_n,$$

and $\sum_{i=1}^{n}{}_{\oplus'} u_i$ to denote the sum:

$$u_1 \oplus' u_2 \oplus' \ldots \oplus' u_n$$

when the dual addition is defined.

The notation is of course a shorthand rather than an essential part of the algebra, and will be used without comment in various familiar ways. For example, in a division belt, we can use induction on (2-19) to prove:

$$\left(\sum_{j=1}^{n}{}_{\oplus} x_j \right)^{-1} = \sum_{j=1}^{n}{}_{\oplus'} x_j^{-1} \tag{3-1}$$

when multiplication is self-dual.

Any of the following notations may occur, with meanings which will be evident from the context:

$$\sum_{i \in J}{}_{\oplus} x_i \; ; \; \sum_{u \in T}{}_{\oplus'} u \; ; \; \prod_{\substack{r=1 \\ r \neq s}}^{t}{}_{\otimes} w_{rs} \; ; \; \prod_{r<s}{}_{\otimes'} t_{rs} \qquad \text{etc. etc.}$$

We now discuss some of the basic formal properties of these operators. Many of these properties are familiar from the theory of lattice polynomials, but (trivial though they are) we review them in order to establish certain general manipulation principles, to which we can appeal in later proofs.

If $y_1, \ldots, y_n$ are all equal to a constant y, then by an obvious induction using the idempotency under $\oplus$ and $\oplus'$ we have:

$$\sum_{i=1}^{n}{}_{\oplus} y_i = \sum_{i=1}^{n}{}_{\oplus'} y_i = y \tag{3-2}$$

Suppose now that $x_1, \ldots, x_n$ and $y_1, \ldots, y_n$ are (not necessarily distinct) elements of a commutative band, such that:

$$\forall i = 1, \ldots, n, \quad x_i \geqslant y_i \tag{3-3}$$

Using (2-6) we obtain by an obvious induction that:

$$\sum_{i=1}^{n}{}_{\oplus} x_i \geqslant \sum_{i=1}^{n}{}_{\oplus} y_i \tag{3-4}$$

If a dual addition is defined, we induce from (2-13) that:

$$\sum_{i=1}^{n}{}_{\oplus'} x_i \geq \sum_{i=1}^{n}{}_{\oplus'} y_i \tag{3-5}$$

If the commutative band is a belt, we induce from (2-9) that:

$$\prod_{i=1}^{n}{}_{\otimes} x_i \geq \prod_{i=1}^{n}{}_{\otimes} y_i \tag{3-6}$$

And if a dual multiplication is defined, we induce from the dual of (2-9) that:

$$\prod_{i=1}^{n}{}_{\otimes'} x_i \geq \prod_{i=1}^{n}{}_{\otimes'} y_i \tag{3-7}$$

Suppose S and T are finite subsets of a commutative band, say $S = \{x_1, \ldots, x_m\}$ and $T = \{y_1, \ldots, y_n\}$, such that $x \geq y$ for all $x \in S$, $y \in T$. If $m \leq n$, then we may "fill" S with $(n-m)$ further copies of x_1; if $m \geq n$ then we may "fill" T with $(m-n)$ further copies of y_1. Applying (3-4) and (3-5), we have:

If $x \geq y$ for all $x \in S$, $y \in T$, then $\sum_{x \in S}{}_{\oplus} x \geq \sum_{y \in T}{}_{\oplus} y$ (3-8)

and if the dual addition is defined, then also $\sum_{x \in S}{}_{\oplus'} x \geq \sum_{y \in T}{}_{\oplus'} y$. (3-9)

Further, taking $y_1 = \ldots = y_m = y$ and using (3-2):

If $x \geq y$ for all $x \in S$, then $\sum_{x \in S}{}_{\oplus} x \geq y$ (3-10)

and if the dual addition is defined, then also $\sum_{x \in S}{}_{\oplus'} x \geq y$. (3-11)

Similarly, if for some fixed x we have $x \geq y$ for all $y \in T$, then:

$$x \geq \sum_{y \in T}{}_{\oplus} y \tag{3-12}$$

and if the dual addition is defined, then also $x \geq \sum_{y \in T}{}_{\oplus'} y$. (3-13)

We remark on the following obvious but important fact. If the finite set T is itself a commutative band, then $\sum_{y \in T}{}_{\oplus} y \in T$, and if T is also closed under the dual addition then $\sum_{y \in T}{}_{\oplus'} y \in T$. In particular, these relations hold for any finite subset T of a given linear commutative band (or belt). In other words, the maximum $\sum_{\oplus}$ and the minimum $\sum_{\oplus'}$, are "attained" on a finite linear set, and the attained value is equal to one of the summands. (We shall use the expressions <u>attained maximum</u> and <u>attained minimum</u> in later proofs.)

We stress that results (3-4) to (3-13) have meaning for <u>finite</u> sets only. Extension of the results of the present chapter to infinite sets requires the introduction of some kind of completeness axiom, and will not be undertaken in the present memorandum.

3-2. The Principle of Closing

Relations (3-4) to (3-13) are of course all trivial, and (3-8) and (3-9) are weak as well as trivial. For the principal interpretation, (3-8) says, for example, that if all x's are greater than all y's then the maximum x is greater than the maximum y. Obviously, we can derive the stronger result that the minimum x is greater than the maximum y, and it is instructive to do this using the principle embodied in the following proposition, which is a verbal restatement of the inequalities (3-3) to (3-7) inclusive.

Proposition 3-1 If V is a commutative band (resp. commutative band with duality, belt or belt with duality), then an inequality, governed by a universal quantifier, may be replaced by an inequality having the operator $\sum_{\oplus}$ (resp. $\sum_{\oplus'}$, $\Pi_{\otimes}$ or $\Pi_{\otimes'}$) governing each side of the original inequality, with the index in both cases running over the same finite subset of V as for the universal quantifier. ●

We shall call this manipulation principle closing with respect to the index in question. It will of course be obvious from the context which sigma or pi we use to do the closing.

In this connection, we may note that (3-2) implies the following:

In closing an inequality with respect to a given index, using either of the operators $\sum_{\oplus}$ or $\sum_{\oplus'}$, it is only necessary to apply the operator to a given side of the inequality if the given index actually occurs therein (other than as dummy variable).

We consider first some very simple examples designed to illustrate these manipulation principles. Some much less trivial applications will then be given.

Accordingly, let S, T be finite subsets of a commutative band in which a dual addition is defined, and suppose:

$$\forall\ x \in S,\ y \in T, \qquad x \geq y$$

For each fixed $x \in S$, we may close with respect to y, to obtain:

$$\forall\ x \in S,\ x \geq \sum_{y \in T}{}_{\oplus}\ y$$

Since $\sum_{y \in T}{}_{\oplus} y$ is some fixed element, we may now close with respect to x to obtain the intuitive result:

$$\sum_{x \in S}{}_{\oplus'}\ x \geq \sum_{y \in T}{}_{\oplus}\ y \qquad (3\text{-}14)$$

As a further example, let S be any finite set. We obtain from (2-1), using the definition of $\sum_{\oplus}$:

$$\forall\ x \in S,\ \sum_{x \in S}{}_{\oplus}\ x \geq x \qquad (3\text{-}15)$$

Closing with respect to x:

$$\sum_{x \in S}{}_{\oplus}\ x \geq \sum_{x \in S}{}_{\oplus'}\ x \qquad (3\text{-}16)$$

In (3-15), we used a notational trick which we had, for clarity, hitherto avoided, but which we shall use extensively from now on - namely the same letter 'x' was used as dummy variable both for the algebraic operator $\sum_{\oplus}$ and for the universal quantifier $\forall$. This is, of course, quite permissible and enables results such as (3-16) to be directly derived in a tidy form using only one dummy variable.

The dual of (3-15) obviously holds:

$$\forall\, x \in S, \quad x \geqslant \sum_{x \in S}{}_{\oplus'}\, x \tag{3-17}$$

and combining (3-15) and (3-17):

$$\forall\, x \in S, \quad \sum_{x \in S}{}_{\oplus}\, x \geqslant x \geqslant \sum_{x \in S}{}_{\oplus'}\, x \tag{3-18}$$

which expresses the intuitive fact that S is bounded by its maximum and minimum.

The following theorems illustrate the use of the principle of closing. They also illustrate the use of a double suffix notation.

<u>Theorem 3-2</u>. <u>Let</u> w_{ij} $(i=1,\ldots,m;\ j=1,\ldots,n)$ <u>be mn arbitrary elements of a belt. Then</u> $P \geqslant Q$, <u>where</u> $P = \prod_{i=1}^{m}{}_{\otimes}\left(\sum_{j=1}^{n}{}_{\oplus}\, w_{ij}\right)$ <u>and</u> $Q = \sum_{j=1}^{n}{}_{\oplus}\left(\prod_{i=1}^{m}{}_{\otimes}\, w_{ij}\right)$. <u>Moreover, if a duality is defined, we have:</u>

$$P \geqslant Q \geqslant S \geqslant T \quad \underline{\text{and}} \quad P' \geqslant Q' \geqslant S' \geqslant T'$$

<u>where</u> $S = \sum_{j=1}^{n}{}_{\oplus'}\left(\prod_{i=1}^{m}{}_{\otimes}\, w_{ij}\right)$ <u>and</u> $T = \prod_{i=1}^{m}{}_{\otimes}\left(\sum_{j=1}^{n}{}_{\oplus'}\, w_{ij}\right)$

<u>and where P', Q', S', T' are derived from P, Q, S, T respectively by replacing $\otimes$ by $\otimes'$.</u>

<u>Proof</u>. From (3-15):

$$\forall,\ i=1,\ldots,m;\ j=1,\ldots,n, \quad \sum_{j=1}^{n}{}_{\oplus}\, w_{ij} \geqslant w_{ij}$$

Closing w.r.t i :

$$\forall\ j=1,\ldots,n, \quad \prod_{i=1}^{m}{}_{\otimes}\left(\sum_{j=1}^{n}{}_{\oplus}\, w_{ij}\right) \geqslant \prod_{i=1}^{m}{}_{\otimes}\, w_{ij}$$

Closing w.r.t.j :

$$\prod_{i=1}^{m}{}_{\otimes}\left(\sum_{j=1}^{n}{}_{\oplus}\, w_{ij}\right) \geqslant \sum_{j=1}^{n}{}_{\oplus}\left(\prod_{i=1}^{m}{}_{\otimes}\, w_{ij}\right)$$

This is exactly $P \geqslant Q$, and $S \geqslant T$ is proved similarly.

Now in (3-16) take S to be the set of products $\prod_{i=1}^{m}{}_{\otimes} w_{ij}$ (j=1,...,n), giving:

$$\sum_{j=1}^{n}{}_{\oplus} \left(\prod_{i=1}^{m}{}_{\otimes} w_{ij} \right) \geq \sum_{j=1}^{n}{}_{\oplus'} \left(\prod_{i=1}^{m}{}_{\otimes} w_{ij} \right)$$

This is exactly $Q \geq S$. Similarly, we obtain $P' \geq Q' \geq S' \geq T'$

If m=2, we can in particular cases interpolate a term between Q and S (Q' and S'), as we now show.

<u>Theorem 3-3. Suppose m=2 in Theorem 3-2. Then $Q \geq R$ and $R' \geq S'$, where:</u>

<u>either</u> $R = \left(\sum_{j=1}^{n}{}_{\oplus} w_{1j}\right) \otimes \left(\sum_{j=1}^{n}{}_{\oplus'} w_{2j}\right)$ <u>or</u> $R = \left(\sum_{j=1}^{n}{}_{\oplus'} w_{1j}\right) \otimes \left(\sum_{j=1}^{n}{}_{\oplus} w_{2j}\right)$

<u>and either</u> $R' = \left(\sum_{j=1}^{n}{}_{\oplus'} w_{1j}\right) \otimes' \left(\sum_{j=1}^{n}{}_{\oplus} w_{2j}\right)$ <u>or</u> $R' = \left(\sum_{j=1}^{n}{}_{\oplus} w_{1j}\right) \otimes' \left(\sum_{j=1}^{n}{}_{\oplus'} w_{2j}\right)$

<u>Furthermore, if the elements w_{ij} are drawn from a belt with duality which either is linear, or has self-dual multiplication, then $R \geq S$ and $Q' \geq R'$.</u>

<u>Proof</u>: There are two interpretations for R, and two for R'. Let us first consider the case:

$$R = \left(\sum_{j=1}^{n}{}_{\oplus} w_{1j}\right) \otimes \left(\sum_{j=1}^{n}{}_{\oplus'} w_{2j}\right)$$

$$= \sum_{j=1}^{n}{}_{\oplus} \left(w_{1j} \otimes \sum_{j=1}^{n}{}_{\oplus'} w_{2j}\right) \qquad \text{(using induction on } X_6\text{)} \qquad (3\text{-}19)$$

Now by (3-17):

$$\forall\, j=1,\ldots,n, \qquad w_{2j} \geq \sum_{j=1}^{n}{}_{\oplus'} w_{2j}$$

Left multiplying by w_{1j} (an isotone operation):

$$\forall\, j=1,\ldots,n, \quad w_{1j} \otimes w_{2j} \geq w_{1j} \otimes \sum_{j=1}^{n}{}_{\oplus'} w_{2j}$$

Closing w.r.t.j, we obtain $Q \geq R$. We must now prove $R \geq S$, under the given extra assumptions.

If multiplication is self dual, then a dual form of the foregoing argument will suffice. So let us assume now that the belt is linear. Then there are indices $J(1\leq J\leq n)$ and $K(1\leq K\leq n)$ such that:

$$w_{1J} = \sum_{j=1}^{n}{}_{\oplus} w_{1j} \text{ and } w_{2K} = \sum_{j=1}^{n}{}_{\oplus'} w_{2j}, \text{ so } R = w_{1J} \otimes w_{2K}.$$

Evidently $w_{1J} \geq w_{1j}$ $(j=1,\ldots,n)$, so in particular:

$$w_{1J} \geq w_{1K}$$

Hence, using (2-9):

$$w_{1J} \otimes w_{2K} \geq w_{1K} \otimes w_{2K}$$

$$\geq \sum_{j=1}^{n}{}_{\oplus'} (w_{1j} \otimes w_{2j}) \qquad \text{(by (3-17)).}$$

But this is: $R \geq S$. Hence $Q \geq R \geq S$. The other case for R is handled similarly, and similarly $Q' \geq R' \geq S'$ for both interpretations for R' ●

Recall now from Proposition 2-9 that a division belt can always be given a duality with self-dual multiplication.

<u>Theorem 3-4</u>. <u>In a division belt, we have for all</u> a_j, b_j <u>$(j=1,\ldots,n)$</u>:

$$\sum_{j=1}^{n}{}_{\oplus'} (a_j \otimes b_j^{-1}) \leq (\sum_{j=1}^{n}{}_{\oplus} a_j) \otimes (\sum_{j=1}^{n}{}_{\oplus} b_j)^{-1} \leq \sum_{j=1}^{n}{}_{\oplus} (a_j \otimes b_j^{-1})$$

<u>and</u>

$$\sum_{j=1}^{n}{}_{\oplus'} (a_j^{-1} \otimes b_j) \leq (\sum_{j=1}^{n}{}_{\oplus} a_j)^{-1} \otimes (\sum_{j=1}^{n}{}_{\oplus} b_j) \leq \sum_{j=1}^{n}{}_{\oplus} (a_j^{-1} \otimes b_j)$$

<u>Proof</u>. The first pair of relations follow from Theorem 3-3 on writing $w_{1j} = a_j$ and $w_{2j} = b_j^{-1}$ and using (3-1). The second pair are proved similarly. ●

<u>3-3. The Principle of Opening</u>. Relation (3-15) suggests another principle of manipulation: <u>an expression governed by an operator $\sum_{\oplus}$ is decreased if the operator is removed</u>. Similarly, from (3-17): <u>an expression governed by an operator $\sum_{\oplus'}$ is increased if the operator is removed</u>.

We may take this further. Suppose in (3-15) we apply separately to the two expressions x and $\sum_{x\in S}{}_{\oplus} x$ identical sequences of translations, left and right multiplications, dual translations and dual left and right multiplications.

The validity of the inequality $\geq$ is preserved because, as observed in Proposition 2-8, multiplications and translations and their duals are isotone. We arrive in each case at an inequality $\alpha \geq \beta$ between two expressions such that β can be formally obtained from α by deleting the operator $\sum_{\oplus}$. We infer the manipulation principles embodied in the following proposition.

Proposition 3-5. Let an algebraic expression β be obtained from a (well-formed) algebraic expression α by deleting an operator $\sum_{\oplus}$ which does not form a part of the argument of any inversion operation $(\ )^{-1}$. Write an inequality: $\alpha \geqslant \beta$, preceded by a universal quantifier whose index runs over the same (finite) set as for the operator $\sum_{\oplus}$. Then this inequality is valid for the commutative band (resp. belt, with or without duality) for which the expression α was written.

Dually, let an algebraic expression β be obtained from a (well-formed) algebraic expression α by deleting an operator $\sum_{\oplus'}$, which does not form a part of the argument of any inversion operation $(\ )^{-1}$. Write an inequality: $\alpha \leqslant \beta$ preceded by a universal quantifier whose index runs over the same (finite) set as for the operator $\sum_{\oplus'}$. Then this inequality is valid for the commutative band (resp. belt) with duality, for which the expression α was written. ●

For example, suppose in (3-15) that we add a given element a, using $\oplus$, left-multiply by an element b, using $\otimes$, and add an element c, using $\oplus'$. We get:

$$\forall\, x \in S,\ (b \otimes [(\sum_{x\in S}{}_{\oplus}\, x) \oplus a]) \oplus' c \geqslant (b \otimes [x \oplus a]) \oplus' c$$

In the practical application of these principles, we must take note of the proviso that the operator does not fall within the scope of some inversion operation, because inversion is not an isotone operation, as (2-10) shows.

Thus:

$$\forall\, x \in S,\ (a^{-1} \oplus' b)^{-1} \oplus (c^{-1} \otimes \sum_{x\in S}{}_{\oplus} x) \geqslant (a^{-1} \oplus' b)^{-1} \oplus (c^{-1} \otimes x)$$

represents a legitimate application of the principle of opening but:

$$\forall\, x \in S,\ (a^{-1} \oplus' b)^{-1} \oplus (c^{-1} \otimes \sum_{x\in S}{}_{\oplus} x)^{-1} \geqslant (a^{-1} \oplus' b)^{-1} \oplus (c^{-1} \otimes x)^{-1}$$

certainly does not.

(However, the dual translation $x \to a \oplus' x$ is of course isotone even if the definition of $\oplus'$ explicitly involves inversion, as in (2-19)).

Application of either half of Proposition 3-5 will be called opening with respect to the index in question. A fully rigorous discussion of these principles obviously demands a metamathematical apparatus, but this seems unnecessarily tedious in the present context. In the situations where we use the principles, it will always be clear that the procedures are well-defined and effective. The following theorem contains a typical application of Proposition 3-5.

Theorem 3-6. Let a_{ij} $(i=1,\ldots,m;\ j=1,\ldots,n)$ and b_i $(i=1,\ldots,m)$ be $m(n+1)$ given elements from a division belt. Then $P \leqslant \emptyset$ where:

$$P = \sum_{i=1}^{m}{}_{\oplus} \left(\sum_{j=1}^{n}{}_{\oplus} \left[a_{ij} \otimes \sum_{k=1}^{m}{}_{\oplus'} (a_{kj}^{-1} \otimes b_k)\right] \otimes b^{-1}\right)$$

<u>If the division belt is linear, then $P = \emptyset$.</u>

<u>Proof</u>. Consider $$\left[a_{ij} \otimes \sum_{k=1}^{m}{}_{\oplus'} (a_{kj}^{-1} \otimes b_k)\right] \otimes b_i^{-1}$$

Opening w.r.t. k:

$$\forall\ i=1,\ldots, m;\ j=1,\ldots, n;\ k=1,\ldots, m,\ \left[a_{ij} \otimes \sum_{k=1}^{m}{}_{\oplus'} (a_{kj}^{-1} \otimes b_k)\right] \otimes b_i^{-1} \leqslant a_{ij} \otimes a_{kj}^{-1} \otimes b_k \otimes b_i^{-1}$$

Taking in particular k=i:

$$\forall\ i=1,\ldots, m\ j=1,\ldots, n:\ \left[a_{ij} \otimes \sum_{k=1}^{m}{}_{\oplus'} (a_{kj}^{-1} \otimes b_k)\right] \otimes b_i^{-1} \leqslant \emptyset$$

Closing w.r.t. i and j, we obtain $P \leqslant \emptyset$.

On the other hand, opening P with respect to i and j we have:

$$\forall\ i=1,\ldots,m;\ j=1,\ldots,n,\ P \geqslant \left[a_{ij} \otimes \sum_{k=1}^{m}{}_{\oplus'} (a_{kj}^{-1} \otimes b_k)\right] \otimes b_i^{-1}.$$

Now, for a <u>linear</u> division belt, the sum in the last expression is equal to one of the summands. Hence in this case:

$$\forall\ i=1,\ldots,m;\ j=1,\ldots,n,\ P \geqslant a_{ij} \otimes a_{k(j)j}^{-1} \otimes b_{k(j)} \otimes b_i^{-1}$$

where k(j) is the index for which $a_{kj}^{-1} \otimes b_k$ achieves its minimum for given j. Choosing i=k(j) for each j, we have $P \geqslant \emptyset$, so $P = \emptyset$ in this case. ●

4. THE PRINCIPAL INTERPRETATION

4-1. Blogs

In Chapter 2, we defined the algebraic structures called belts and considered a number of different particular forms of such structures, in particular the division belts. We now define a related type of belt, which we shall call a blog. Division belts and blogs will be the two most frequently occurring types of belt in the sequel, embracing as they do the real numbers and extended real numbers respectively.

Accordingly, let $(G,\oplus,\otimes)$ be a division belt. We now progressively extend $(G,\oplus,\otimes)$ as follows. Firstly, by use of (2-19) we introduce a dual addition $\oplus'$, so that as discussed in Section 2-8, G becomes a lattice-ordered group, or equivalently a division belt having a duality with self-dual multiplication.
We now define a system W, constructed by adjoining universal bounds to G, i.e. $W = G \cup \{-\infty\} \cup \{+\infty\}$ where $-\infty < x < +\infty$ for all $x \in G$, and $+\infty$, $-\infty$ are the adjoined elements.

As remarked in Section 2-8, G is a distributive lattice, and it is easily verified that W is then a distributive lattice also. We now introduce an operation $\otimes$ as follows:

$$\left.\begin{array}{l} \text{If } x, y \in G \text{ then } x \otimes y \text{ is already defined in } (G,\otimes) \\ \text{The products } -\infty \otimes +\infty \text{ and } +\infty \otimes -\infty \text{ both equal } -\infty \\ \text{If } x \in G \cup \{-\infty\} \text{ then } x \otimes -\infty = -\infty \otimes x = -\infty \text{ by definition} \\ \text{If } x \in G \cup \{+\infty\} \text{ then } x \otimes +\infty = +\infty \otimes x = +\infty \text{ by definition} \end{array}\right\} \qquad (4\text{-}1)$$

If we call the elements of G finite elements, then definitions (4-1) extend the scope of the operation $\otimes$ by means of the rule "finite element times infinite element = infinite element" together with the conventional rule" $-\infty$ times $+\infty$ = $-\infty$". This last ensures that $-\infty$ acts as a null element in the entire system $(W,\oplus,\otimes)$ but constitutes an asymmetry between $-\infty$ and $+\infty$ which we redress by introducing a dual multiplication $\otimes'$ which acts exactly like $\otimes$ except that we stipulate: $-\infty \otimes' +\infty = +\infty \otimes' -\infty = +\infty$. (See [37] where this idea is presented, but in a different notation).
For convenience we set out all the definitions in tabular form below.

Fig. 4-1 A Blog

x	y	$x \oplus y$	$x \oplus' y$	$x \otimes y$	$x \otimes' y$
$-\infty$	$-\infty$	$-\infty$	$-\infty$	$-\infty$	$-\infty$
$-\infty$	$y \in G$	y	$-\infty$	$-\infty$	$-\infty$
$-\infty$	$+\infty$	$+\infty$	$-\infty$	$-\infty$	$+\infty$
$x \in G$	$-\infty$	x	$-\infty$	$-\infty$	$-\infty$
$x \in G$	$y \in G$	$x \oplus y$	$(x^{-1} \oplus y^{-1})^{-1}$	$x \otimes y$	$x \otimes y$
$x \in G$	$+\infty$	$+\infty$	x	$+\infty$	$+\infty$
$+\infty$	$-\infty$	$+\infty$	$-\infty$	$-\infty$	$+\infty$
$+\infty$	$y \in G$	$+\infty$	y	$+\infty$	$+\infty$
$+\infty$	$+\infty$	$+\infty$	$+\infty$	$+\infty$	$+\infty$

We call a system $(W,\oplus,\otimes,\oplus',\otimes')$ constructed this way a blog (acronym for Bounded Lattice-Ordered Group) and we refer to G as the group of the blog W. Thus the finite elements in W are just the elements of the group of W.

As usual, we shall denote the identity element in the group G by $\emptyset$. From Fig. 4-1 we see that $\emptyset$ is an identity element for the whole of W with respect to both $\otimes$ and $\otimes'$. The inverse of an element $x \in G$ is as usual written x^{-1}. In many practical applications, multiplication in G will be commutative, but we do not need to assume this yet in the abstract argument. Most of our results are essentially extendable to any lattice-ordered semigroup which satisfies any of the criteria for embeddability in a group (see e.g. [27]).

By a direct verification of the relevant axioms, and an appeal to the discussion of Section 2-8, we can confirm the following:

Proposition 4-1. The system $(W, \oplus, \otimes, \oplus', \otimes')$ is a belt having a duality, and containing as sub-belts the systems $(G \cup \{-\infty\}, \oplus, \otimes, \oplus')$ and $(G \cup \{+\infty\}, \oplus, \otimes, \oplus')$, which are belts having a duality with self-dual multiplication, and the system $(G, \oplus, \otimes, \oplus')$ which is a division belt having a duality (with self-dual multiplication). In all these systems, $\emptyset$ acts as identity element with respect to both $\otimes$ and $\otimes'$, whilst $+\infty$ and $-\infty$ (when present) acts as null-elements with respect to $\otimes$ and $\otimes'$ respectively. ●

Notice that we do not exclude the possibility that G may be the trivial group $\{\emptyset\}$, in which case the belts mentioned in Proposition 4-1 may well be degenerate, or have trivial dualities.

By a direct consideration of cases, we can confirm that a blog satisfies the following "associative inequalities". These will be important in our later discussions, when we shall often refer to them as axiom X_{12}.

Proposition 4-2. Let $W(\oplus, \otimes, \oplus', \otimes')$ be a blog. Then for all $x, y, z \in W$ we have:

$$X_{12}: \quad \left.\begin{aligned} x \otimes (y \otimes' z) &\leq (x \otimes y) \otimes' z \\ x \otimes' (y \otimes z) &\geq (x \otimes' y) \otimes z \end{aligned}\right\} \qquad (4\text{-}2)$$ ●

(Inequalities X_{12} occur, in another notation, in [37])

For future convenience, we record the following trivial results also.

Proposition 4-3. Let $(W, \oplus, \otimes, \oplus', \otimes')$ be a blog and let $x, y \in W$. Then: If $x \otimes y = +\infty$ or $x \oplus y = +\infty$ then either $x = +\infty$ or $y = +\infty$ (or both). If $x \otimes y = -\infty$ then either $x = -\infty$ or $y = -\infty$ (or both). If $x \oplus y = -\infty$ then both $x = -\infty$ and $y = -\infty$ ●

4-2. The Principal Interpretation

The blog which arises when G is taken as the linearly ordered arithmetically additive group F of finite real numbers will be called the principal interpretation (of axioms X_1 to X_6, X_1' to X_6' and L_3 of Fig 2-1). Thus the principal interpretation is $(R, \max, \otimes, \min, \otimes')$ where $R = F \cup \{-\infty\} \cup \{+\infty\}$ (the set of extended real numbers). Evidently the following holds.

Proposition 4-4. The principal interpretation is a linear belt and satisfies X_7, X_8,

No.	Element-set	Type	Duality	X_7	X_8	X_9	X_{10}	X_{11}	X_{12}
1	Reals with $\pm\infty$								
2	Rationals with $\pm\infty$	Blog	Yes	✓	✓		✓		✓
3	Integers with $\pm\infty$								
4	Reals	Division Belt	Yes, with self-dual multiplication						
5	Rationals			✓	✓	✓			✓
6	Integers								
7	Neg.reals with $-\infty$								
8	Neg.rationals with $-\infty$	Belt	Yes	✓			✓		✓
9	Neg.integers with $-\infty$								
10	Non-pos.reals with $-\infty$								
11	Non-pos.rationals with $-\infty$	Belt	Yes	✓	✓		✓	✓	✓
12	Non-pos.integers with $-\infty$								
13	Pos.reals with $+\infty$								
14	Pos.rationals with $+\infty$	Belt	Yes	✓					✓
15	Pos.integers with $+\infty$								
16	Non-neg.reals with $+\infty$								
17	Non-neg.rationals with $+\infty$	Belt	Yes	✓	✓				✓
18	Non-neg.integers with $+\infty$								

Fig 4-2. Some sub-belts of the Principal Interpretation

X_{10}, X'_{10} and X_{12}

If we take a particular subgroup of the arithmetically additive real numbers, we can create a blog which is a sub-blog of R (i.e. a sub-belt which is itself a blog). Or we may take a subsemigroup of (F,+), say the non-negative integers, extend this to a subgroup of (F,+) and then to a sub-blog of R. It is clear that the principal interpretation has a number of naturally occurring sub-belts ([17],[21]), some of which are listed in Fig 4-2. All these systems take $\oplus$ as max, $\oplus'$ as min, and $\otimes$, $\otimes'$ as defined from arithmetical addition following Fig 4-1. System 1 in the table is the principal interpretation itself. Systems 7 to 12 obviously have a special relationship with Systems 13 to 18 respectively, to which we shall return when we discuss conjugacy in Chapter 7. Evidently we could extend the table by listing all the systems which arise by adjoining, and/or deleting, $\pm\infty$ from Systems 7 to 18.

The practical problems discussed in Section 1-2 can all be formulated using an appropriate system from Fig 4-2. It will often be natural to start with a system other than System 1, if the physical nature of the problem implies a restriction to whole numbers, or positive numbers, but since all these systems can be embedded in (a sub-blog of) R, it is clear that the properties of blogs in general, and R in particular, are of central importance in the theory.

4-3. The 3-element Blog ③

In section 1-2.6, we considered some Boolean problems, which do not receive a natural expression in any of the systems in Fig 4-2. However, the 2-element Boolean algebra ② can be embedded in any blog W since it is evidently isomorphic to the system consisting of the two elements $-\infty$, $\emptyset \in W$ together with the functions $\oplus$, $\otimes$ restricted to these elements.

We may thus embed the Boolean algebra ② in the principal interpretation, for example, but a rather more convenient choice is the 3-element blog ③, which is by definition a system having three elements $-\infty$, $\emptyset$, $+\infty$, with the operations specified in Fig 4-3.

x	y	$x \oplus y$	$x \oplus' y$	$x \otimes y$	$x \otimes' y$
$-\infty$	$-\infty$	$-\infty$	$-\infty$	$-\infty$	$-\infty$
$-\infty$	$\emptyset$	$\emptyset$	$-\infty$	$-\infty$	$-\infty$
$-\infty$	$+\infty$	$+\infty$	$-\infty$	$-\infty$	$+\infty$
$\emptyset$	$-\infty$	$\emptyset$	$-\infty$	$-\infty$	$-\infty$
$\emptyset$	$\emptyset$	$\emptyset$	$\emptyset$	$\emptyset$	$\emptyset$
$\emptyset$	$+\infty$	$+\infty$	$\emptyset$	$+\infty$	$+\infty$
$+\infty$	$-\infty$	$+\infty$	$-\infty$	$-\infty$	$+\infty$
$+\infty$	$\emptyset$	$+\infty$	$\emptyset$	$+\infty$	$+\infty$
$+\infty$	$+\infty$	$+\infty$	$+\infty$	$+\infty$	$+\infty$

Fig 4-3

Evidently, the group of the blog ③ is the trivial group $\{\emptyset\}$.

Theorem 4-5. The 3-element blog ③ can be embedded isomorphically in any blog.

Proof. If W is a blog then W contains at least three distinct elements $-\infty$, ϕ, $+\infty$. By comparing the tables of Figs 4-1 and 4-3, we see that the system which arises when the operations in W are restricted to these three elements, is isomorphic to ③. ●

We show now that ③ is the only blog having a finite number of elements. First we record a simple lemma, for convenience in later arguments.

Lemma 4-6. If u, v are elements of a division belt, or finite elements of a blog, then $u < v$ if and only if $u^{-1} > v^{-1}$.

Proof. If $u < v$, then by definition $u \neq v$, but $u^{-1} \geq v^{1}$ by (2-10). However $u^{-1} \neq v^{-1}$ since $u \neq v$. Hence $u^{-1} > v^{-1}$. Similarly for the converse. ●

Theorem 4-7. Let G be a division belt other than the one-element division belt $\{\phi\}$. Then for each $u \in G$ we can find $v \in G$ such that $v > u$.

Proof. By hypothesis we can find $\alpha \in G$ with $\alpha \neq \phi$. Now $\alpha \oplus \phi \geq \phi$, giving two possibilities:

(i) $\alpha \oplus \phi > \phi$

(ii) $\alpha \oplus \phi = \phi$, so $\phi \geq \alpha$. But $\phi \neq \alpha$, whence $\phi > \alpha$.

In case (i), define $\beta = \alpha \oplus \phi$; in case (ii) define $\beta = \alpha^{-1}$. Then in either case (using Lemma 4-6 for case (ii)) we have found $\beta > \phi$. If now $u \in G$ is arbitrary we have by (2-9):

$$\beta \otimes u \geq \phi \otimes u = u$$

But we cannot have $\beta \otimes u = u$, else:

$$\beta = (\beta \otimes u) \otimes u^{-1} = \phi, \text{ contradicting } \beta > \phi.$$

Hence $\beta \otimes u > u$. ●

Corollary 4-8. The only division belt having a finite number of elements is the one-element division belt $\{\phi\}$ and the only blog having a finite number of elements is the 3-element blog ③ .

Proof. If G is a division belt having a finite number of elements, define:

$$u = \sum_{x \in G} {}_{\oplus}\, x.$$

So $u \geq x$ for all $x \in G$. Hence we cannot find $v \in G$ with $v > u$, so by Theorem 4-7, G cannot be other than $\{\phi\}$. Finally if G is the group of a blog W having a finite number of elements, then G is $\{\phi\}$, so W is ③. ●

Recall that axiom X_{10} (Section 2-6) asserts the existence of a null element in a given belt $(V, \oplus, \otimes)$:

X_{10}: $\exists\ \theta \in V$ such that for all $x \in V$ there holds:

$$x \oplus \theta = x \text{ and } x \otimes \theta = \theta \otimes x = \theta$$

Corollary 4-9. The only division belt satisfying axiom X_{10} is the one-element division belt $\{\phi\}$.

Proof. If $(G, \oplus, \otimes)$ is a division belt with a null element u^{-1}, then we cannot find $v^{-1} \varepsilon G$ such that $v^{-1} < u^{-1}$. Hence we cannot find $v \varepsilon G$ such that $v > u$, whence G cannot be other than $\{\emptyset\}$. And clearly $\{\emptyset\}$ has the null element $\emptyset$. ●

4-4. Further Properties of Blogs

Suppose we write a triple product $x \otimes y \otimes z$. By introducing brackets, we can make two different expressions, namely $(x \otimes y) \otimes z$ and $x \otimes (y \otimes z)$. If we now are at liberty to add a prime to both, to either or to neither occurrence of the symbol $\otimes$, (thus making the symbol $\otimes'$), we arrive at altogether eight different formal triple products. In a division belt, by virtue of the associative law and of the self-duality of multiplication, all eight expressions are equal for any x, y, z. In a blog, however, this is not the case, although a definite structure of inequalities holds among the expressions, as the following result shows.

Theorem 4-10. Let $(W,\oplus,\otimes,\oplus',\otimes')$ be a blog and let $x, y, z \varepsilon W$. Then:

$$\begin{array}{ccccccc} (x \otimes' y) \otimes' z & \geqslant & (x \otimes y) \otimes' z & \geqslant & x \otimes (y \otimes' z) & \geqslant & x \otimes (y \otimes z) \\ \| & & & & & & \| \\ x \otimes' (y \otimes' x) & \geqslant & x \otimes' (y \otimes z) & \geqslant & (x \otimes' y) \otimes z & \geqslant & (x \otimes y) \otimes z \end{array}$$

Proof. Certainly, $(y \otimes' z) \geqslant (y \otimes z)$ since these products are actually equal in all cases except where one of y, z is $+\infty$ and the other is $-\infty$, when we have $(y \otimes' z) = +\infty > -\infty = (y \otimes z)$. Hence by (2-9):

$$x \otimes (y \otimes' z) \geqslant x \otimes (y \otimes z)$$

Similarly $(x \otimes' y) \otimes' z \geqslant (x \otimes y) \otimes' z$

And these results, coupled with inequalities X_{12}, prove the first row of inequalities in the theorem. The second row is proved similarly. The equality of the first terms, and of the last terms, follows from the associative law. ●

Our next two results will be useful in future manipulative arguments.

Lemma 4-11. Let $(W,\oplus,\otimes,\oplus',\otimes')$ be a blog. If $u_1, u_2, v_1, v_2 \varepsilon W$ are all finite and are such that either $u_1 \geqslant v_1$ and $u_2 > v_2$, or $u_1 > v_1$ and $u_2 \geqslant v_2$, then $u_1 \otimes u_2 > v_1 \otimes v_2$.

Proof. Suppose $u_1 \geqslant v_1$ and $u_2 > v_2$. Evidently $u_1 \otimes u_2 \geqslant v_1 \otimes v_2$. Suppose in fact that $u_1 \otimes u_2 = v_1 \otimes v_2$. From the fact that $u_1 \geqslant v_1$ we have, on left-multiplying by u_1^{-1}, that $\emptyset \geqslant u_1^{-1} \otimes v_1$. Hence:

$$u_2 = u_1^{-1} \otimes (u_1 \otimes u_2) = u_1^{-1} \otimes (v_1 \otimes v_2) = (u_1^{-1} \otimes v_1) \otimes v_2 \leqslant \emptyset \otimes v_2 = v_2$$

But this contradicts $u_2 > v_2$. Hence $u_1 \otimes u_2$ cannot equal $v_1 \otimes v_2$ so the only possiblity is that $u_1 \otimes u_2 > v_1 \otimes v_2$. The other case is proved similarly. ●

Corollary 4-12. Let $(W,\oplus,\otimes,\oplus',\otimes')$ be a blog. If $u_1,\ldots, u_n, v_1,\ldots, v_n \in W$ are all finite and $u_i \geq v_i$ $(i=1,\ldots, n)$ with $u_j > v_j$ for some j $(1 \leq j \leq n)$, then:

$$\prod_{i=1}^{n}{}_{\otimes} u_i > \prod_{i=1}^{n}{}_{\otimes} v_i$$

Hence if $v_i \leq \emptyset$ $(i=1,\ldots, n)$ and $\prod_{i=1}^{n}{}_{\otimes} v_i = \emptyset$ then $v_i = \emptyset$ $(i=1,\ldots, n)$

Proof. The first assertion follows from Lemma 4-11 by an obvious iteration, and the second follows by then taking $u_1 = \ldots = u_n = \emptyset$. ●

Finally, the following result will be useful when we discuss conjugation later.

Lemma 4-13. Let $(W,\oplus,\otimes,\oplus',\otimes')$ be a blog. For each $x \in W$, define $x^* \in W$ as follows:

If x is finite then $x^* = x^{-1}$

If $x = -\infty$ then $x^* = +\infty$

If $x = +\infty$ then $x^* = -\infty$

Then: (i) $\forall\, x \in W$, $x \otimes x^* = x^* \otimes x$
$= \emptyset$ if x is finite and $-\infty$ otherwise.

(ii) $\forall\, x \in W$, $x \otimes' x^* = x^* \otimes' x$
$= \emptyset$ if x is finite and $+\infty$ otherwise.

(iii) $\forall\, x, y \in W$, the following relations are all equivalent:

$x \geq y$, $x^* \otimes y \leq \emptyset$; $y \otimes x^* \leq \emptyset$; $x \otimes' y^* \geq \emptyset$; $y^* \otimes' x \geq \emptyset$

Proof. Relations (i) and (ii) follow directly from the definition of a blog. Hence we may write: $x \otimes x^* = x^* \otimes x \leq \emptyset \leq x \otimes' x^* = x^* \otimes' x$ (4-3)

Now if $x \geq y$, we may left-multiply by x^* to give:

$$x^* \otimes y \leq x^* \otimes x$$
$$\leq \emptyset \qquad \text{(by (4-3))}$$

Conversely, if $x^* \otimes y \leq \emptyset$ we may left-multiply by x to give:

$$x \otimes'(x^* \otimes y) \leq x \tag{4-4}$$

However, $x \otimes'(x^* \otimes y) \geq (x \otimes' x^*) \otimes y$ (by Proposition 4-2)
$\geq \emptyset \otimes y$ (by (4-3) and (2-9))
$= y$ (4-5)

Evidently, (4-4) and (4-5) imply $x \geq y$. Hence $x \geq y$ if and only if $x^* \otimes y \leq \emptyset$. The equivalence of the other relations is proved similarly. ●

5. THE SPACES E_n AND $\mathcal{M}_{mn}$

5-1. Band-Spaces

Let $V = (V, \oplus, \otimes)$ be a belt. Let $T = (T, \oplus)$ be a given system and let there be defined a right-multiplication of elements of T by elements of V, that is to say a binary combining operation $\otimes$ such that $x \otimes \lambda \in T$ for each pair x, λ with $x \in T$ and $\lambda \in V$. We shall say that $(T, \oplus)$ (or just T) is a right band-space, or briefly a space (over $(V, \oplus, \otimes)$ or briefly over V) if the following axioms hold:

S_1: $(T, \oplus)$ is a commutative band

S_2: $\forall\, x \in T$ and $\forall\, \lambda, \mu \in V$, $\quad (x \otimes \lambda) \otimes \mu = x \otimes (\lambda \otimes \mu)$

S_3: $\forall\, x, y \in T$ and $\forall\, \lambda \in V$, $\quad (x \oplus y) \otimes \lambda = (x \otimes \lambda) \oplus (y \otimes \lambda)$

S_4: $\forall\, x \in T$ and $\forall\, \lambda, \mu \in V$, $\quad x \otimes (\lambda \oplus \mu) = (x \otimes \lambda) \oplus (x \otimes \mu)$

Such spaces, then, play the role of vector-spaces in our theory. We have followed the usual algebraic practice of using the same symbols $\oplus$ and $\otimes$ to denote operations within V, within T and between T and V. We assume further:

S_5: If V has an identity element ϕ then:

$$\forall\, x \in T, \quad x \otimes \phi = x$$

A subspace is of course a subset of a space which is itself a space over the same belt V under the same combining operations. The following results are immediate deductions from axioms S_1 to S_4 and the general arguments of Section 2-5.

Proposition 5-1. Let $(V, \oplus, \otimes)$ be a belt. Then $(V, \oplus)$ is a space over $(V, \oplus, \otimes)$. If $(T, \oplus)$ is any space over $(V, \oplus, \otimes)$ then $(V, \oplus, \otimes)$ has a representation as a belt of endomorphisms of $(T, \oplus)$. Finally if $(W, \oplus, \otimes)$ is a sub-belt of $(V, \oplus, \otimes)$ then $(T, \oplus)$ is also a space over $(W, \oplus, \otimes)$. ●

As a non-trivial example of a space, let V receive the principal interpretation: $(V, \oplus, \otimes)$ is $(R, \max, +)$. Let $(T, \oplus)$ be $(R^R, \max)$, ie the set of all extended real-valued functions of one extended real variable. For $f \in T$ and $\lambda \in E_1$ define $f \otimes \lambda$ as the "vertical translation" of f by λ, ie. $(f \otimes \lambda)(x) = f(x) + \lambda$. Then $(T, \oplus)$ is a space over $(V, \oplus, \otimes)$.

If $(S, \oplus)$ and $(T, \oplus)$ are both spaces over V, we shall call a homomorphism $g: (S, \oplus) \to (T, \oplus)$ right-linear (over V) if it satisfies:

$$\forall\, x \in S \text{ and } \forall\, \lambda \in V, \quad g(x \otimes \lambda) = g(x) \otimes \lambda \tag{5-1}$$

The set of all right-linear homomorphisms of $(S, \oplus)$ to $(T, \oplus)$ over V will be called $\text{Hom}_V(S, T)$. Using familiar arguments, we easily establish the following.

Proposition 5-2. Let $(S, \oplus)$, $(T, \oplus)$ be given spaces over a belt $(V, \oplus, \otimes)$. Then $(\text{Hom}_V(S, T), \oplus)$ is a commutative band when $\oplus$ is defined as in (2-2). ●

5-2. Two-Sided Spaces

If $(V, \oplus, \otimes)$ is a belt and $(T, \oplus)$ a given system, let there be defined a left-multiplication of elements of T by elements of V, that is to say a binary combining operation $\otimes$ such that $\lambda \otimes x \in T$ for each pair $\lambda \in V$ and $x \in T$. It is clear that we may formulate "left" variants of S_1 to S_5 and define thereby the axioms of a left band-space (over V). All the arguments of Section 5-1 obviously then go through mutatis mutandis. In particular, we may define left-linear homomorphisms of left band spaces by analogy with (5-1).

As an example of a left band space, let $(T, \oplus)$ be the set of all extended real-valued functions of one extended real variable, under max, and let $(V, \oplus, \otimes)$ be the belt of all monotone non-decreasing extended real-valued functions of one extended real variable, under max and composition. For $\lambda \in V$ and $f \in T$ define $\lambda \otimes f \in T$ as a function such that $(\lambda \otimes f)(x) = \lambda(f(x))$ for all extended reals x. Then $(T, \oplus)$ is a left band-space over V.

The following easily established result gives a further example.

Proposition 5-3. Let $(T, \oplus)$ be a right band-space over a belt $(V, \oplus, \otimes)$. Then $(\text{Hom}_V(T,T), \oplus, \otimes)$ is a belt when $\oplus$ is defined as in (2-2) and $\otimes$ denotes composition; and $(T, \oplus)$ is a left band-space over this belt when for each $g \in \text{Hom}_V(T, T)$ and $x \in T$ we define $g \otimes x$ to be $g(x)$. ●

Notice that we shall allow only the epithet "right" to be tacit: "space", without further modifiers, will always mean "right band space", never "left band space"

The foregoing result leads us naturally to the following structure.

A 2-sided space is a triple (L, T, R) such that the following axioms hold:

T_1: L is a belt and T is a left band-space over L.

T_2: R is a belt and T is a right band-space over R.

T_3: $\forall \lambda \in L$, $\forall x \in T$ and $\forall \mu \in R$:

$$\lambda \otimes (x \otimes \mu) = (\lambda \otimes x) \otimes \mu \tag{5-2}$$

Evidently (5-1) is the particular form of (5-2) for the 2-sided space

$$(\mathrm{Hom}_V\,(S,\,S),\,S,\,V).$$

As an easy consequence of the axioms T_1 to T_3 and the general arguments of Section 2-5, we record the following result.

<u>Proposition 5-4.</u> <u>Let L, R be belts and T a commutative band. Then the following statements are equivalent:</u>

(i) <u>(L, T, R) is a 2-sided space.</u>

(ii) <u>T is a right band-space over R and L has a representation as a sub-belt of $(\mathrm{Hom}_R\,(T,T),\,\oplus,\otimes)$.</u> ●

For a given 2-sided space (L, T, R), axiom S_3 implies that right-multiplication by an element of R constitutes an endomorphism of T, and so is isotone. Similarly, left-multiplication by an element of L is an isotone function on T to T. Then the following <u>principle of isotonicity for 2-sided spaces</u> follows.

<u>Proposition 5-5.</u> <u>Let (L, T, R) be a 2-sided space. Let K consist of the identity mapping i_T together with the addition $\oplus$ (defined on T^2 to T), all translations of T, all left multiplications of T by elements of L and all right multiplications of T by elements of R. Then $\hat{K}$, the composition algebra generated by K, consists entirely of isotone functions. Moreover, if T is actually a belt then we may adjoin its multiplication $\otimes$ to K and $\hat{K}$ will still consist entirely of isotone functions.</u> ●

We remark that any right band-space or left band-space may be construed as a 2-sided space by supplying the missing belt in the form of a one-element belt containing only the identity mapping. Hence the above proposition implies an isotonicity principle for these structures also.

<u>5-3. Function Spaces.</u>

Let $E_1 = (E_1, \oplus, \otimes)$ be a given belt. An important class of spaces over E_1 is the class of <u>function spaces</u>, that is to say spaces in which the commutative band $(T, \oplus)$ is actually the commutative band $((E_1)^U, \oplus)$ of all functions from some given set U to E_1, with addition defined as in (2-2). Such spaces are naturally 2-sided since for $f \in (E_1)^U$ and $\lambda \in E_1$ we may defined $f \otimes \lambda$, $\lambda \otimes f \in (E_1)^U$ respectively as functions such that:

$$\forall\, x \in U, \quad \begin{cases} (f \otimes \lambda)\,(x) = f(x) \otimes \lambda \\ (\lambda \otimes f)\,(x) = \lambda \otimes f(x) \end{cases}$$

Axioms T_1, T_2, T_3 are readily verified for the 2-sided space $(E_1,(E_1)^U,E_1)$ and so the representation and isotonicity properties implied by Propositions 5-4 and 5-5 hold good.

When discussing function spaces over a belt E_1, we shall often call the elements $\lambda \in E_1$ <u>scalars</u>. The multiplications $f \mapsto f \otimes \lambda$ and $f \mapsto \lambda \otimes f$ will then be called <u>(right) scalar multiplications</u> and <u>left scalar multiplications</u> respectively.

The system $(R^R, \max)$ considered as an example in Section 5-1 is of course a function space and function spaces of this sort arise in the applications discussed in Section 1-3; indeed our aim in developing a basic formalism in such general terms is to embrace these applications in future publications. In the present publication, however, we shall be concentrating upon function spaces where the set U is finite. Such a space $(T, \oplus)$ thus takes T as $(E_1)^n$ for some integer $n \geq 1$, but for convenience we shall write E_n instead of $(E_1)^n$. Evidently E_1 itself is formally identical with $(E_1)^1$. Our spaces E_n are thus <u>spaces of n-tuples</u>.

Elements of E_n will generally be denoted by underlined lower-case letters, thus: $\underline{x} \in E_n$. The value taken by the function $\underline{x} \in E_n$ at a particular integer $j(1 \leq j \leq n)$ is of course the jth component of $\underline{x}$, denoted by $\{\underline{x}\}_j$. This latter notation is useful for extracting the jth component of an n-tuple defined by a formula as in eg.: $\{\underline{a} \oplus (\lambda \otimes \underline{b})\}_j$.

Related to the spaces of n-tuples over E_1 are the <u>spaces of matrices</u>, defined as follows. Let $m, n \geq 1$ be given integers. By an <u>$(m \times n)$ matrix (over E_1)</u> we understand (as usual) a rectangular array of elements of E_1 arranged in m rows and n columns:

$$\begin{bmatrix} a_{11}, & \cdots, & a_{1n} \\ & \cdots & \\ a_{m1}, & \cdots, & a_{mn} \end{bmatrix}$$

If $m = n$, we shall speak of a <u>square matrix</u> (of <u>order</u> n).

Notations such as $\underline{A}$ or $[a_{ij}]$ will be used to denote matrices in the usual way, and the notation $\{\underline{A}\}_{ij}$ will mean: the element in row i and column j of matrix $\underline{A}$. Thus $\underline{A} = [\{\underline{A}\}_{ij}]$ and $\{[a_{ij}]\}_{ij} = a_{ij}$.

By $\mathcal{M}_{mn}$ we denote the set of all (m x n) matrices over E_1. If $\underline{A}\varepsilon\mathcal{M}_{mn}$ and $\underline{B}\varepsilon\mathcal{M}_{pq}$ then $\underline{A} = \underline{B}$ if and only if $m = p$, $n = q$ and $\{\underline{A}\}_{ij} = \{\underline{B}\}_{ij}$ for all $i=1,\ldots, m$ and $j=1,\ldots, n$.

Evidently if $\underline{A}\varepsilon\mathcal{M}_{mn}$ and U is the set of mn index-pairs $(1,1),\ldots, (m,n)$ then $\underline{A}$ can be considered to be formally identical with the function $h_{\underline{A}}:U \to E_1$ defined by:

$$h_{\underline{A}}: (i,j) \to \{\underline{A}\}_{ij}.$$

In this way $\mathcal{M}_{mn}$ can be considered formally identical with $(E_1)^U$.

The algebraic operations $\oplus$, $\otimes$ can be extended from E_1 to matrices over E_1 by analogues of the usual definitions for matrix sums and products, as proposed independently by quite a number of authors publishing in different contexts and notations (eg. [6], [11], [17], [22],):

$$\forall\ [a_{ij}],\ [b_{ij}]\ \varepsilon\mathcal{M}_{mn},\ [a_{ij}] \oplus [b_{ij}] \quad \text{is by definition:}$$

$$[a_{ij} \oplus b_{ij}]\ \varepsilon\mathcal{M}_{mn} \tag{5-3}$$

$$\forall\ [a_{ij}]\ \varepsilon\mathcal{M}_{mp},\ [b_{ij}]\ \varepsilon\mathcal{M}_{pn},\ [a_{ij}] \otimes [b_{ij}] \quad \text{is by definition}$$

$$[\sum_{r=1}^{p}{}_{\oplus}(a_{ir} \otimes b_{rj})]\ \varepsilon\mathcal{M}_{mn} \tag{5-4}$$

Evidently, by identifying E_n as $\mathcal{M}_{n1}$, we obtain from (5-4) a definition of left-multiplication of n-tuples by matrices. As for conventional matrix algebra, multiplication of matrices is in general not commutative even if multiplication is commutative for E_1.

If $\underline{A}\ \varepsilon\mathcal{M}_{mn}$ and $\underline{B}\ \varepsilon\mathcal{M}_{pq}$ we shall say that $\underline{A}$ and $\underline{B}$ are <u>conformable for addition</u> whenever both $m = p$ and $n = q$; and <u>conformable for $\underline{A} \otimes \underline{B}$</u> whenever $n = p$. The use of the same symbols for operations in both E_1 and $\mathcal{M}_{mn}$ parallels standard algebraic practice.

Throughout the rest of the present publication, the notations E_n and $\mathcal{M}_{mn}$ will be used with the meanings defined above, and the definitions will not be repeated. When E_1 is a blog, we shall adapt the use of the word "finite" to apply to E_n and $\mathcal{M}_{mn}$ also - specifically, an n-tuple or matrix will be called finite if all its elements are finite.

5-4. Matrix Algebra

We have not yet formally identified the structures formed by the sets $\mathcal{M}_{mn}$ and before doing so it is useful to establish a particular form of words in order to economise on verbiage when in future chapters we examine other matrix structures. So let the letters $\mathcal{G}, \mathcal{B}$ be used in a given context in such a way that for given integers m, n $\geqslant$ 1 the notations $\mathcal{G}_{mn}, \mathcal{B}_{mn}$ represent classes of (m x n) matrices over a given belt V. (Thus in the present chapter the use of the letter $\mathcal{M}$ in this way has been introduced with $V = E_1$). By the $(\mathcal{G}, \mathcal{B}, V)$ - hom - rep statement we shall mean the following statement:

For all integers m, n, p $\geqslant$ 1: $(\mathcal{G}_{mn}, \oplus)$ is a commutative band and $(\mathcal{B}_{np}, \oplus)$ is a function space over $(V, \oplus, \otimes)$; for each $\underline{A} \in \mathcal{G}_{mn}$ the left-multiplication $g_{\underline{A}} : \underline{B} \mapsto \underline{A} \otimes \underline{B}$ is a right-linear homomorphism of $(\mathcal{B}_{np}, \oplus)$ into $(\mathcal{B}_{mp}, \oplus)$ over V and is therefore isotone; the mapping τ: $\underline{A} \mapsto g_{\underline{A}}$ is a representation of $(\mathcal{G}_{mn}, \oplus)$ as a commutative sub-band $(\mathcal{L}\mathcal{G}_{mn}, \oplus)$ of $(\text{Hom}_V(\mathcal{B}_{np}, \mathcal{B}_{mp}), \oplus)$; in particular if m = n then $(\mathcal{G}_{nn}, \mathcal{B}_{np}, V)$ is a 2-sided space and τ is a representation of the belt $(\mathcal{G}_{nn}, \oplus, \otimes)$ as a sub-belt $(\mathcal{L}\mathcal{G}_{nn}, \oplus, \otimes)$ of $(\text{Hom}_V(\mathcal{B}_{np}, \mathcal{B}_{np}), \oplus, \otimes)$.

Theorem 5-6. Let E_1 be a belt. Then the $(\mathcal{M}, \mathcal{M}, E_1)$ hom-rep statement is true.

Proof. The demonstrations that $(\mathcal{M}_{nn}, \oplus, \otimes)$ is a belt, that $(\mathcal{M}_{mn}, \oplus)$ is a commutative band and that $(\mathcal{M}_{np}, \oplus)$ is a left band-space over the belt $(\mathcal{M}_{nn}, \oplus, \otimes)$, consist of direct verifications of the relevant axioms, using the definitions (5-3) and (5-4). Apart from axiom X_3 (whose verification is trivial) the verifications are classical. For example, associativity of matrix multiplication follows from distributivity, associativity and additive commutativity of E_1.

To construe $(\mathcal{M}_{np}, \oplus)$ as a function space over E_1 we notice that if $\mathcal{M}_{np}$ is considered formally identical with the set $(E_1)^U$ of functions from the index-set U = ((1,1),..., (n,p)) to E_1, then the operation $\oplus$ defined as in (2-2) coincides with that in (5-3), so the commutative bands $(\mathcal{M}_{np}, \oplus)$ and $((E_1)^U, \oplus)$ may be considered formally identical also. We then define scalar multiplications by:

$$\left.\begin{aligned} [a_{ij}] \otimes \lambda &= [a_{ij} \otimes \lambda] \\ \lambda \otimes [a_{ij}] &= [\lambda \otimes a_{ij}] \end{aligned}\right\} \quad (\forall [a_{ij}] \in \mathcal{M}_{np}, \ \forall \lambda \in E_1)$$

The fact that $g_{\underline{A}}$ is a right-linear homomorphism follows from the classical verification of the following identities:

$$\left.\begin{array}{l}\underline{A} \otimes (\underline{B} \oplus \underline{C}) = (\underline{A} \otimes \underline{B}) \oplus (\underline{A} \otimes \underline{C}) \\ \\ \underline{A} \otimes (\underline{B} \otimes \lambda) = (A \otimes B) \otimes \lambda\end{array}\right\} (\forall\ \underline{A} \in \mathcal{M}_{mn}, \forall\ \underline{B}, \underline{C} \in \mathcal{M}_{np}, \forall \lambda \in E_1)$$

Finally the fact that τ is a representation follows by the general arguments of Section 2-5.

Let us define a left $\mathcal{M}_{qm}$-multiplication to be a left-multiplication using a matrix from $\mathcal{M}_{qm}$, and similarly, we define a right-$\mathcal{M}_{nr}$-multiplication. An $\mathcal{M}_{mn}$-translation is a function $f: \mathcal{M}_{mn} \mapsto \mathcal{M}_{mn}$ such that there exists $\underline{A} \in \mathcal{M}_{mn}$ whereby $f(\underline{X}) = \underline{A} \oplus \underline{X}$ for all $\underline{X} \in \mathcal{M}_{mn}$.

The following result follows immediately from Theorems 5-5 and 5-6 and Proposition 2-3.

Proposition 5-7. Let $(E_1, \oplus, \otimes)$ be a belt and let K consist of the following functions (as applied to $\mathcal{M}_{mn}$): $\oplus$, $i_{\mathcal{M}_{mn}}$, together with all left and right scalar multiplications, all left $\mathcal{M}_{mm}$-multiplications, all right $\mathcal{M}_{nn}$-multiplications and all $\mathcal{M}_{mn}$-translations. Then $\hat{K}$, the composition algebra generated by K, consists entirely of isotone functions. Moreover, if $m = n$, we may adjoin the function $\otimes$ to K and the foregoing statement remains true. ●

Propostion 5-7 embodies what we may call the principle of isotonicity for matrix operations.

The commutative band $(\mathcal{M}_{n1}, \oplus)$ is of course formally identical with the commutative band $(E_n, \oplus)$. According to Theorem 5-6, E_n is not only a function space over E_1, but also a space over $\mathcal{M}_{nn}$. We see here, of course, the classical role of matrices as linear transformations (endomorphisms) of spaces of n-tuples, and we shall be looking more closely later at the properties of these transformations.

Finally it is clear that the system $(\mathcal{M}_{11}, \oplus, \otimes)$ is formally identical with $(E_1, \oplus, \otimes)$, for any belt E_1. This follows directly from (5-3), (5-4) for $m = n = p = 1$.

5-5. The Identity Matrix

The results of the previous section show that there are extensive formal similarities between the properties of matrices in minimax algebra and those in conventional matrix theory.

However, the existence of a multiplicative identity element for E_1 (axiom X_8) does not yet imply the existence of such an element for the belt $\mathcal{M}_{nn}$. As the following theorem shows, we must also require E_1 to possess a null element, i.e. to satisfy:

$$X_{10} : \exists\, \theta \varepsilon E_1 \text{ such that for all } x \varepsilon E_1:$$
$$x \oplus \theta = x \text{ and } x \otimes \theta = \theta \otimes x = \theta.$$

Theorem 5-8. Let $(E_1, \oplus, \otimes)$ be a belt. Then a necessary and sufficient condition that the belt $(\mathcal{M}_{nn}, \oplus, \otimes)$ $(n > 1)$ satisfy axiom X_8 is that $(E_1, \oplus, \otimes)$ satisfy both axiom X_8 and axiom X_{10}; and then $(\mathcal{M}_{nn}, \oplus, \otimes)$ satisfies axiom X_{10} also.

Proof. Sufficiency is straightforward: if E_1 satisfies X_8 and X_{10} then the $(n \times n)$ "identity matrix" $\underline{I}_n$ whose diagonal elements all equal $\emptyset$ and whose off-diagonal elements all equal θ, is easily verified (from 5-4) to act as a left and right multiplicative identity in $\mathcal{M}_{nn}$; and the "null matrix" $\underline{\Phi}_n$, all of whose elements equal θ, has the properties of a null element (axiom X_{10}).

Conversely, suppose $\underline{X} = [x_{ij}] \varepsilon \mathcal{M}_{nn}$ is a left and right identity element in $\mathcal{M}_{nn}$. Let $\underline{U} \varepsilon \mathcal{M}_{nn}$ have all its elements equal to a given arbitrary $u \varepsilon E_1$. Then:

$$u = \{\underline{X} \otimes \underline{U}\}_{ij} = \sum_{r=1}^{n}{}_{\oplus} (x_{ir} \otimes u) = (\sum_{r=1}^{n}{}_{\oplus} x_{ir}) \otimes u \quad (i,j=1,\ldots,n)$$

Since u is arbitrary, this demonstrates the existence of a left identity $= \sum_{r=1}^{n}{}_{\oplus} x_{ir}$ in E_1, and the existence of a right identity follows similarly. Hence E_1 has a (necessarily unique) two-sided identity element $\emptyset$ (say) and X_8 is verified.

Now let $\underline{V} \varepsilon \mathcal{M}_{nn}$ have diagonal elements all equal to $\emptyset$ and off-diagonal elements all equal to a given arbitrary $u \varepsilon E_1$. For $i \neq j$ we have:

$$u = \{\underline{X} \otimes \underline{V}\}_{ij} = \sum_{r=1}^{n}{}_{\oplus} (x_{ir} \otimes \{\underline{V}\}_{rj}) = x_{ij} \oplus (\sum_{\substack{r=1 \\ r\neq j}}^{n}{}_{\oplus} x_{ir}) \otimes u$$

whence by (2-1), $x_{ij} \leqslant u$

But u was arbitrary. Hence every off-diagonal x_{ij} is equal to (necessarily the same, unique) θ such that $\theta \oplus u = u$ for all $u \varepsilon E_1$. Hence for all $i=1,\ldots,n$:

$$\sum_{r=1}^{n}{}_{\oplus}\, x_{ir} = x_{ii} \oplus \sum_{\substack{r=1\\ r\neq i}}^{n}{}_{\oplus}\, x_{ir} = x_{ii} \oplus \Theta = x_{ii}$$

But we had already shown that $\sum_{r=1}^{n}{}_{\oplus}\, x_{ir} = \emptyset$. Hence $x_{ii} = \emptyset$ for all $i=1,\ldots, n$, so $\underline{X}$ has diagonal elements equal to $\emptyset$ and off-diagonal elements equal to Θ

Now let $\underline{W} \in \mathcal{M}_{nn}$ have, for given i and j, the element in row i, column j equal to a given arbitrary $u \in E_1$ and all other elements equal to a given arbitrary $v \in E_1$. Then:

$$u = \{\underline{X} \otimes \underline{W}\}_{ij} = (\sum_{\substack{r=1\\ r\neq i}}^{n}{}_{\oplus}\, \Theta \otimes v) \oplus (\emptyset \otimes u) = (\Theta \otimes v) \oplus u$$

Hence $(\theta \otimes v) \oplus u = u$ for arbitrary u, and so $(\theta \otimes v)$ equals Θ by the uniqueness of θ. Similarly $v \otimes \theta = \theta$ for arbitrary $v \in E_1$. Hence X_{10} is verified for E_1.

<u>Corollary 5-9. If E_1 is a blog, then the belt $\mathcal{M}_{nn} (n \geqslant 1)$ satisfies X_8. If E_1 is a division belt then $\mathcal{M}_{nn} (n \geqslant 1)$ does not satisfy X_8 unless either $n = 1$ or $E_1 = \{\emptyset\}$.</u>

<u>Proof</u>. If E_1 is a blog then by Proposition 4-1, E_1 satisfies X_8 and X_{10} and hence by Theorem 5-8, so does $\mathcal{M}_{nn}$. If E_1 is a division belt then $\mathcal{M}_{11}$ is isomorphic to E_1 and hence satisfies X_8. If $E_1 = \{\emptyset\}$ then $\mathcal{M}_{nn}$ trivially satisfies X_8 and X_{10} for all $n \geqslant 1$.

If $E_1 \neq \{\emptyset\}$ and $n > 1$ then $\mathcal{M}_{nn}$ cannot satisfy X_8. For by Theorem 5-8 this would imply that E_1 satisfied axiom X_{10}, contradicting Corollary 4-9. ●

The notation $\underline{I}_n$ for the $(n \times n)$ "unit matrix" or "identity matrix" used in the above proof will be standard in subsequent contexts. (If, $n=1$, define $\underline{I}_n = [\emptyset]$).

<u>5-6 . Matrix Transformations</u>.

We can now discuss a question which arises naturally in connection with the concepts introduced earlier in the chapter. In conventional linear algebra, we can characterise linear transformations of vector spaces entirely in terms of matrices: is the same true in the present theory? In other words, is $\mathcal{M}_{mn}$ isomorphic to $\mathrm{Hom}_{E_1} (E_n, E_m)$ for all integers $m, n \geqslant 1$? The following results give necesarry and sufficient conditions for this to be the case.

<u>Theorem 5-10. Let E_1 be a belt which satisfies axioms X_8 and X_{10}. Then for all integers $m, n \geqslant 1$, $\mathcal{M}_{mn}$ is isomorphic to $\mathrm{Hom}_{E_1}(E_n, E_m)$, and the representation</u>

$\tau: \underline{A} \mapsto g_{\underline{A}}$ of $\mathcal{M}_{mn}$ as a commutative sub-band $\mathcal{L}\mathcal{M}_{mn}$ of $\mathrm{Hom}_{E_1}(\mathcal{M}_{np}, \mathcal{M}_{mp})$ is faithful for all integers $m, n, p \geqslant 1$, and surjective for $p = 1$.

Proof. First let $p = 1$, and let $\underline{u}(j) \in \mathcal{M}_{n1}$ $(j=1,\ldots,n)$ be the jth column of $\underline{I}_n$ the $(n \times n)$ identity matrix. Arguing exactly as in elementary linear algebra we can show that the action of any $g \in \mathrm{Hom}_{E_1}(\mathcal{M}_{n1}, \mathcal{M}_{m1})$ on $\mathcal{M}_{n1}$ coincides with the action of the left-multiplication $g_{\underline{A}}$ where $\underline{A}$ is the matrix given by:

$$\{\underline{A}\}_{ij} = \{g(\underline{u}(j))\}_{i1} \quad (i=1,\ldots,m;\ j=1,\ldots,n) \tag{5-5}$$

Hence $\tau : \underline{A} \mapsto g_{\underline{A}}$ is surjective of $\mathcal{M}_{mn}$ onto $\mathrm{Hom}_{E_1}(\mathcal{M}_{n1}, \mathcal{M}_{m1})$.

Moreover, if $\underline{A}, \underline{B} \in \mathcal{M}_{mn}$ are different then their action on some $\underline{u}(j)$ will be different, i.e. $g_{\underline{A}} \neq g_{\underline{B}}$, showing that τ is injective. Hence $\mathcal{M}_{mn}$ is isomorphic to $\mathrm{Hom}_{E_1}(\mathcal{M}_{n1}, \mathcal{M}_{m1})$ and so to $\mathrm{Hom}_{E_1}(E_n, E_m)$.

If $p > 1$ and $\underline{A}, \underline{B} \in \mathcal{M}_{mn}$ are different, let $\underline{C} \in \mathcal{M}_{np}$ have as its first column some $\underline{u}(j)$ for which $\underline{A} \otimes \underline{u}(j) \neq \underline{B} \otimes \underline{u}(j)$. Then $\underline{A} \otimes \underline{C} \neq \underline{B} \otimes \underline{C}$, ie. $g_{\underline{A}} \neq g_{\underline{B}}$ and so τ is injective, i.e. faithful. ●

Corollary 5-11. Let E_1 be a belt, and let $n > 1$ be a given integer. Then a necessary and sufficient condition that $\mathcal{M}_{mn}$ be isomorphic to $\mathrm{Hom}_{E_1}(E_n, E_m)$ for all integers $m \geqslant 1$ is that E_1 satisfy axioms X_8 and X_{10}.

Proof. The identity mapping i_{E_n} belongs to, and acts as multiplicative identity element in, the belt $\mathrm{Hom}_{E_1}(E_n, E_n)$. So for the isomorphism to hold when $m = n$ in particular, it is necessary that the belt $\mathcal{M}_{nn}$ also have a multiplicative identity element which implies by Theorem 5-8 that E_1 should satisfy axioms X_8 and X_{10}.

The sufficiency follows from Theorem 5-10. ●

In the light of Corollaries 5-9 and 5-11, we see that the isomorphism of $\mathcal{M}_{mn}$ and $\mathrm{Hom}_{E_1}(E_n, E_m)$ holds when E_1 is a blog but fails in general when E_1 is a division belt. In the latter case τ is not surjective since no matrix can supply the identity mapping. However, we shall see later when we consider residuomorphisms in Chapter 10 that τ is faithful for a class of belts which includes the division belts.

5-7. Further Notions

Let $(V, \oplus, \otimes, \oplus', \otimes')$ be a belt with duality. We shall say that a space $(T, \oplus)$ over V has a duality if:

D_1: A dual addition $\oplus'$ is defined whereby $(T, \oplus, \oplus')$ is a commutative band with duality.

D_2: $(T, \oplus')$ is a space over the belt $(V, \oplus', \otimes')$

Further definitions may be made in the obvious way of left-band-space with duality, two-sided space with duality and so on, and dual forms obtained of the results of the present chapter.

If V is a belt with duality then the duality can be extended to any function space over V by dualising the definitions of $\oplus$ and scalar multiplication in the obvious way.

Another topic in the classical spirit which we shall consider later concerns characterising of those matrix transformations which hold certain spaces fixed. We lay the groundwork for this now. Accordingly, let $(S, \oplus)$ and $(T, \oplus)$ be given spaces over a given belt $(V, \oplus, \otimes)$. Let $\dot{S} \subset S$ and $\dot{T} \subset T$ be given. Then we define:

$$\mathrm{Stab}_V(\dot{S}, \dot{T}) = \{g \mid g \in \mathrm{Hom}_V(S, T) \text{ and } g(s) \in \dot{T} \text{ for all } s \in \dot{S}\} \qquad (5\text{-}6)$$

It is straighforward to verify the following results, which extend Propositions 5-2, 5-3 and 5-6, and Theorem 5-10. (The isomorphisms noted at the end of Proposition 5-13 follow from Theorem 5-10).

Proposition 5-12. Let $(S, \oplus)$, $(T, \oplus)$ be given spaces over a given belt $(V, \oplus, \otimes)$, let $(\dot{T}, \oplus)$ be a commutative sub-band of $(T, \oplus)$, and let $\dot{S}$ be a subset of S. Then $(\mathrm{Stab}_V(\dot{S}, \dot{T}), \oplus)$ is a commutative sub-band of $(\mathrm{Hom}_V(S, T), \oplus)$ when $\oplus$ is defined as in (2-2). If moreover $S = T$ and $\dot{S} = \dot{T}$ then $(\mathrm{Stab}_V(\dot{T}, \dot{T}), \oplus, \otimes)$ is a sub-belt of $(\mathrm{Hom}_V(T, T), \oplus, \otimes)$ when $\otimes$ denotes composition; and $(\dot{T}, \oplus)$ is a left-band-space over this belt when for each $g \in \mathrm{Stab}_V(\dot{T}, \dot{T})$ and each $x \in \dot{T}$ we define $g \otimes x$ to be $g(x)$. ●

Proposition 5-13. Let $(E_1, \oplus, \otimes)$ be a belt with a sub-belt $(F_1, \oplus, \otimes)$. For any integers $m, n, p \geq 1$ let $\mathcal{O}_{np} \subset \mathcal{M}_{np}$ consist of all $(n \times p)$ matrices $\underline{A}$ such that $\{\underline{A}\}_{ij} \in F_1$ $(i=1,\ldots, n;\ j=1,\ldots, p)$ and let $\mathcal{N}_{mn} \subset \mathcal{M}_{mn}$ consist of all $(m \times n)$ matrices $\underline{A}$ such that the left-multiplication $g_{\underline{A}} : \mathcal{M}_{np} \to \mathcal{M}_{mp}$, defined by $g_{\underline{A}} : \underline{B} \mapsto \underline{A} \otimes \underline{R}$, satisfies:

$$\underline{g_{\underline{A}} \in \mathrm{Stab}_{E_1}(\mathcal{O}_{np}, \mathcal{O}_{mp})}$$

<u>Then the $(\mathcal{N}, \mathcal{O}, F_1)$-hom-rep statement is true. Furthermore, if E_1 satisfies axioms X_8 and X_{10} then for all integers $m, n > 1$:</u>

<u>$(\mathcal{N}_{mn}, \oplus)$ is isomorphic to $(\mathrm{Stab}_{E_1}(F_n, F_m), \oplus)$</u>

<u>$(\mathcal{N}_{nn}, \oplus, \otimes)$ is isomorphic to $(\mathrm{Stab}_{E_1}(F_n, F_n), \oplus, \otimes)$.</u>

<u>where F_n, F_m are respectively the spaces of n-tuples and m-tuples over F_1.</u> ●

6. DUALITY FOR MATRICES

6-1. The Dual Operations

If $(E_1, \oplus, \otimes, \oplus', \otimes')$ is a belt with duality then the duality may be extended to the algebra of matrices over E_1 by using the dual operations $\oplus'$, $\otimes'$ instead of $\oplus$, $\otimes$ in (5-3) and (5-4). Thus:

$$\forall\ [a_{ij}],\ [b_{ij}] \in \mathcal{M}_{mn},\ [a_{ij}] \oplus' [b_{ij}] \quad \text{is by definition}$$

$$[a_{ij} \oplus' b_{ij}] \in \mathcal{M}_{mn} \tag{6-1}$$

$$\forall\ [a_{ij}] \in \mathcal{M}_{mp},\ [b_{ij}] \in \mathcal{M}_{pn},\ [a_{ij}] \otimes' [b_{ij}] \quad \text{is by definition}$$

$$\Big[\sum_{r=1}^{p}{}_{\oplus'} (a_{ir} \otimes' b_{rj})\Big] \in \mathcal{M}_{mn} \tag{6-2}$$

The expressions <u>conformable for $\oplus'$</u> and <u>conformable for $A \otimes' B$</u> will be used in the obvious way.

Evidently, the dual of Theorem 5-6 holds, which together with Theorem 5-6 itself yields the following:

<u>Proposition 6-1. Let $(E_1, \oplus, \otimes, \oplus', \otimes')$ be a belt with duality. Then $(\mathcal{M}_{nn}, \oplus, \otimes, \oplus', \otimes')$ is a belt with duality, for any integer $n \geqslant 1$. Moreover $(\mathcal{M}_{mn}, \oplus, \oplus')$ is a function space with duality over E_1, and also a space with duality over the belt $(\mathcal{M}_{nn}, \oplus, \otimes, \oplus', \otimes')$ for any integers $m, n \geqslant 1$.</u> ●

The duals of the various statements in Proposition 5-7 evidently hold, and with the obvious appropriate definitions of <u>left dual $\mathcal{M}_{mm}$-multiplication</u>, <u>right dual $\mathcal{M}_{nn}$-multiplication</u> and <u>dual $\mathcal{M}_{mn}$-translation</u>, we can develop the <u>principle of isotonicity for matrix operations with duality</u> embodied in the following result.

<u>Proposition 6-2. Let $(E_1, \oplus, \otimes, \oplus', \otimes')$ be a belt with duality and let K consist of the following functions (as applied to $\mathcal{M}_{mn}$): $\oplus$, $\oplus'$, $i_{\mathcal{M}_{mn}}$, together with all left and right scalar multiplications and dual scalar multiplications, all left $\mathcal{M}_{mm}$-multiplications and left dual $\mathcal{M}_{mm}$-multiplications, all right $\mathcal{M}_{nn}$-multiplications and right dual $\mathcal{M}_{nn}$-multiplications and all $\mathcal{M}_{mn}$-translations and dual $\mathcal{M}_{mn}$-translations. Then $\hat{K}$, the composition algebra generated by K, consists entirely of isotone functions. Moreover, if $m = n$, we may adjoin the functions $\otimes$ and $\otimes'$ to K and the foregoing statement remains true.</u> ●

6-2. Some Matrix Inequalities

The following theorem lists some basic "mixed inequalities" which will be useful in matrix manipulative work. We use the notation $\{\underline{X}\}_{ij}$, with i, j undefined, to denote any typical element of a given matrix $\underline{X}$, and the notation '$\dashv$' means "when the matrices are conformable for the relevant operations".

Theorem 6-3. Let E_1 be a belt with duality. Then the following inequalities hold for the matrices over E_1:

(i) $\underline{X} \oplus (\underline{Y} \oplus' \underline{Z}) \leq (\underline{X} \oplus \underline{Y}) \oplus' (\underline{X} \oplus \underline{Z}) \dashv$

(ii) $\underline{X} \oplus'(\underline{Y} \oplus \underline{Z}) \geq (\underline{X} \oplus' \underline{Y}) \oplus (\underline{X} \oplus' \underline{Z}) \dashv$

(iii) $\underline{X} \otimes (\underline{Y} \oplus' \underline{Z}) \leq (\underline{X} \otimes \underline{Y}) \oplus' (\underline{X} \otimes \underline{Z}) \dashv$

(iv) $(\underline{Y} \oplus' \underline{Z}) \otimes \underline{X} \leq (\underline{Y} \otimes \underline{X}) \oplus' (\underline{Z} \otimes \underline{X}) \dashv$

(v) $\underline{X} \otimes' (\underline{Y} \oplus \underline{Z}) \geq (\underline{X} \otimes' \underline{Y}) \oplus (\underline{X} \otimes' \underline{Z}) \dashv$

(iv) $(\underline{Y} \oplus \underline{Z}) \otimes' \underline{X} \geq (\underline{Y} \otimes' \underline{X}) \oplus (\underline{Z} \otimes' \underline{X}) \dashv$

Moreover, associative inequalities X_{12} hold in general for matrices over E_1, if they hold for $\mathcal{M}_{11}$:

$$X_{12}: \quad \begin{array}{l} \underline{X} \otimes (\underline{Y} \otimes' \underline{Z}) \leq (\underline{X} \otimes \underline{Y}) \otimes' \underline{Z} \dashv \\ \underline{X} \otimes'(\underline{Y} \otimes \underline{Z}) \geq (\underline{X} \otimes' \underline{Y}) \otimes \underline{Z} \dashv \end{array}$$

In particular, inequalities X_{12} hold if E_1 is a division belt or a blog.

Proof. Relations (i) to (vi) are direct consequences of Lemma 2-7 and the fact that multiplications, translations and their duals act as homomorphisms.

To prove inequalities X_{12}, assume that the inequalities X_{12} hold in $\mathcal{M}_{11}$ (i.e. in E_1) and consider:

$$\{\underline{X} \otimes (Y \otimes' Z)\}_{ij} = \sum_{r}{}_{\oplus}(\{\underline{X}\}_{ir} \otimes \{\underline{Y} \otimes' \underline{Z}\}_{rj}) = \sum_{r}{}_{\oplus} (\{\underline{X}\}_{ir} \otimes \sum_{s}{}_{\oplus'}(\{\underline{Y}\}_{rs} \otimes' \{\underline{Z}\}_{sj}))$$

We now apply the principles of opening and closing developed in Chapter 3. Opening w.r.t. the index s:

$$\forall s, \{\underline{X} \otimes (\underline{Y} \otimes' \underline{Z})\}_{ij} \leq \sum_{r}{}_{\oplus} (\{\underline{X}\}_{ir} \otimes (\{\underline{Y}\}_{rs} \otimes' \{\underline{Z}\}_{sj})) \qquad (6\text{-}3)$$

Now, X_{12} applied to $\mathcal{M}_{11}$ gives for all r, s that:

$$\{\underline{X}\}_{ir} \otimes (\{\underline{Y}\}_{rs} \otimes' \{\underline{Z}\}_{sj}) \leq (\{\underline{X}\}_{ir} \otimes \{\underline{Y}\}_{rs}) \otimes' \{\underline{Z}\}_{sj}$$

Closing w.r.t. the index r:

$$\forall s,\ \sum_{r\oplus} (\{\underline{X}\}_{ir} \otimes (\{\underline{Y}\}_{rs} \otimes' \{\underline{Z}\}_{sj})) \leq \sum_{r\oplus} ((\{\underline{X}\}_{ir} \otimes \{\underline{Y}\}_{rs}) \otimes' \{\underline{Z}\}_{sj}) \qquad (6\text{-}4)$$

But from (2-17):

$$(\sum_{r\oplus}(\{\underline{X}\}_{ir} \otimes \{\underline{Y}\}_{rs})) \otimes' \{\underline{Z}\}_{sj} \geq \sum_{r\oplus} ((\{\underline{X}\}_{ir} \otimes \{\underline{Y}\}_{rs}) \otimes' \{\underline{Z}\}_{sj}) \qquad (6\text{-}5)$$

But the expression on the left-hand side of (6-5) is just:

$$\{\underline{X} \otimes \underline{Y}\}_{is} \otimes' \{\underline{Z}\}_{sj}.$$

Hence, from (6-3), (6-4) and (6-5):

$$\forall s,\ \{\underline{X} \otimes (\underline{Y} \otimes' \underline{Z})\}_{ij} \leq \{\underline{X} \otimes \underline{Y}\}_{is} \otimes' \{\underline{Z}\}_{sj}.$$

Closing w.r.t. the index s:

$$\{\underline{X} \otimes (\underline{Y} \otimes' \underline{Z})\}_{ij} \leq \sum_{s\oplus'} (\{\underline{X} \otimes \underline{Y}\}_{is} \otimes' \{\underline{Z}\}_{sj}) = \{(\underline{X} \otimes \underline{Y}) \otimes' \underline{Z}\}_{ij}$$

This implies the first half of X_{12} and the second half is proved similarly. Now axiom X_{12} holds for E_1 if E_1 is blog (by Proposition 4-2) or if E_1 is a division belt (which has a self-dual associative multiplication operation). Hence in both these cases axiom X_{12} holds for E_1 and so for $\mathcal{M}_{11}$ which is isomorphic to E_1. ●

In manipulative work, it is useful to have a mnemonic to save looking up each relation in Theorem 6-3 separately. Here is one:

Let α,β be two algebraic expressions such that $\alpha = \beta$ would hold as an identity valid for any belt, if all accents were dropped, i.e. if in both α and β all occurrences of $\oplus'$, $\otimes'$ were replaced by $\oplus$, $\otimes$ respectively. If the occurrences of $\oplus'$ (resp. $\otimes'$) lie deeper in α than in β, then $\alpha \leq \beta$ identically.

By saying than an occurrence of e.g. $\oplus'$ lies deeper in α than in β, we mean that there are more unmated left-hand brackets to the left of it in α than in β. As an illustration, the expression $\underline{X} \otimes (\underline{Y} \otimes' \underline{Z})$ and $(\underline{X} \otimes \underline{Y}) \otimes' \underline{Z}$ would (by the associative law) become identically equal in any belt if the accents were dropped. But in the expression $\underline{X} \otimes (\underline{Y} \otimes' \underline{Z})$, the occurrence of $\otimes'$ is at "depth one", whilst in the expression $(\underline{X} \otimes \underline{Y}) \otimes' \underline{Z}$ it is at "depth zero". Hence the first half of X_{12} holds.

Again:

$$\underline{X} \otimes (\underline{Y} \oplus \underline{Z}) = (\underline{X} \otimes \underline{Y}) \oplus (\underline{X} \otimes \underline{Z})$$

holds identically in any belt. In the expression $\underline{X} \otimes' (\underline{Y} \oplus \underline{Z})$, the occurrence of $\otimes'$ is at "depth zero" whereas in the expression

$$(\underline{X} \otimes' \underline{Y}) \oplus (\underline{X} \otimes' \underline{Z})$$

both occurrences of $\otimes'$ are at "depth one". Hence inequality (v) holds in Theorem 6-3.

We put the above rule forward purely as a mnemonic. Obviously, in more formal dress, it could be promoted to a metamathematical theorem, but there seems little motivation for this in the present context. As it stands, the mnemonic is intended for application to Theorem 6-3 only and cannot, without further refinement, be applied generally.

7. CONJUGACY

7-1. Conjugacy for Belts

In conventional linear operator theory, there are two common starting points for a theory of conjugacy. One approach, which one might call the functional conjugacy approach, defines the conjugate of a given space S as a certain set S^* of (linear, scalar-valued) functions defined on S. The other, which one might call the axiomatic conjugacy approach, defines a certain involution $x \to x^*$ on the elements of S, subject to certain axioms. It is characteristic of the classical theory of linear operators that these two definitions give rise to the same structures in certain important cases.

A similar, if rather more complicated, situation arises in minimax algebra. In the present chapter we make our first excursion into the theory of conjugacy for minimax algebra, beginning with a discussion of axiomatic conjugacy for belts and matrices over a belt.

Accordingly, let $(V, \oplus)$ and $(W, \oplus')$ be given commutative bands.

We shall say that $(W, \oplus')$ is conjugate to $(V, \oplus)$ if there is a function $g: V \to W$ which satisfies the following axioms:

N_1: g is bijective

N_2: $\forall\, x, y \in V$, $g(x \oplus y) = g(x) \oplus' g(y)$

In particular, if $(V, \oplus, \oplus')$ is a commutative band with duality, we shall say that $(V, \oplus, \oplus')$ is self-conjugate if $(V, \oplus')$ is conjugate to $(V, \oplus)$.

In the language of lattice theory, conjugate commutative bands V and W are just semi-lattices and g is a semi-lattice anti-isomorphism. Evidently, V and W are, as partially ordered sets, isomorphic to each others dual, and the inverse bijection g^{-1} is also an anti-isomorphism. Hence conjugacy is a symmetric relation, and if U and W are both conjugate to V, then U and W are isomorphic.

We are, of course, less interested in this completely trivial situation than in that in which some algebraic structure is present. So suppose now that the systems V and W have the further binary combining operations $\otimes$ and $\otimes'$ respectively, such that $(V, \oplus, \otimes)$ and $(W, \oplus' \otimes')$ are belts. We shall say that $(W, \oplus', \otimes')$ is conjugate to $(V, \oplus, \otimes)$ if axioms N_1, N_2 and N_3 hold, where axiom N_3 is:

N_3: $\forall\, x, y \in V$, $\qquad g(x \otimes y) = g(y) \otimes' g(x)$.

In particular, if $(V, \oplus, \otimes, \oplus', \otimes')$ is a belt with duality, we shall say that $(V, \oplus, \otimes, \oplus', \otimes')$ is self-conjugate if $(V, \oplus', \otimes')$ is conjugate to $(V, \oplus, \otimes)$.

Evidently the bijection g satisfies N_3 if and only if its inverse g^{-1} satisfies N_3' (with g^{-1} in the role of g). And since it is entirely a matter of convention which of the additions $\oplus$ and $\oplus'$ we call "dual" addition, it is evident that conjugacy remains a symmetric relation when its meaning is extended to belts.

Furthermore, if $(U, \oplus', \otimes')$ and $(W, \oplus', \otimes')$ are both conjugate to $(V, \oplus, \otimes)$ there exist bijections $g: V \to W$ and $h: V \to U$ such that

$g(x \otimes y) = g(y) \otimes' g(x)$ and $h(x \otimes y) = h(y) \otimes' h(x)$

from which it follows that the bijection $goh^{-1}: U \to W$ defined by $goh^{-1}(z)=g(h^{-1}(z))$ satisfies:

$$\forall\, u, v \in U, \quad goh^{-1}(u \otimes' v) = goh^{-1}(u) \otimes' goh^{-1}(v).$$

Hence goh^{-1} is an algebraic, as well as an order, isomorphism. Summarising:

Proposition 7-1. Conjugacy is a symmetric relation, and two commutative bands or belts which are conjugate to the same are algebraically and order-isomorphic.

In view of the isomorphism just discussed, we can say, if W is conjugate to V, that W is the conjugate of V. We shall denote the conjugate of a given system V (if it has one) by the notation V^*. Hence $(V^*)^* = V$. Examples of belts having a conjugate are given by Systems 7 to 12 from the table of Fig 4-2, which are conjugate to Systems 13 to 18 respectively.

Let V, V^* be conjugate, and let $x \in V$, $y \in V$. Just as in conventional linear operator theory, we shall henceforth employ the notation x^* (rather than $g(x)$) to denote the element of V^* corresponding to x under the bijection, and the notation y^* (rather than $g^{-1}(y)$ to denote the element of V corresponding to y under the bijection. Thus $(x^*)^* = x$. Future references to axioms N_1 to N_3 will assume the axioms have been rewritten in this more symmetric notation. We shall call the element x^* the conjugate of x.

An immediate consequence of the definitions is the following:

Proposition 7-2. If V, V^* are conjugate systems, then $x \geqslant y$ in V if and only if $x^* \leqslant y^*$ in V^*. Similarly, $x > y$ in V if and only if $x^* < y^*$ in V^*.

Finally, a straightforward verification yields the following result.

Proposition 7-3. Every division belt is self-conjugate, under the bijection:

$$x^* = x^{-1} \tag{7-1}$$

and every blog is self-conjugate under the bijection:

$$\left.\begin{aligned} (-\infty)^* &= +\infty \\ (+\infty)^* &= -\infty \\ x^* &= x^{-1} \text{ if } x \text{ is finite.} \end{aligned}\right\} \tag{7-2}$$

7-2. Conjugacy for Matrices

Given systems E_1, E_1^* which are conjugate, we may extend the conjugacy to matrices over E_1 and E_1^* as follows. If $\underline{A}$ is an (mxn) matrix over E_1, then $\underline{A}^*$ is by definition an (nxm) matrix over E_1^* such that:

$$\{\underline{A}^*\}_{ij} = (\{\underline{A}\}_{ji})^* \qquad (1 \leqslant i \leqslant m;\ 1 \leqslant j \leqslant n) \tag{7-3}$$

In other words, $\underline{A}^*$ is obtained by "transposing and conjugating".

Then if V is any collection of matrices over E_1, we define V^* to be

$$V^* = \{\underline{A}^* \mid \underline{A} \in V\} \tag{7-4}$$

Evidently, $\mathcal{M}_{mn}^*$ is the set of all (nxm) matrices over E_1^*.

The following theorem shows that the "conjugacy" of sets of matrices defined by (7-3) and (7-4) is a consistent extension of the notion of conjugacy defined in

Section 7-1, justifying the use of the notations $\underline{A}^*$ and V^*.

<u>Theorem 7-4. Let $(E_1,\oplus)$ and $(E_1^*,\oplus')$ be conjugate. Then so are $(\mathcal{M}_{mn},\oplus)$ and $(\mathcal{M}_{mn}^*,\oplus')$, for any integers m, n $\geq$ 1. If moreover multiplications $\otimes$, $\otimes'$ are defined for E_1, E_1^* respectively such that $(E_1, \oplus, \otimes)$, and $(E_1^*, \oplus', \otimes')$ are conjugate, then multiplications $\otimes$, $\otimes'$ are also defined for $\mathcal{M}_{nn}$, $\mathcal{M}_{nn}^*$ respectively, for each integer n $\geq$ 1, such that $(\mathcal{M}_{nn}, \oplus, \otimes)$ and $(\mathcal{M}_{nn}^*, \oplus', \otimes')$ are conjugate. In all cases, the conjugate of a given matrix $\underline{A}$ is the matrix $\underline{A}^*$ defined in (7-3).</u>

<u>Proof.</u> Let V be the set of all matrices over E_1, and let $g:V \to V^*$ be defined by $g: \underline{A} \mapsto \underline{A}^*$ according to (7-3). It is immediately clear that the restriction of g to $\mathcal{M}_{mn}$ (for given m, n $\geq$ 1) is a bijection. Moreover, for all i=1,..., m; j=1,..., n; $\underline{A}, \underline{B} \in \mathcal{M}_{mn}$:

$$\{g(\underline{A} \oplus \underline{B})\}_{ij} = \{(\underline{A} \oplus \underline{B})^*\}_{ij}$$

$$= (\{\underline{A} \oplus \underline{B}\}_{ji})^* \qquad \text{(by (7-3))}$$

$$= (\{\underline{A}\}_{ji} \oplus \{\underline{B}\}_{ji})^* \qquad \text{(by (5-3))}$$

$$= (\{\underline{A}\}_{ji})^* \oplus' (\{\underline{B}\}_{ji})^* \qquad \text{(by } N_1 \text{ for } E_1\text{)}$$

$$= \{\underline{A}^*\}_{ij} \oplus' \{\underline{B}^*\}_{ij} \qquad \text{(by (7-3))}$$

$$= \{\underline{A}^* \oplus' \underline{B}^*\}_{ij} \qquad \text{(by dual of (5-3))}$$

$$= \{g(\underline{A}) \oplus' g(\underline{B})\}_{ij}$$

Hence $g(\underline{A} \oplus \underline{B}) = g(\underline{A}) \oplus' g(\underline{B})$, so N_2 is satisfied for $(\mathcal{M}_{mn}, \oplus)$, $(\mathcal{M}_{mn}^*, \oplus')$.

Now if multiplications $\otimes$, $\otimes'$ are defined for E_1, E_1^* respectively, then they are also defined for $\mathcal{M}_{nn}$, $\mathcal{M}_{nn}^*$ respectively, for each integer n $\geq$ 1, via (5-4) and its dual. So if N_3 holds for E_1, E_1^*, we have for all i=1,..., n; $\underline{A},\underline{B} \in \mathcal{M}_{nn}$:

$$\{g(\underline{A}\otimes\underline{B})\}_{ij} = \{(\underline{A}\otimes\underline{B})^*\}_{ij}$$

$$= (\{\underline{A}\otimes\underline{B}\}_{ji})^* \qquad \text{(by (7-3))}$$

$$= \Big(\sum_{k=1}^{n}{}_{\oplus}(\{\underline{A}\}_{jk} \otimes \{\underline{B}\}_{ki})\Big)^* \qquad \text{(by (5-4))}$$

$$= \sum_{k=1}^{n}{}_{\oplus'} (\{\underline{A}\}_{jk} \otimes \{\underline{B}\}_{ki})^* \qquad \text{(by induction on } N_2\text{)}$$

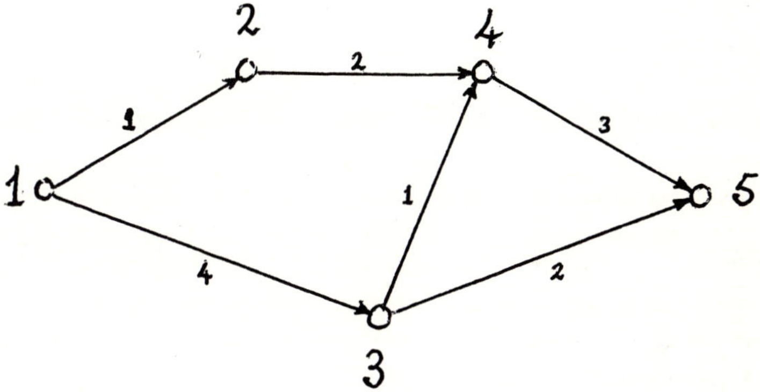

Fig 7-1: A network.

$$= \sum_{k=1}^{n} {}_{\oplus'} ((\{\underline{B}\}_{ki})^{*} \otimes' (\{\underline{A}\}_{jk})^{*}) \qquad \text{(by } N_3)$$

$$= \sum_{k=1}^{n} {}_{\oplus'} (\{\underline{B}^{*}\}_{ik} \otimes' \{\underline{A}^{*}\}_{kj}) \qquad \text{(by (7-3)}$$

$$= \{\underline{B}^{*} \otimes' \underline{A}^{*}\}_{ij} \qquad \text{(by dual of (5-4))}$$

$$= \{g(\underline{B}) \otimes' g(\underline{A})\}_{ij}$$

Hence $g(\underline{A}\otimes\underline{B}) = g(\underline{B}) \otimes' g(\underline{A})$

So N_3 is satisfied for $(\mathcal{M}_{nn}, \oplus, \otimes)$, $(\mathcal{M}_{nn}, \oplus', \otimes')$. ●

Proposition 7-5. If $(E_1, \oplus, \otimes, \oplus', \otimes')$ is a self-conjugate belt, then $\mathcal{M}^{*}_{mn} = \mathcal{M}_{nm}$ for all integers $m, n \geq 1$, and $(\mathcal{M}_{nn}, \oplus, \otimes, \oplus', \otimes')$ is a self-conjugate belt. ●

7-3 . Two Examples

We conclude this chapter with a couple of examples, for the principal interpretation, showing a possible physical interpretation of the conjugate of a matrix.

The diagram of Fig 7-1 is intended to suggest a (highly simplified!) activity network for building a hospital. Nodes 2 , 3 and 4 of the network denote construction activities, and nodes 1 and 5 are the "start" and "finish" activity respectively. An arrow connecting two nodes indicates that the activity at the head of the arrow cannot be started before the activity at the tail of the arrow has finished. The number associated with each arrow indicates that at least that amount of time must elapse between the starting of the tail-activity and the starting of the head-activity (the number thus includes an allowance for the activity-time of the tail-activity itself, as well as the delay exerted on the head-activity by the tail-activity).
We may now analyse this network by forward recursion or backward recursion.

Under forward recusion, we assume that the date of admission of the first patient has been specified as a target, and, in order not to tie up money unnecessarily in a wholly or partly finished hospital standing idle, we wish to start each preceding activity as late as possible, consistent with meeting that target.

Following the analysis of Section 1-2.3, we set up the matrix $\underline{A}$:

$$\underline{A} = \begin{bmatrix} 0 & -1 & -4 & \infty & \infty \\ \infty & 0 & \infty & -2 & \infty \\ \infty & \infty & 0 & -1 & -2 \\ \infty & \infty & \infty & 0 & -3 \\ \infty & \infty & \infty & \infty & 0 \end{bmatrix} \qquad (7\text{-}5)$$

and are led to the problem: $\underline{t} = \underline{A} \otimes' \underline{t}$ (7-6)

Under backward recursion, we assume that the starting date of the project is given, and that we wish to start each subsequent activity as early as possible. We wish to know just how early that may be - in particular, when can we admit the first patient?

Arguing along the general lines of Section 1-2.1, we evidently have to satisfy:

$$t_i = \max_{1 \leq j \leq 5} (a_{ij} + t_j) \qquad (i=1,\ldots, 5)$$

where t_i $(i=1,\ldots, 5)$ are the relevant start-times, and a_{ij} now represents the irreducible amount of time by which the start of activity j must precede the start of activity i (and $a_{ij} = -\infty$ if activity j does not constrain activity i in this way).

In other words we have the problem:

$$\underline{t} = \underline{A} \otimes \underline{t}, \tag{7-7}$$

where:

$$\underline{A} = \begin{bmatrix} 0 & -\infty & -\infty & -\infty & -\infty \\ 1 & 0 & -\infty & -\infty & -\infty \\ 4 & -\infty & 0 & -\infty & -\infty \\ -\infty & 2 & 1 & 0 & -\infty \\ -\infty & -\infty & 2 & 3 & 0 \end{bmatrix} \tag{7-8}$$

Evidently the $\underline{A}$ of (7-6) is the conjugate of the $\underline{A}$ of (7-8). To rationalise the notation, let us now call the matrix in (7-6) $\underline{A}^*$. Then we can say that the matrices $\underline{A}$ and $\underline{A}^*$ define the structure of the system as we analyse it respectively forward or backward in time.

We may generalise this. If we have a repetitive cyclic process as considered in 1-2.1, we may specify the earliest start-times for cycles 1,2,..., n, given the start-times $\underline{x}$ for cycle zero, by the sequence:

$$\underline{A} \otimes \underline{x}, \quad \underline{A}^2 \otimes \underline{x}, \ \ldots, \ \underline{A}^n \otimes \underline{x} \tag{7-9}$$

Dually, if we are given the start-times $\underline{y}$ for cycle n, we may specify the latest start-times for the preceding cycles, consistent with the given information $\underline{y}$, by the sequence:

$$\underline{A}^* \otimes' \underline{y}, \ \underline{A}^{2*} \otimes' \underline{y}, \ \ldots, \ \underline{A}^{n*} \otimes' \underline{y} \tag{7-10}$$

(where the notation $\underline{A}^{2*}$ denotes $\underline{A}^* \otimes' \underline{A}^*$ and in general $\underline{A}^{r*} = \underline{A}^{(r-1)*} \otimes' \underline{A}^*$).

The sequence (7-10) runs backward in time. The conjugacy $\underline{A} \leftrightarrow \underline{A}^*$ thus has an interesting physical interpretation in this situation, and is related to the concept of the inverse problem in the theory of machine-scheduling [43].

We shall consider the relationship between (7-9) and (7-10) further later.

8. AA^* RELATIONS

8-1. Pre-residuation

Let E_1 be a self-conjugate belt. We shall say that E_1 is pre-residuated if it satisfies the following axiom X_{13}:

X_{13}: If $J \subset E_1$ is a finite set of elements, then:

$$\left.\begin{aligned} \forall x \in E_1, \ [\textstyle\sum_{a \in J}{}_{\oplus'} (a^* \otimes' a)] \otimes x \geq x \\ x \otimes [\textstyle\sum_{a \in J}{}_{\oplus'} (a \otimes' a^*)] \geq x \end{aligned}\right\} \tag{8-1}$$

The reason for choosing the term pre-residuated will emerge in Chapter 9.

Theorem 8-1. Let E_1 be a pre-residuated belt. Then E_1 satisfies:

X'_{13}: If $J \subset E_1$ is a finite set of elements, then:

$$\left.\begin{aligned} \forall x \in E_1, \ [\textstyle\sum_{a \in J}{}_{\oplus} (a \otimes a^*)] \otimes' x \leq x \\ x \otimes' [\textstyle\sum_{a \in J}{}_{\oplus} (a^* \otimes a)] \leq x \end{aligned}\right\} \tag{8-2}$$

Proof: Applying Proposition 7-2 to the second half of axiom X_{13}, we have:

$$(x \otimes [\textstyle\sum_{a \in J}{}_{\oplus'} (a \otimes' a^*)])^* \leq x^*$$

i.e. $[\sum_{a \in J}{}_{\oplus} (a \otimes a^*)] \otimes' x^* \leq x^*$ (by N'_2, N_3, N'_3)

But x was arbitrary, so (since * is a bijection) we may write x^* for x, to give the first half of X'_{13}. The second half is proved similarly. ●

We remark that, since * is a bijection, we may interchange a and a^* throughout X_{13}, X'_{13}.

Theorem 8-2. All division belts, and all blogs, are pre-residuated. In particular, the principal interpretation, and the 3-element blog ③, are pre-residuated.

Proof. Proposition 7-3 has already established that all division belts, and all blogs, are self-conjugate. So let J be a finite subset of a division belt or blog E_1, and let $a \in J$, $x \in E_1$ be arbitrary.

If E_1 is a division belt, then $a^* \otimes' a = \emptyset$ by (7-1).

If E_1 is a blog, then $a^* \otimes' a = \emptyset$ or $+\infty$ by (7-2) and Lemma 4-13.

Hence in all cases $\sum_{a \in J}{}_{\oplus'} (a^* \otimes' a) \geq \emptyset$ and the first half of axiom X_{13} follows by (2-9). The second half is proved similarly. ●

In the following result, we recall that the symbol '⊣' appended to a matrix identity or relation, means "when the matrices are conformable for the relevant operations".

Lemma 8-3. Let E_1 be a pre-residuated belt and J a given finite set of matrices over E_1. Let X be an arbitrary matrix over E_1. Then:

$$\underline{P} \otimes' \underline{X} \leq \underline{X} \leq \underline{Q} \otimes \underline{X} \;\dashv$$

$$\underline{X} \otimes' \underline{P} \leq \underline{X} \leq \underline{X} \otimes \underline{Q} \;\dashv$$

where
$$\underline{P} = \sum_{\underline{A}\varepsilon J}{}_{\oplus}(\underline{A} \otimes \underline{A}^*) \dashv$$

$$\underline{Q} = \sum_{A\varepsilon J}{}_{\oplus'}(\underline{A}^* \otimes' \underline{A}) \dashv$$

Proof. We prove first $\underline{Q} \otimes \underline{X} \geq \underline{X} \dashv$ Now $\{\underline{Q} \otimes \underline{X}\}_{ij}$ is by definition:

$$\sum_{s}{}_{\oplus}\left(\left[\sum_{\underline{A}\varepsilon J}{}_{\oplus'}\left(\sum_{r_{\underline{A}}}{}_{\oplus'}(\{\underline{A}^*\}_{ir_{\underline{A}}} \otimes' \{\underline{A}\}_{r_{\underline{A}}s})\right)\right] \otimes \{\underline{X}\}_{sj}\right) \dashv$$

where r_A, s are row indices for $\underline{A}$, $\underline{X}$ respectively.

Opening w.r.t. the index s we have:

$$\forall s: \{\underline{Q} \otimes \underline{X}\}_{ij} \geq \left[\sum_{\underline{A}\varepsilon J}{}_{\oplus'}\left(\sum_{r_{\underline{A}}}{}_{\oplus'}(\{\underline{A}^*\}_{ir_{\underline{A}}} \otimes' \{\underline{A}\}_{r_{\underline{A}}s})\right)\right] \otimes \{\underline{X}\}_{sj} \quad \dashv$$

Taking in particular s=i:

$$\{\underline{Q} \otimes \underline{X}\}_{ij} \geq \left[\sum_{\underline{A}\varepsilon J}{}_{\oplus'} \sum_{r_{\underline{A}}}{}_{\oplus'}(\{\underline{A}^*\}_{ir_{\underline{A}}} \otimes' \{\underline{A}\}_{r_{\underline{A}}i})\right] \otimes \{\underline{X}\}_{ij} \quad \dashv$$

$$= \left[\sum_{\underline{A}\varepsilon J}{}_{\oplus'} \sum_{r}{}_{\oplus'}((\{\underline{A}\}_{r_{\underline{A}}i})^* \otimes' \{\underline{A}\}_{r_{\underline{A}}i})\right] \otimes \{\underline{X}\}_{ij} \dashv \qquad \text{(by (7-3))}$$

$$\geq \{\underline{X}\}_{ij} \quad \text{by } X_{13}$$

This proves $\underline{Q} \otimes \underline{X} \geq \underline{X} \dashv$ and the remaining cases are proved similarly. ●

Corollary 8-4. If E_1 is a pre-residuated belt, then $\mathcal{M}_{nn}$ is a pre-residuated belt for each integer $n \geq 1$.

Proof. Follows immediately from Proposition 7-5 and Theorem 8-3. ●

8-2 . Alternating AA^* Products

We now prove a number of identities and inequalities involving an arbitrary matrix $\underline{A}$ and its conjugate $\underline{A}^*$, over a pre-residuated belt in which axiom X_{12} (Section 4-1) is also satisfied.

Theorem 8-5. Let E_1 be a pre-residuated belt satisfying axiom X_{12} and $\underline{A}$ an arbitrary matrix over E_1. Then the products:

$\underline{A} \otimes (\underline{A}^* \otimes' \underline{A})$; $\underline{A} \otimes'(\underline{A}^* \otimes \underline{A})$; $(\underline{A} \otimes' \underline{A}^*) \otimes \underline{A}$; $(\underline{A} \otimes \underline{A}^*) \otimes' \underline{A}$

<u>always exist, and are all equal to $\underline{A}$. Similarly the products</u>

$$\underline{A}^* \otimes (\underline{A} \otimes' \underline{A}^*);\quad \underline{A}^* \otimes' (\underline{A} \otimes \underline{A}^*);\quad (\underline{A}^* \otimes' \underline{A}) \otimes \underline{A}^*;\quad (\underline{A}^* \otimes \underline{A}) \otimes' \underline{A}^*$$

<u>always exist, and are all equal to $\underline{A}^*$.</u>

<u>Proof</u>. It is evident that the matrices $\underline{A}$, $\underline{A}^*$ are conformable for all these products.

Moreover: $\underline{A} \otimes (\underline{A}^* \otimes' \underline{A}) \geq \underline{A}$ (by Lemma 8-3)

But $\underline{A} \otimes (\underline{A}^* \otimes' \underline{A}) \leq (\underline{A} \otimes \underline{A}^*) \otimes' \underline{A}$ (by axiom X_{12} in Theorem 6-3)

$\leq \underline{A}$ (by Lemma 8-3)

Hence $\underline{A} \otimes (\underline{A}^* \otimes' \underline{A}) = \underline{A}$, and the remaining results follow similarly and dually. ●

<u>Theorem 8-6. Let E_1 be a pre-residuated belt satisfying axiom X_{12}, and $\underline{A}$ an arbitrary matrix over E_1. Then:</u>

$$(\underline{A} \otimes \underline{A}^*) \otimes' (\underline{A} \otimes \underline{A}^*) = \underline{A} \otimes \underline{A}^*$$

$$(\underline{A}^* \otimes \underline{A}) \otimes' (\underline{A}^* \otimes \underline{A}) = \underline{A}^* \otimes \underline{A}$$

$$(\underline{A} \otimes' \underline{A}^*) \otimes (\underline{A} \otimes' \underline{A}^*) = \underline{A} \otimes' \underline{A}^*$$

$$(\underline{A}^* \otimes' \underline{A}) \otimes (\underline{A}^* \otimes' \underline{A}) = \underline{A}^* \otimes' \underline{A}$$

<u>The relevant products always exist.</u>

<u>Proof</u>. It is evident that $\underline{A}$, $\underline{A}^*$ are always conformable for the given multiplications.

Now, $(\underline{A} \otimes \underline{A}^*) \otimes' (\underline{A} \otimes \underline{A}^*) \leq (\underline{A} \otimes \underline{A}^*)$ (by Lemma 8-3)

On the other hand:

$(\underline{A} \otimes \underline{A}^*) \otimes' (\underline{A} \otimes \underline{A}^*) \geq ((\underline{A} \otimes \underline{A}^*) \otimes' \underline{A}) \otimes \underline{A}^*$ (by axiom X_{12} in Theorem 6-3)

$= \underline{A} \otimes \underline{A}^*$ (by Theorem 8-5)

Hence $(\underline{A} \otimes \underline{A}^*) \otimes' (\underline{A} \otimes \underline{A}^*) = \underline{A} \otimes \underline{A}^*$ and the remaining results follow similarly and dually. ●

Let us now write a "word" consisting of alternating occurrences of the letters $\underline{A}$ and $\underline{A}^*$, starting with either, the case of one single letter being allowed. For example:

$$\underline{A}\,\underline{A}^*\ ;\ \underline{A}^*\,\underline{A}\,\underline{A}^*\ ;\ \underline{A}^*\ ;\ \underline{A}\,\underline{A}^*\,\underline{A}\,\underline{A}^*\,\underline{A}\,\underline{A}^*$$

If the word consist of $k > 1$ letters, let us insert $(k-1)$ alternating occurrences of copies of the symbols $\otimes$ and $\otimes'$, starting with either, the case of one single insertion being (necessarily) allowed. For example:

$$\underline{A} \otimes' \underline{A}^*\ ;\ \underline{A}^* \otimes \underline{A} \otimes' \underline{A}^*\ ;\ \underline{A}^*\ ;\ \underline{A} \otimes \underline{A}^* \otimes' \underline{A} \otimes \underline{A}^* \otimes' \underline{A} \otimes \underline{A}^*$$

Let us finally insert brackets in any arbitrary way so as to make a well-formed algebraic expression. For example:

$$\underline{A} \otimes' \underline{A}^*\ ;\ \underline{A}^* \otimes (\underline{A} \otimes' \underline{A}^*)\ ;\ \underline{A}^*\ ;\ ((\underline{A} \otimes (\underline{A}^* \otimes' \underline{A})) \otimes (\underline{A}^* \otimes' \underline{A})) \otimes \underline{A}^*$$

Any algebraic expression which can be constructed in this way will be called an alternating $\underline{A}\ \underline{A}^*$-product. Let us classify these expressions as follows.

If an expression contains an odd number of letters ($\underline{A}$ or $\underline{A}^*$) we shall say that it is of type $\underline{A}$ if it begins (and ends) with an $\underline{A}$, and that it is of type $\underline{A}^*$ if it begins (and ends) with an $\underline{A}^*$. If an expression contains an even number of letters ($\underline{A}$ or $\underline{A}^*$) we shall say that it is of type

$$\underline{A} \otimes \underline{A}^* \text{ or } \underline{A} \otimes' \underline{A}^* \quad\text{or}\quad \underline{A}^* \otimes \underline{A} \quad\text{or}\quad \underline{A}^* \otimes' \underline{A}$$

exactly according to the first two letters with separating operator, regardless of how the brackets lie in the total expression.

For example:

$\underline{A} \otimes' \underline{A}^*$	is of type $\underline{A} \otimes' \underline{A}^*$
$\underline{A}^* \otimes (\underline{A} \otimes' \underline{A}^*)$	is of type $\underline{A}^*$
$\underline{A}^*$	is of type $\underline{A}^*$
$((\underline{A} \otimes (\underline{A}^* \otimes' \underline{A})) \otimes (\underline{A}^* \otimes' \underline{A})) \otimes \underline{A}^*$	is of type $\underline{A} \otimes \underline{A}^*$

Theorem 8-7. Let E_1 be a pre-residuated belt satisfying axiom X_{12}, and $\underline{A}$ an arbitrary matrix over E_1. Then every alternating $\underline{A}\ \underline{A}^*$-product $\underline{P}$ is well-defined, and if $\underline{P}$ is of type $\underline{Q}$, then $\underline{P} = \underline{Q}$.

Proof. We proceed by induction on the number k of letters in the word from which $\underline{P}$ was produced. If k=1 or 2 the theorem is trivial. If k=3 then Theorem 8-5 applies. We now assume $k \geq 4$. Now $\underline{P}$ is a product of two $\underline{A}\ \underline{A}^*$ products $\underline{P}_1$, $\underline{P}_2$ each having less than k letters. A little deliberation shows that the sixteen cases exhibited in Fig 8-1 may arise.

For example, $\underline{P}_1$ must be one of the six types defined above. Suppose $\underline{P}_1$ is of type $\underline{A}$. Then $\underline{P}_1$ ends with an $\underline{A}$, so $\underline{P}_2$ must begin with an $\underline{A}^*$. Thus $\underline{P}_2$ is of type $\underline{A}^*$, type $\underline{A}^* \otimes \underline{A}$ or type $\underline{A}^* \otimes' \underline{A}$. Suppose (to take the third case from the table for illustration) that $\underline{P}_2$ is of type $\underline{A}^* \otimes \underline{A}$. Then the first operator in $\underline{P}_2$ is $\otimes$ so the operator between $\underline{P}_1$ and $\underline{P}_2$ must be $\otimes'$. The number of letters in $\underline{P}$ = number in $\underline{P}_1$ + number in $\underline{P}_2$ = an odd number, so $\underline{P}$ is of type $\underline{A}$, since $\underline{P}_1$, and thus $\underline{P}$ itself, begins with an $\underline{A}$.

Now by an obvious induction hypothesis we can assume (for this same case) that $\underline{P}_1$ and $\underline{P}_2$ are well-defined matrix products and that $\underline{P}_1 = \underline{A}$ and $\underline{P}_2 = \underline{A}^* \otimes \underline{A}$. Hence by Theorem 8-5, $\underline{P} = \underline{P}_1 \otimes' \underline{P}_2$ is a well-defined matrix product and $\underline{P} = \underline{A}$. All the other cases are treated analogously, using Theorem 8-5 or 8-6 as appropriate to complete the induction. ●

8-3. Modified $\underline{A}\ \underline{A}^*$ Products

We examine now some matrix products which arise when an $\underline{A}$ or an $\underline{A}^*$ in an alternating $\underline{A}\ \underline{A}^*$ product is replaced by another matrix $\underline{X}$. These results will be important later in our discussion of residuation.

$\underline{P}_1$ of type	$\underline{P}_2$ of type	$\underline{P}$ is	of type
$\underline{A}$	$\underline{A}^*$	$\underline{P}_1 \otimes \underline{P}_2$	$\underline{A} \otimes \underline{A}^*$
$\underline{A}$	$\underline{A}^*$	$\underline{P}_1 \otimes' \underline{P}_2$	$\underline{A} \otimes' \underline{A}^*$
$\underline{A}$	$\underline{A}^* \otimes \underline{A}$	$\underline{P}_1 \otimes' \underline{P}_2$	$\underline{A}$
$\underline{A}$	$\underline{A}^* \otimes' \underline{A}$	$\underline{P}_1 \otimes \underline{P}_2$	$\underline{A}$
$\underline{A}^*$	$\underline{A}$	$\underline{P}_1 \otimes \underline{P}_2$	$\underline{A}^* \otimes \underline{A}$
$\underline{A}^*$	$\underline{A}$	$\underline{P}_1 \otimes' \underline{P}_2$	$\underline{A}^* \otimes' \underline{A}$
$\underline{A}^*$	$\underline{A} \otimes \underline{A}^*$	$\underline{P}_1 \otimes' \underline{P}_2$	$\underline{A}^*$
$\underline{A}^*$	$\underline{A} \otimes' \underline{A}^*$	$\underline{P}_1 \otimes \underline{P}_2$	$\underline{A}^*$
$\underline{A} \otimes \underline{A}^*$	$\underline{A}$	$\underline{P}_1 \otimes' \underline{P}_2$	$\underline{A}$
$\underline{A} \otimes \underline{A}^*$	$\underline{A} \otimes \underline{A}^*$	$\underline{P}_1 \otimes' \underline{P}_2$	$\underline{A} \otimes \underline{A}^*$
$\underline{A} \otimes' \underline{A}^*$	$\underline{A}$	$\underline{P}_1 \otimes \underline{P}_2$	$\underline{A}$
$\underline{A} \otimes' \underline{A}^*$	$\underline{A} \otimes' \underline{A}^*$	$\underline{P}_1 \otimes \underline{P}_2$	$\underline{A} \otimes' \underline{A}^*$
$\underline{A}^* \otimes \underline{A}$	$\underline{A}^*$	$\underline{P}_1 \otimes' \underline{P}_2$	$\underline{A}^*$
$\underline{A}^* \otimes \underline{A}$	$\underline{A}^* \otimes \underline{A}$	$\underline{P}_1 \otimes' \underline{P}_2$	$\underline{A}^* \otimes \underline{A}$
$\underline{A}^* \otimes' \underline{A}$	$\underline{A}^*$	$\underline{P}_1 \otimes \underline{P}_2$	$\underline{A}^*$
$\underline{A}^* \otimes' \underline{A}$	$\underline{A}^* \otimes' \underline{A}$	$\underline{P}_1 \otimes \underline{P}_2$	$\underline{A}^* \otimes' \underline{A}$

Fig 8-1 Types of $\underline{AA}^*$ - Product

<u>Theorem 8-8. Let E_1 be a pre-residuated belt satisfying axiom X_{12}. The following inequalities apply for arbitrary matrices $\underline{A}$, $\underline{X}$ over E_1:</u>

$$\left.\begin{array}{l} \underline{A}^* \otimes (\underline{A} \otimes' \underline{X}) \\ \underline{A} \otimes (\underline{A}^* \otimes' \underline{X}) \\ (\underline{X} \otimes' \underline{A}) \otimes \underline{A}^* \\ (\underline{X} \otimes' \underline{A}^*) \otimes \underline{A} \end{array}\right\} \leq \underline{X} \leq \left\{\begin{array}{l} \underline{A}^* \otimes' (\underline{A} \otimes \underline{X}) \\ \underline{A} \otimes' (\underline{A}^* \otimes \underline{X}) \\ (\underline{X} \otimes \underline{A}) \otimes' \underline{A}^* \\ (\underline{X} \otimes \underline{A}^*) \otimes' \underline{A} \end{array}\right.$$

<u>Proof</u>. Let $\underline{A} \in \mathcal{M}_{mn}$ and $\underline{X} \in \mathcal{M}_{nr}$. We have:

$$\begin{aligned} \underline{A}^* \otimes' (\underline{A} \otimes \underline{X}) &\geq (\underline{A}^* \otimes' \underline{A}) \otimes \underline{X} && \text{(by axiom } X_{12}) \\ &\geq \underline{X} && \text{(by Lemma 8-3)} \end{aligned}$$

The other relations are proved similarly and dually. ●

Theorem 8-8 can in fact be regarded as the first of a hierarchy of relationships in which the letter $\underline{X}$ is substituted for the first or last letter in an alternating

$\underline{A}\ \underline{A}^*$ product. The theorem sets the pattern for products involving an odd number of letters, and Theorem 8-9 (below) sets the pattern for products involving an even number. An evident inductive argument provides an analogue of Theorem 8-7, but we shall not carry this out because it involves yet another tedious examination of cases, and does not contribute much to the general theory. The following theorem on quadruple products, however, does have practical application later.

For a given alternating $\underline{A}\ \underline{A}^*$ product $\underline{P}$ with $k > 1$ letters, let us define $\underline{P}(\underline{X})$ to be the formal product obtained when the last (rightmost) letter ($\underline{A}$ or $\underline{A}^*$) is replaced by an $\underline{X}$, and $\underline{P}[\underline{X}]$ to be the formal product obtained when the first (leftmost) letter ($\underline{A}$ or $\underline{A}^*$) is replaced by an $\underline{X}$.

Theorem 8-9. Let E_1 be a pre-residuated belt satisfying axiom X_{12}, and $\underline{A}, \underline{X}$ be arbitrary matrices over E_1. If $\underline{P}$ is an alternating $\underline{A}\ \underline{A}^*$ product containing 4 letters and $\underline{P}$ is of type $\underline{Q}$, then:

$$\underline{P}(\underline{X}) = \underline{Q}(\underline{X}) \dashv \quad ; \quad \text{and} \quad \underline{P}[\underline{X}] = \underline{Q}[\underline{X}] \dashv .$$

Proof. Suppose first that $\underline{P}$ is of type $\underline{A} \otimes \underline{A}^*$. Then $\underline{Q}(\underline{X})$ is $\underline{A} \otimes \underline{X}$, and $\underline{P}$ is:

$$\underline{P}_1 = ((\underline{A} \otimes \underline{A}^*) \otimes' \underline{A}) \otimes \underline{A}^* \quad ; \quad \text{or } \underline{P}_2 = (\underline{A} \otimes \underline{A}^*) \otimes' (\underline{A} \otimes \underline{A}^*) \ ;$$

$$\text{or} \quad \underline{P}_3 = \underline{A} \otimes (\underline{A}^* \otimes' (\underline{A} \otimes \underline{A}^*))$$

So $\underline{P}(\underline{X})$ is:

$$\underline{P}_1(\underline{X}) = ((\underline{A} \otimes \underline{A}^*) \otimes' \underline{A}) \otimes \underline{X} \quad ; \quad \text{or } \underline{P}_2(\underline{X}) = (\underline{A} \otimes \underline{A}^*) \otimes' (\underline{A} \otimes \underline{X}) \ ;$$

$$\text{or } \underline{P}_3(\underline{X}) = \underline{A} \otimes (\underline{A}^* \otimes' (\underline{A} \otimes \underline{X})).$$

We now suppose $\underline{A} \in \mathcal{M}_{mn}$ and $\underline{X} \in \mathcal{M}_{nr}$, so all relevant products exist. We have:

	$\underline{P}_1(\underline{X})$	$=$	$\underline{A} \otimes \underline{X}$	(by Theorem 8-5)
And	$\underline{P}_2(\underline{X})$	$\geq$	$\underline{P}_1(\underline{X})$	(by axiom X_{12})
But	$\underline{P}_2(\underline{X})$	$\leq$	$\underline{A} \otimes \underline{X}$	(by Lemma 8-3)
Therefore	$\underline{P}_2(\underline{X})$	$=$	$\underline{A} \otimes \underline{X}$ also.	
Furthermore	$\underline{P}_3(\underline{X})$	$\leq$	$\underline{P}_2(\underline{X})$	(by axiom X_{12})

But from Theorem 8-8 we have

$$\underline{A}^* \otimes' (\underline{A} \otimes \underline{X}) \geq \underline{X}$$

and since multiplications are isotone we infer:

$$\underline{P}_3(\underline{X}) = \underline{A} \otimes (\underline{A}^* \otimes' (\underline{A} \otimes \underline{X})) \geq \underline{A} \otimes \underline{X}$$

Hence we have $\underline{P}_3(\underline{X}) = \underline{A} \otimes \underline{X} = \underline{Q}(\underline{X}) = \underline{P}_2(\underline{X}) = \underline{P}_1(\underline{X})$, for all cases when $\underline{Q} = \underline{A} \otimes \underline{A}^*$. The case of $\underline{P}[\underline{X}]$ for this same $\underline{Q}$, and of $P(\underline{X})$ and $\underline{P}[\underline{X}]$ for the three other possible formulae for $\underline{Q}$, are handed similarly. ●

We remark in conclusion that the results of Sections 8-2 and 8-3 apply in particular when E_1 is a division belt or a blog, since by Theorem 8-2, division belts and blogs are pre-residuated; and axiom X_{12} holds for division belts (which have a self-dual associative multiplication) and for blogs (by Proposition 4-2). In particular, the results of Sections 8-2 and 8-3 apply if E_1 is given the principal interpretation, or is the 3-element blog ③.

Moreover, since E_1 is formally identical to $\mathcal{M}_{11}$, the results of Sections 8-2 and 8-3 apply to E_1 itself if we write scalars in place of matrices throughout.

8-4. Some Bijections

We might summarise our results on alternating $\underline{A}\ \underline{A}^*$ products (Section 8-2) by saying that adjacent letters $\underline{A}$ and $\underline{A}^*$ in such an expression may be cancelled without changing the value of the expression. In other words, $\underline{A}$ and $\underline{A}^*$ act a little as one another's inverse. We shall return to this idea when we discuss the solution of linear equations in Chapter 14 , but first we must make it more precise.

We shall find it useful in the subsequent discussion to use the notations Ran f and Dom f to denote respectively the range and domain of a given function f. The restriction of f to a given set $U \subset \text{Dom} f$, written $f \backslash U$, is a function h such that Dom h = U and h(x) = f(x) for all $x \in U$.

It will also be convenient, in discussing systems of 1-ary functions, to use the notation o to denote composition. Thus, if f, g are given 1-ary functions such that $\text{Ran } g \cap \text{Dom } f$ is not empty then f o g is a function such that:

$$\text{Dom}(f \circ g) = \{x \mid x \in \text{Dom } g \text{ and } g(x) \in \text{Dom } f\}$$

and

$$(f \circ g)(x) = f(g(x)) \text{ for all } x \in \text{Dom}(f \circ g).$$

The following result, which is virtually self-evident, will be used to make precise our foregoing remarks about $\underline{A}$ and $\underline{A}^*$.

Proposition 8-10. Let S, T be given sets and $f: S \to T$ and $g: T \to S$ given functions such that:

$$\left.\begin{aligned} f \circ g \circ f &= f \\ g \circ f \circ g &= g \end{aligned}\right\} \quad (8\text{-}3)$$

Then $f \backslash \text{Ran } g$ and $g \backslash \text{Ran } f$ are mutually inverse bijections. ●

Our matrix algebra abounds with such pairs of functions. For example, let $\underline{A} \in \mathcal{M}_{mn}$ and define $f: \mathcal{M}_{np} \to \mathcal{M}_{mp}$ and $g: \mathcal{M}_{mp} \to \mathcal{M}_{np}$ by: $f(\underline{X}) = \underline{A} \otimes \underline{X}$ and $g(\underline{Y}) = \underline{A}^* \otimes' \underline{Y}$. Then Theorem 8-9 implies that relations (8-3) are satisfied. Similarly if we define f and g by $f(\underline{X}) = \underline{X}^* \otimes' \underline{A}^*$ and $g(\underline{Y}) = \underline{A}^* \otimes' \underline{Y}^*$, relations (8-3) are again satisfied.

The interaction of left-right symmetry with duality enables us to deduce many pairs of such functions from Theorem 8-9, and it is convenient to have a notation in which to express this succinctly.

Consider the diagram of Fig 8-2. We shall say that such a diagram is valid if the following are true:

Z_1 : S and T are given sets

Z_2 : $f : S \to T$ and $g : T \to S$ are given mappings

Z_3 : U = Ran g and V = Ran f.

Z_4 : f\U and g\V are mutually inverse bijections.

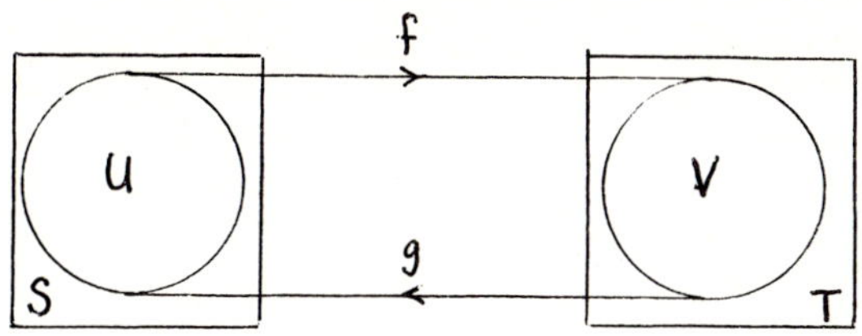

Fig 8-2 Valid Diagrams

Furthermore, if $\underline{A} \in \mathcal{M}_{mn}$, and $\mathcal{W}_{np} \subset \mathcal{M}_{np}$ is a set of matrices, we define

$$\underline{A} \otimes \mathcal{W}_{np} = \{\underline{A} \otimes \underline{X} \mid \underline{X} \in \mathcal{W}_{np}\}$$

with similar definitions for $\underline{A} \otimes' \mathcal{W}_{np}$, $\mathcal{W}_{pm} \otimes \underline{A}$ and $\mathcal{W}_{pm} \otimes' \underline{A}$.

Then the following result is an immediate deduction from Theorem 8-9 and Proposition 8-10. (The notation $\mathcal{W}_{mn}$, etc., is used in Figs 8-3 and 8-4 in order that these figures may be used again later in another context.)

Proposition 8-11. Let E_1 be a pre-residuated belt satisfying axiom X_{12}. Then the diagrams of Figs 8-3 and 8-4 are valid for arbitrary $\underline{A} \in \mathcal{M}_{mn}$, when the notation $\mathcal{W}_{np}$ (etc) is taken to be synonymous with the notation $\mathcal{M}_{np}$ (etc).

8-5. A Worked Example

It will perhaps be helpful to illustrate the results of the present chapter with a numerical example for the principal interpretation. Accordingly, let:

$$\underline{A} = \begin{bmatrix} 2 & 0 & +\infty \\ -\infty & -\infty & -1 \\ 3 & -3 & 0 \end{bmatrix}, \qquad \underline{X} = \begin{bmatrix} -1 & +\infty \\ -1 & -2 \\ 0 & -\infty \end{bmatrix}$$

Define $\underline{Y} = \underline{A} \otimes \underline{X}$. Then $\underline{Y} \in$ Ran f, where $f: \mathcal{M}_{32} \to \mathcal{M}_{32}$ is defined by: $f(\underline{X}) = \underline{A} \otimes \underline{X}$.

$$\underline{Y} = \begin{bmatrix} +\infty & +\infty \\ -1 & -\infty \\ 2 & +\infty \end{bmatrix} \quad ; \quad \underline{A}^* = \begin{bmatrix} -2 & +\infty & -3 \\ 0 & +\infty & 3 \\ -\infty & 1 & 0 \end{bmatrix}$$

$$\underline{A}^* \otimes' \underline{Y} = \begin{bmatrix} -1 & +\infty \\ 5 & +\infty \\ 0 & -\infty \end{bmatrix} \quad ; \quad \underline{A} \otimes (\underline{A}^* \otimes' \underline{Y}) = \begin{bmatrix} +\infty & +\infty \\ -1 & -\infty \\ 2 & +\infty \end{bmatrix}$$

W mp
X ↦ X* ⊗' A
W pn
A⊗'W np
W pm ⊗'A
A ⊗' Y* ↤ Y
W ↦ A⊗'W
X ↦ A* ⊗ X
Z ↦ Z ⊗' A
Y ↦ Y ⊗ A*
W ↦ W* ⊗ A*
A*⊗W mp
W pn⊗A*
W np
W pm
A* ⊗ Z* ↤ Z
W mp
X ↦ X* ⊗ A
W pn
A⊗W np
W pm ⊗A
A ⊗ Y* ↤ Y
W ↦ A ⊗ W
X ↦ A* ⊗' X
Z ↦ Z ⊗ A
Y ↦ Y ⊗' A*
W ↦ W* ⊗' A*
A*⊗'W mp
W pn ⊗'A*
W np
W pm
A* ⊗' Z* ↤ Z

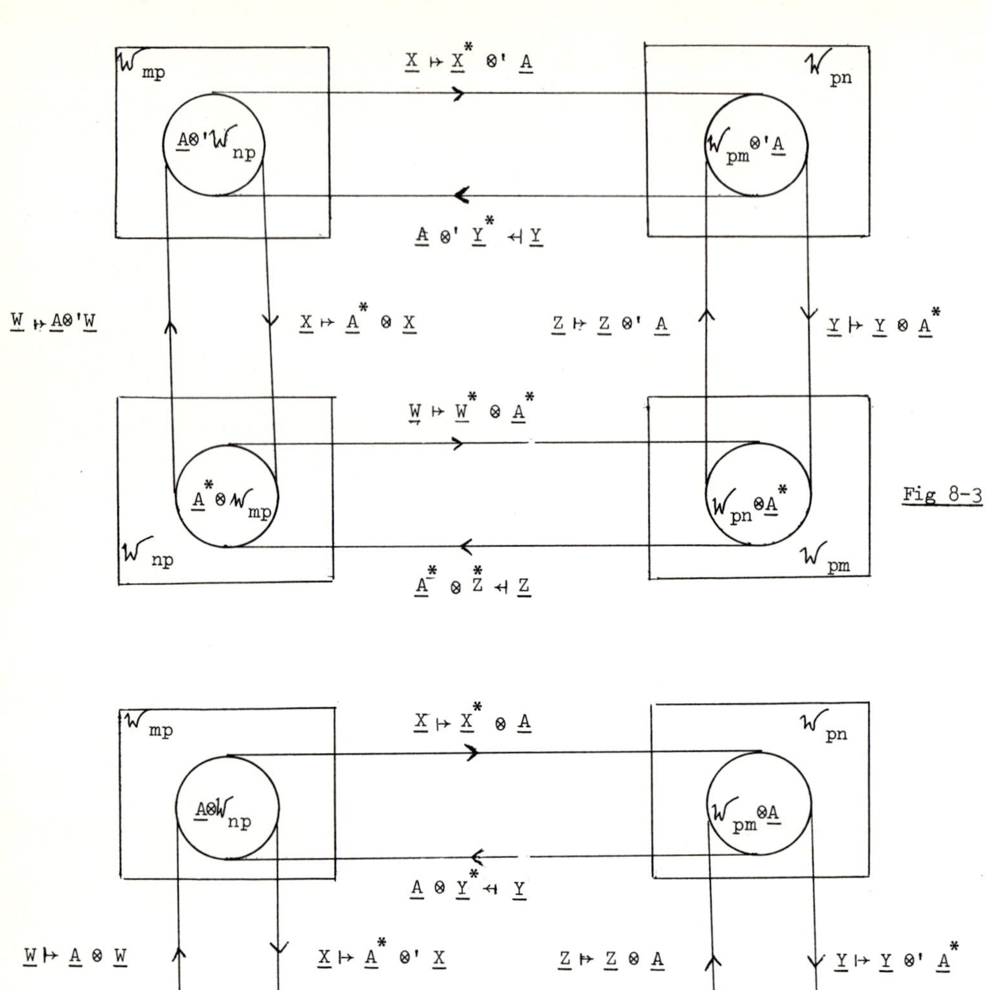

Fig 8-3

Fig 8-4

Hence we confirm that, as stated in Theorem 8-8:

$$\underline{A}^* \otimes' \underline{Y} = A^* \otimes' (\underline{A} \otimes \underline{X}) \geq \underline{X}$$

and as stated in Proposition 8:

$$(f \circ g)(\underline{Y}) = \underline{Y}$$

where $g: \mathcal{M}_{32} \to \mathcal{M}_{32}$ is defined by $g(\underline{Y}) = \underline{A}^* \otimes' \underline{Y}$

9. SOME SCHEDULE ALGEBRA

9-1. Feasibility and Compatibility

In Section 7-3, we observed how the relationship between forward planning and backward planning in a class of scheduling problems was reflected by the relation of conjugacy $\underline{A} \leftrightarrow \underline{A}^*$. With the notation of that section, suppose that the (nxn) matrix $\underline{A}$ defines a process for which the vector of earliest permissible start-times for cycle nought is $\underline{x} \in E_n$. Then the vector of actual start-times for cycle nought will be some $\underline{t}(o) \geq \underline{x}$. So the vector of earliest permissible start-times for cycle one will be $\underline{A} \otimes \underline{t}(o)$, and the vector of actual start-times for cycle one will be $\underline{t}(1) \geq \underline{A} \otimes \underline{x}$. By an obvious induction, the vector of actual start-times of cycle r will be $\underline{t}(r)$ where:

$$\underline{t}(r) \geq \underline{A} \otimes \underline{t}(r-1) \geq \ldots \geq \underline{A}^r \otimes \underline{t}(o) \geq \underline{A}^r \otimes \underline{x} \qquad (9\text{-}1)$$

Suppose the process is due to run up to and including cycle N, and must terminate at or before a given vector of finishing-times $\underline{z}$. Evidently, this requires that $\underline{A}^{N+1} \otimes \underline{x} \leq \underline{z}$ and we shall call the pair $\underline{x},\underline{z}$ of vectors <u>compatible for ($\underline{A}$, N)</u> if this holds. A sequence $\{\underline{t}(r)\}$ (r=0,..., N) will be called <u>feasible for ($\underline{x},\underline{z}$,N)</u> if we have:

$$\left.\begin{array}{lll}\text{(i)} & \underline{t}(r+1) \geq \underline{A} \otimes \underline{t}(r) & (r=0,\ldots, N)\\ \text{(ii)} & \underline{t}(N+1) \leq \underline{z} & \\ \text{(iii)} & \underline{t}(0) \geq \underline{x} & \end{array}\right\} \qquad (9\text{-}2)$$

We shall require the following simple but important lemma.

<u>Lemma 9-1. Let E_1 be a pre-residuated belt which satisfies axiom X_{12}. If $\underline{A} \in \mathcal{M}_{mn}$, $\underline{x} \in E_n$ and $\underline{z} \in E_m$, we have $\underline{z} \geq \underline{A} \otimes \underline{x}$ if and only if $\underline{x} \leq \underline{A}^* \otimes' \underline{z}$.</u>

<u>Hence, if m=n and N≥0, the following are all equivalent:</u>

<u>$\underline{z} \geq \underline{A}^{N+1} \otimes \underline{x}$; $\underline{A}^* \otimes' \underline{z} \geq \underline{A}^N \otimes \underline{x}$; ... ; $\underline{A}^{N*} \otimes' \underline{z} \geq \underline{A} \otimes \underline{x}$; $\underline{A}^{(N+1)*} \otimes' \underline{z} \geq \underline{x}$.</u>

<u>Proof.</u> If $\underline{z} \geq \underline{A} \otimes \underline{x}$ we have:

$$\underline{A}^* \otimes' \underline{z} \geq \underline{A}^* \otimes' (\underline{A} \otimes \underline{x}) \qquad \text{(by isotonicity)}$$

$$\geq \underline{x} \qquad \text{(by Theorem 8-8)}$$

And dually, the converse holds.

Hence if $\underline{z} \geq \underline{A}^{N+1} \otimes \underline{x}$ with $N \geq 0$ we have:

$$\underline{z} \geq \underline{A} \otimes (\underline{A}^N \otimes \underline{x})$$

Hence $\underline{A}^* \otimes' \underline{z} \geq \underline{A}^N \otimes x$, and by iteration: $\underline{A}^{r*} \otimes' \underline{z} \geq \underline{A}^{N+1-r} \otimes \underline{x}$, (r=1,..., N),

and $\underline{A}^{(N+1)*} \otimes' \underline{z} \geq \underline{x}$

And, dually, the converse sequence of implications holds. ●

We may now prove the following theorems in the theory of scheduling.

Theorem 9-2. Let E_1 receive the principal interpretation. Let $\underline{A} \varepsilon \mathcal{M}_{nn}$ and $\underline{x}, \underline{z} \varepsilon E_n$ for some integer $n \geq 1$. Let $N \geq 0$ be a given integer. Then the following conditions are equivalent:

(i) $\underline{x}, \underline{z}$ are compatible for $(\underline{A}, N)$.

(ii) There exists a sequence $\{\underline{t}(r)\}$ which is feasible for $(\underline{x},\underline{z},N)$.

(iii) For all $r=0,1,\ldots, (N+1)$. there holds

$$\underline{A}^r \otimes x \leq \underline{A}^{(N-r+1)*} \otimes' \underline{z}$$

(where $\underline{A}^r \otimes \underline{x} = \underline{x}$ and $\underline{A}^{s*} \otimes' \underline{z} = \underline{z}$ by definition when $r=0$, $s=0$.

Proof. Conditions (i) and (iii) are equivalent by Lemma 9-1. Also (9-2) clearly implies $\underline{z} \geq \underline{t}(N+1) \geq \underline{A} \otimes \underline{t}(N) \geq \ldots \geq \underline{A}^{N+1} \otimes \underline{t}(o) \geq \underline{A}^{N+1} \otimes \underline{x}$, so condition (ii) implies condition (i). Finally, if $\underline{z} \geq \underline{A}^{N+1} \otimes \underline{x}$ then the sequence $\{\underline{t}(r)\}$ defined by: $\underline{t}(r) = \underline{A}^r \otimes \underline{x}$ $(r=0,\ldots, (N+1))$ is a feasible sequence for $(\underline{x},\underline{z},N)$, so condition (i) implies condition (ii). ●

Theorem 9-3. With the notation of Theorem 9-2, any sequence $\underline{t}(r)$ which is feasible for $(\underline{x},\underline{z},N)$ satisfies:

$$\underline{A}^r \otimes \underline{x} \leq \underline{t}(r) \leq \underline{A}^{(N-r+1)*} \otimes' \underline{z} \qquad (r=0,\ldots, (N+1)).$$

Proof. From (i) of (9-2) we have for $0 \leq s < r \leq (N+1)$:

$$\underline{t}(r) \geq \underline{A}^{(r-s)} \otimes \underline{t}(s)$$

So $$\underline{t}(r) \geq \underline{A}^r \otimes \underline{t}(o) \qquad (9\text{-}3)$$

And $$\underline{t}(N+1) \geq \underline{A}^{N-r+1} \otimes \underline{t}(r)$$

By Lemma 9-1, this last relation gives:

$$\underline{t}(r) \leq \underline{A}^{(N-r+1)*} \otimes' \underline{t}(N+1) \quad , \qquad (9\text{-}4)$$

We may now use (ii) and (iii) of (9-2), in (9-4) and (9-3) respectively, to give (by isotonicity):

$$\underline{A}^r \otimes \underline{x} \leq \underline{t}(r) \leq \underline{A}^{(N-r+1)*} \otimes' \underline{z}$$ ●

9-2 . The Float

Suppose at a certain moment that the first-following cycle which we must begin is cycle s $(1 \leq s \leq N)$, and that because of given circumstances the vector of earliest permissible start times for this cycle is $\underline{y}$. The question arises: is the target vector of completion times $\underline{z}$ still achievable? Formally, let $\{\underline{t}(r)\}$ $(r=0,\ldots,(s-1))$ be a sequence satisfying:

$$\left.\begin{array}{l} \text{(i) } \underline{t}(o) \geq \underline{x} \\ \text{(ii) If } s \geq 2 \text{ then } \underline{t}(r+1) \geq \underline{A} \otimes \underline{t}(r) \quad (r=0,\ldots, (s-2)) \end{array}\right\} \qquad (9\text{-}5)$$

By an $(\underline{x},\underline{z},N)$ - feasible y-continuation of the given sequence we mean a sequence $\{t(\underline{r})\}$ $(r=0,\ldots,(N+1))$ which is feasible for $(\underline{x},\underline{z},N)$, which coincides with the given sequence for $r=0,\ldots,(s-1)$, and which satisfies $t(\underline{s}) = \underline{y}$

Theorem 9-4. With the foregoing notation, a sequence $\{\underline{t}(r)\}$ $(r=0,\ldots,(s-1))$ satisfying (9-5) has an $(\underline{x},\underline{z},N)$ - feasible $\underline{y}$-continuation if and only if:

$$\underline{A} \otimes \underline{t}(s-1) \leq \underline{y} \leq \underline{A}^{(N-s+1)*} \otimes' \underline{z} \qquad (9\text{-}6)$$

Proof. The "only if" follows from Theorem 9-3 and (9-2). Conversely, if (9-6) holds, then by Lemma 9-1, $\underline{A}^{(N-s+1)} \otimes \underline{y} \leq \underline{z}$.

It follows, using (9-5) that the sequence:

$$\underline{t}(o),\ldots, \underline{t}(s-1), \underline{y}, \underline{A} \otimes \underline{y}, \ldots, \underline{A}^{(N-s+1)} \otimes \underline{y}$$

is feasible for $(\underline{x},\underline{z},N)$.

Suppose, then, that $\underline{y}$ satisfies (9-6). The earliest start-times for cycle s are given to be $\underline{y}$, but how much later might we begin without prejudice to the target vector of completion times $\underline{z}$? From Theorem 9-4, we see that the vector $\underline{t}(s)$ of start-times for cycle s must satisfy:

$$\underline{y} \leq \underline{t}(s) \leq \underline{A}^{(N-s+1)*} \otimes' \underline{z} \qquad (9\text{-}7)$$

and that there will then be an $(\underline{x},\underline{z},N)$ feasible $\underline{t}(s)$-continuation (if $s < N + 1$). Hence a componentwise comparison of $\underline{y}$ with $\underline{A}^{(N-s+1)*} \otimes' \underline{z}$ tells us, for each activity, what the further delays are which we can tolerate in initiating cycle s, beyond the $\underline{y}$ which is given. The vector of componentwise differences between $\underline{y}$ and $\underline{A}^{(N-s+1)*} \otimes' \underline{z}$ is known in scheduling theory as the total float (vector).

However, we may not wish to delay $\underline{t}(s)$ until equal to $\underline{A}^{(N-s+1)*} \otimes' \underline{z}$, thereby using up all the total float. If $s < N$, therefore, we may ask: what delays can we tolerate in initiating cycle s without prejudice to subsequent cycles - i.e. without reducing the total float which will be available at cycle (s+1)? Now, if we begin cycle s as early as possible, namely at times $\underline{y}$, the total float vector for cycle (s+1) will be obtained by a comparison of $\underline{A} \otimes \underline{y}$ with $\underline{A}^{(N-s)*} \otimes' \underline{z}$. Hence, if we actually begin at times $\underline{t}(s)$ instead of $\underline{y}$, this total float vector will not be diminished provided $\underline{A} \otimes \underline{t}(s)$ does not exceed $A \otimes \underline{y}$. But by Lemma 9-1,

$$\underline{A} \otimes \underline{t}(s) \leq \underline{A} \otimes \underline{y} \text{ if and only if } \underline{t}(s) \leq \underline{A}^{*} \otimes' (\underline{A} \otimes \underline{y}).$$

Hence if we are not to reduce the total float available at cycle (s+1), we must choose $\underline{t}(s)$ to satisfy

$$\underline{y} \leq \underline{t}(s) \leq \underline{A}^{*} \otimes' (\underline{A} \otimes \underline{y}) \qquad (9\text{-}8)$$

rather than (9-7). And (9-8) is indeed a tighter inequality than (9-7) since:

$$\begin{aligned} A^{*} \otimes' (\underline{A} \otimes \underline{y}) &\leq \underline{A}^{*} \otimes' (\underline{A} \otimes (\underline{A}^{(N-s+1)*} \otimes' \underline{z})) \qquad \text{(by isotonicity and (9-7))} \\ &= \underline{A}^{*} \otimes' (\underline{A} \otimes (\underline{A}^{*} \otimes' (\underline{A}^{(N-s)*} \otimes' \underline{z}))) \end{aligned}$$

$$= \underline{A}^* \otimes' (\underline{A}^{(N-s)*} \otimes' \underline{z}) \qquad \text{(by Theorem 8-9)}$$

$$= \underline{A}^{(N-s+1)*} \otimes' \underline{z}$$

So, from (9-8), a comparison of $\underline{y}$ with $A^* \otimes' (A \otimes y)$ tells us for each activity how much further delay we can tolerate without changing the total float vector for cycle s+1. The vector of componentwise differences between $\underline{y}$ and $\underline{A}^* \otimes' (\underline{A} \otimes \underline{y})$ is called the <u>free float (vector)</u>.

9-3. A Worked Example

By way of illustration, let us assume that the matrix $\underline{A}$ is given by:

$$\underline{A} = \begin{bmatrix} 3 & -\infty & 2 \\ -\infty & 1 & 1 \\ 2 & 3 & 2 \end{bmatrix}$$

By direct calculation:

$$\underline{A}^2 = \begin{bmatrix} 6 & 5 & 5 \\ 3 & 4 & 3 \\ 5 & 5 & 4 \end{bmatrix} \quad ; \quad \underline{A}^4 = \begin{bmatrix} 12 & 11 & 11 \\ 9 & 8 & 8 \\ 11 & 10 & 10 \end{bmatrix}$$

Suppose now that:

$$\underline{x} = \begin{bmatrix} 0 \\ 1 \\ 2 \end{bmatrix} \quad ; \quad \underline{z} = \begin{bmatrix} 17 \\ 11 \\ 15 \end{bmatrix} \quad ; \quad N = 3$$

And suppose further that cycle nought has been completed, cycle one initiated, and that delays and other practical problems dictate that cycle 2 cannot now be initiated earlier than the start-times given by:

$$\underline{y} = \begin{bmatrix} 8 \\ 6 \\ 6 \end{bmatrix}$$

Can we still achieve the target completion times $\underline{z}$? We have:

$$\underline{A}^2 \otimes \underline{x} = \begin{bmatrix} 6 & 5 & 5 \\ 3 & 4 & 3 \\ 5 & 5 & 4 \end{bmatrix} \otimes \begin{bmatrix} 0 \\ 1 \\ 2 \end{bmatrix} = \begin{bmatrix} 7 \\ 5 \\ 6 \end{bmatrix}$$

$$\underline{A}^{2*} \otimes' \underline{z} = \begin{bmatrix} -6 & -3 & -5 \\ -5 & -4 & -5 \\ -5 & -3 & -4 \end{bmatrix} \otimes' \begin{bmatrix} 17 \\ 11 \\ 15 \end{bmatrix} = \begin{bmatrix} 8 \\ 7 \\ 8 \end{bmatrix}$$

Hence $\underline{A}^2 \otimes \underline{x} \leq \underline{y} \leq \underline{A}^{2*} \otimes' \underline{z}$, so by Theorem 9-4, the completion times $\underline{z}$ are still achievable. The total float vector is:

$$\begin{bmatrix}8\\7\\8\end{bmatrix} - \begin{bmatrix}8\\6\\6\end{bmatrix} = \begin{bmatrix}0\\1\\2\end{bmatrix}$$

Hence cycle 2 of activities 1, 2 and 3 may be delayed by 0, 1 and 2 units respectively, beyond the given times $\underline{y}$, without prejudice to the completion times $\underline{z}$.

Furthermore,

$$\underline{A} \otimes \underline{y} = \begin{bmatrix}3 & -\infty & 2\\ -\infty & 1 & 1\\ 2 & 3 & 2\end{bmatrix} \otimes \begin{bmatrix}8\\6\\6\end{bmatrix} = \begin{bmatrix}11\\7\\10\end{bmatrix}$$

$$\underline{A}^* \otimes' (\underline{A}\otimes\underline{y}) = \begin{bmatrix}-3 & +\infty & -2\\ +\infty & -1 & -3\\ -2 & -1 & -2\end{bmatrix} \otimes \begin{bmatrix}11\\7\\10\end{bmatrix} = \begin{bmatrix}8\\6\\6\end{bmatrix}$$

Hence $\underline{y} = \underline{A}^* \otimes' (\underline{A} \otimes \underline{y})$ and the free float vector is $\begin{bmatrix}0\\0\\0\end{bmatrix}$.

Hence any delay in initiating cycle 2 for any activity, beyond the given times $\underline{y}$, will reduce the total float at the next cycle for at least one activity.

10. RESIDUATION AND REPRESENTATION

10-1. Some Residuation Theory

A very natural way to express the properties of matrices implied by Theorem 8-9 is to use the language of residuation theory, as set out in [42], for example. We recall that a function $f: S \to T$ where S, T are given partially ordered sets, is called residuated if there exists a function $f^*: T \to S$ such that the following hold:

$$R_1 : f \text{ is isotone and } f^* \text{ is isotone}$$

$$\begin{aligned} R_2 : &\text{(i)} \quad f(f^*(t)) \leq t \quad \text{for all } t \in T \\ &\text{(ii)} \quad f^*(f(s)) \geq s \quad \text{for all } s \in S. \end{aligned} \tag{10-1}$$

The function f^* is called the residual of f. It is well-known (and easy to prove) that the residual of f is unique if f is residuated. An isotone function $f^*: T \to S$ for which an isotone function $f: S \to T$ exists such that R_2(i) and (ii) hold, is called dually residuated.

We shall discuss below how our matrix algebra may be represented as an algebra of residuated functions. Before doing so, however, it will be useful to record some more general facts about such functions. We remark at this point that the notation f^* to denote the residual of f is not in conflict with our previous use of * to denote conjugation. We shall see below that the one is a special case of the other.

Now, conditions R_2 above may be rewritten

$$\begin{aligned} R_2: \quad &\text{(i)} \quad f \circ f^* \leq i_T \\ &\text{(ii)} \quad f^* \circ f \geq i_S \end{aligned} \tag{10-2}$$

where the order-relations are respectively that in T^T and that in S^S, and $i_T \in T^T$, $i_S \in S^S$ are the usual identity mappings, and $\circ$ denotes composition.

Since f is isotone, we have from R_2(ii):

$$f \circ (f^* \circ f) \geq f \circ i_S = f$$

But from R_2(i):

$$(f \circ f^*) \circ f \leq i_T \circ f = f$$

But evidently $(f \circ f^*) \circ f = f \circ (f^* \circ f)$. Hence $f \circ f^* \circ f = f$.

From this result and its dual we deduce using Proposition 8-10 the following standard result of residuation theory.

Proposition 10-1. If f is a residuated function with residual f^*, then $f = f \circ f^* \circ f$ and $f^* = f^* \circ f \circ f^*$. Accordingly, $f \setminus \text{Ran } f^*$ and $f^* \setminus \text{Ran } f$ are mutually inverse bijections. ●

We remark that the pairs of functions displayed in Figs 8-3 and 8-4 are by no means all examples of Proposition 10-1, but of the more general Proposition 8-10. Specifically, mappings such as $\underline{X} \mapsto \underline{X}^* \otimes' \underline{A}$ are antitone rather than isotone.

Now let S be a given set, $K \subset S^S$ a set of 1-ary functions from S to S, and $\hat{K}$ the composition algebra generated by K. Since the composition of 1-ary functions is evidently associative, $(\hat{K},o)$ is a semigroup: $\forall f,g,h\varepsilon\hat{K}$, $f \circ (g \circ h) = (f \circ g) \circ h$

We shall call $(\hat{K},o)$ the composition semigroup generated by K.

Lemma 10-2. Let S be a given partially ordered set. Let $K \subset S^S$ be a given set of residuated functions, and K^* the set of residuals. Then $(\hat{K},o)$, the composition semigroup generated by K, consists entirely of residuated functions, the set of residuals being $(\hat{K}^*, o)$, the composition algebra generated by K^*.

Proof. Let $f_1,f_2 \varepsilon K$ with residuals f_1^*, f_2^* respectively. It is easily verified (bearing in mind the isotonicity of f and f*) that:

$$(f_1 \circ f_2) \circ (f_2^* \circ f_1^*) \leq i_S \leq (f_2^* \circ f_1^*) \circ (f_1 \circ f_2)$$

Hence $(f_1 \circ f_2)$ is residuated, with residual $f_2^* \circ f_1^*$, and by an immediate induction, each composition product of a finite number of functions from K is residuated. But $\hat{K}$ consists exactly of those products; and the set of residuals is just the set of finite composition products of elements of K^* - i.e. $\hat{K}^*$. ●

We make now some definitions through which the concepts of residuation theory can interact with those of our minimax algebra.

An ordered semigroup is by definition a semigroup for which all left multiplications and all right multiplications are isotone functions relative to a given partial order on the elements of the semigroup. A residuated semigroup is an ordered semigroup for which all left multiplications and all right multiplications are residuated mappings.

We shall say that a belt is residuated if all left multiplications and all right multiplications are residuated mappings relative to the partial order corresponding to the addition operation. Evidently, if $(S,\oplus,\otimes)$ is a residuated belt then $(S, \otimes)$ is a residuated semigroup.

By a residuated belt with duality, we understand a belt with duality in which all left multiplications and right multiplications are residuated, and all left dual multiplications and all right dual multiplications are dually residuated, relative to the partial order corresponding to the addition operations.

Theorem 10-3. If E_1 is a pre-residuated belt satisfying axiom X_{12}, then E_1 is a residuated belt with duality.

Proof. E_1 is by definition self-conjugate. For arbitrary $a \varepsilon E_1$, let a^* denote the conjugate of a, and define the multiplications g_a, $g_a^* \varepsilon (E_1)^{E_1}$ by:

$$\forall\, x, y \,\varepsilon\, E_1,\ g_a(x) = a \otimes x, \qquad g_a^*(y) = a^* \otimes' y.$$

By Proposition 2-8, the functions g_a and g_a^* are isotone, and by Theorem 8-8, (applied to elements of $E_1 = \mathcal{M}_{11}$):

$$\forall\, x \,\varepsilon\, E_1, \quad a \otimes (a^* \otimes' x) \leq x \leq a^* \otimes' (a \otimes x)$$

i.e.
$$(g_a \circ g_a^*)(x) \leq x \leq (g_a^* \circ g_a)(x) \tag{10-3}$$

Hence g_a is residuated and g_a^* is dually residuated. Similar arguments apply to right multiplications. ●

Corollary 10-4. If E_1 is a pre-residuated belt satisfying axiom X_{12}, then $\mathcal{M}_{nn}$ is a residuated belt with duality for each integer $n \geqslant 1$.

Proof. By Theorem 6-3 and Corollary 8-4, $\mathcal{M}_{nn}$ is a pre-residuated belt satisfying axiom X_{12}. Then Theorem 10-3 (with $\mathcal{M}_{nn}$ in the role of E_1) implies that $\mathcal{M}_{nn}$ is a residuated belt with duality. ●

10-2. Residuomorphisms.

A relation such as: $\lambda \otimes (x \oplus y) = (\lambda \otimes x) \oplus (\lambda \otimes y)$ among certain elements λ, x, y of our algebraic structures, brings together two ideas. One is the idea of linearity, suggesting analogies with the concepts of linear algebra - spaces, matrices, invariant subspaces, etc. The other is the idea of isotonicity, implied by linearity as discussed in Section 2-3, and leading to the algebra of order-relations.

We shall be interested in a class of functions in which these two ideas interact in a very specific way. Accordingly, let $(S, \oplus, \oplus')$, $(T, \oplus, \oplus')$ be commutative bands with duality. By a residuomorphism from S to T we shall mean a function $f \in T^S$ such that there exists a function $f^* \in S^T$ whereby the following hold:

$$\left.\begin{array}{lll} R_2: & & f \circ f^* \leqslant i_T \text{ and } f^* \circ f \geqslant i_S \\ R_3: & (i) & \forall\, x, y \in S,\ f(x \oplus y) = f(x) \oplus f(y) \\ & (ii) & \forall\, a, b \in T,\ f^*(a \oplus' b) = f^*(a) \oplus' f^*(b) \end{array}\right\} \qquad (10\text{-}4)$$

Evidently, R_3 (i) and (ii) imply the isotonicity of f and f^* respectively, which with R_2 shows that f is residuated with residual f^*. Hence f^* is unique.

It is convenient to have a succinct notation for displaying classes of residuomorphisms. Accordingly, the notation:

$$M: S \Leftrightarrow T: N$$

will always mean: $(S, \oplus, \oplus')$, $(T\ \oplus, \oplus')$ are given commutative bands with duality, M is a given set of residuomorphisms from S to T, and N is the set of residuals of elements of M.

If S=T, we shall sometimes call the functions $f \in M$ endo-residuomorphisms (of S).

The following result shows that a set of residuomorphisms can always be embedded in a commutative band of residuomorphisms.

Theorem 10-5. If $M: S \Leftrightarrow T: N$, then we can find V, V^* such that:

(i) $(V,\oplus)$ and $(V^*,\oplus')$ are commutative bands

(ii) $M \subset V$ and $N \subset V^*$

(iii) $V: S \Leftrightarrow T: V^*$

(iv) V and V^* are conjugate, the conjugate of $f \in V$ being its residual $f^* \in V^*$.

(v) If $(M,\oplus)$ and $(N,\oplus')$ are commutative bands, then $(M,\oplus)$ is $(V,\oplus)$ and $(N,\oplus')$ is

$(V^*,\oplus')$.

Proof. Suppose $g, h \varepsilon M$. We may define $g \oplus h$ as in (2- 2), and $g^* \oplus' h^*$ dually. We have for all $t \varepsilon T$:

$$g(g^*(t)\oplus' h^*(t)) \leq (gog^*)(t) \oplus' (goh^*)(t) \qquad \text{(by Lemma 2-7)}$$
$$\leq (gog^*)(t)$$
$$\leq t \qquad \text{(by } R_2\text{)}$$

Similarly, $h(g^*(t) \oplus' h^*(t)) \leq t$, whence:

$$((g\oplus h) \circ (g^*\oplus' h^*))(t) = g(g^*(t)\oplus' h^*(t)) \oplus h(g^*(t) \oplus' h^*(t)) \qquad \text{(by definition)}$$
$$\leq t$$

i.e. $(g\oplus h) \circ (g^*\oplus' h^*) \leq i_T$

Dually, $(g^* \oplus' h^*) \circ (g \oplus h) \geq i_S$

Thus, with $g\oplus h$ in the role of f, and $g^*\oplus' h^*$ in the role of f^*, we can satisfy (10-2) and hence $g\oplus h$ is residuated with residual $(g^*\oplus' h^*)$. Moreover, as in Section 2-3, $g\oplus h$ is a homomorphism, so we can satisfy R_3(i) of (10-4) and, dually, R_3(ii) of (10-4).

Hence, for each $g,h \varepsilon M$, we have that $g \oplus h$ is again a residuomorphism, and by induction we have that each finite sum $\sum_{i=1}^{n}{}_{\oplus} f_i$, where $f_i \varepsilon M$, $(i=1,\ldots, n)$ is again a residuomorphism with residual $\sum_{i=1}^{n}{}_{\oplus'} f_i^*$. Let V be the set of such sums, and V^* the set of residuals of elements of V. Evidently $(V,\oplus)$ and $(V^*,\oplus')$ are commutative bands which are identical to $(M,\oplus)$ and $(N,\oplus')$ if these are already commutative bands. Thus parts (i), (ii), (iii) and (v) of the theorem are established.

Finally, the bijection $f \leftrightarrow f^*$ satisfies N_1 and N_2 of Section 7-1, showing that V and V^* are conjugate.

The following theorem considers the case S=T, i.e.

$$M: S \Leftrightarrow S: N$$

and shows that a set of endo-residuomorphisms can always be embedded in a belt of such mappings.

Theorem 10-6. If $M:S\Leftrightarrow S:N$, then we can find V, V^* such that:

(i) $(V,\oplus,\otimes)$ and $(V^*,\oplus',\otimes')$ are belts

(ii) $M \subset V$ and $N \subset V^*$

(iii) $V:S \Leftrightarrow S:V^*$

(iv) V and V^* are conjugate, the conjugate of $f \varepsilon V$ being its residual $f^* \varepsilon V^*$.

(v) If $(M,\oplus,\otimes)$ and $(N,\oplus',\otimes')$ are belts, then $(V,\oplus,\otimes)$ is $(M,\oplus,\otimes)$ and $(V^*,\oplus',\otimes')$ is $(N,\oplus',\otimes')$.

Proof. Introducing the composition semigroups $(\hat{M},\circ)$, $(\hat{N},\circ)$ we have by Proposition 2-3 and Lemma 10-2 that $\hat{M}: S\Leftrightarrow S: \hat{N}$. Embed $\hat{M}$, $\hat{N}$ in their respective commutative bands V, V^* of finite sums as in the proof of Theorem 10-5 and confirm that $(V,\oplus,\otimes)$, $(V^*, \oplus',\otimes')$ are belts, when $\otimes$, $\otimes'$ denote $\circ$. The bijection $f\leftrightarrow f^*$ satisfies N_3 of Section 7-1, by the proof of Lemma 10-2.

Other details of the proof are as in Theorem 10-5.

Theorems 10-5 and 10-6 show, as remarked above, that the use of the notation f^* to denote the residual of f is fully consistent with the notation of conjugacy.

10-3. Representation Theorems

In abstract algebra, a representation of a given structure V is a morphism of V into a set of endomorphisms of some structure S. The "passive" structure V is thereby, as it were, given an active role as a set of operators. The representation τ is called faithful if it is a bijection. We have seen these ideas at work already in Chapter 2, for example. We now investigate what can be said about such representations in the light of residuation theory.

Accordingly, let $(V,\oplus,\otimes)$ be a given belt, and define Λ_V to be the set of all left-multiplications $g_a \in V^V$, with $a \in V$, as in (2-8). Make Λ_V into a belt as in (2-2) and (2-7). Following the terminology of ring theory, we shall call the mapping $\tau : V \to \Lambda_V$ defined by $\tau : a \mapsto g_a$ the (left) regular representation of V (See Proposition 2-4). Similarly, we may define the right regular representation of V. If now $(V,\oplus,\otimes,\oplus',\otimes')$ is a self-conjugate belt, then we may consider the belt $(\Lambda_V^*,\oplus',\otimes')$ of left dual multiplications $g_a^* \in V^V$ defined, for each $a \in V$ by $g_a^* : x \mapsto a^* \otimes' x$. Then the (left) dual regular representation of V is the mapping:

$$\tau^* : (V,\oplus',\otimes') \mapsto (\Lambda_V^*,\oplus',\otimes')$$

defined by $\tau^* : a \mapsto g^*_{a^*}$.

Similarly, we may define the right dual regular representation of V. The left and right, regular and dual regular, representations of V will be jointly known as the regular representations of V. In the singular, the regular representation of V will always mean the left regular representation of V.

Lemma 10-7. If V is a pre-residuated belt satisfying axiom X_{12}, then, with the foregoing notation:

$$\forall a \in V, \quad (g_a \circ g_a^*)(a) = a \tag{10-5}$$

Proof. Relation (10-5) asserts:

$$\forall a \in V, \quad a \otimes (a^* \otimes' a) = a$$

which is guaranteed by Theorem 8-5 (with $E_1 = \mathcal{M}_{11} = V$).

Theorem 10-8. If V is a pre-residuated belt satisfying axiom X_{12}, then the regular representations of V are faithful.

Proof. With the foregoing notation, if $a,b \in V$ are such that $\tau(a) = \tau(b)$, i.e. $g_a = g_b$, then:

$$a = (g_a \circ g_a^*)(a) \quad \text{(by (10-5))}$$

$$= (g_b \circ g_a^*)(a)$$

$$= b \otimes (a^* \otimes' a)$$

$\geqslant b$ (by (8-1))

Hence $a \geqslant b$ and similarly $b \geqslant a$, so $a = b$ and τ is bijective. Similar arguments hold for the other regular representations.

Theorem 10-9. If V is a pre-residuated belt satisfying axiom X_{12}, then:

$$(\Lambda_V,\oplus,\otimes): (V,\oplus,\oplus') \Leftrightarrow V(\oplus,\oplus'): (\Lambda_V^*,\oplus',\otimes').$$

Proof. From (10-3) we see that $g_a \in \Lambda_V$ and $g_a^* \in \Lambda_V^*$ satisfy R_2 of (10-4), and they certainly satisfy R_3 of (10-4) (by axioms X_5, X_5'). Hence g_a, for each $a \in V$, is a residuomorphism, with residual g_a^*, and the result follows.

Evidently, we may summarise these results in the following way.

Proposition 10-10. A pre-residuated belt satisfying axiom X_{12} has a faithful representation as a belt of endo-residuomorphisms of a commutative band.

10-4. Representation for Matrices.

A classical topic is that of the representation of matrices as linear transformations of finite-dimensional vector spaces. In Chapter 5, we have already broached the analogous topic in minimax algebra.

Now, if $\underline{A} \in \mathcal{M}_{mn}$ and $\underline{X} \in \mathcal{M}_{np}$ are matrices over a given belt E_1, then the product $\underline{A} \otimes \underline{X}$ is known as soon as its columns are known. Hence the effect of left multiplication by $\underline{A}$ is known for all elements of $\mathcal{M}_{np}$ as soon as it is known for all elements of E_n. In considering representations of sets of matrices as residuomorphisms, therefore, it suffices to restrict ourselves to residuomorphisms of E_n.

Accordingly, for given integers $m,n \geqslant 1$, we define $\mathcal{L}\mathcal{M}_{mn}$ to be the set of left multiplications $\{g_{\underline{A}} | \underline{A} \in \mathcal{M}_{mn}\}$ defined by:

$$g_{\underline{A}}: E_n \to E_m, \text{ namely } \forall\, \underline{x} \in E_n, \quad g_{\underline{A}}(\underline{x}) = \underline{A} \otimes \underline{x} \tag{10-6}$$

And, if E_1^* is conjugate to E_1, we define $\mathcal{M}_{mn}^*$ as in Section 7-2, and let $\mathcal{L}^*\mathcal{M}_{mn}^*$ be the set of left dual multiplications $\{g_{\underline{A}}^* | \underline{A} \in \mathcal{M}_{mn}\}$ defined by:

$$g_{\underline{A}}^*: \ E_m \to E_n, \text{ namely } \ \forall\, \underline{y} \in E_m, \ g_{\underline{A}}^*(\underline{y}) = \underline{A}^* \otimes' \underline{y} \tag{10-7}$$

$\mathcal{L}\mathcal{M}_{mn}$ and $\mathcal{L}^*\mathcal{M}_{mn}^*$ may be construed as commutative bands, and, if m=n, as belts, in the usual way.

Theorem 10-11. Let E_1 be a pre-residuated belt satisfying axiom X_{12}, and let $m,n \geqslant 1$ be given integers. Then:

$$\mathcal{L}\mathcal{M}_{mn}: \ E_n \Leftrightarrow E_m : \mathcal{L}^*\mathcal{M}_{mn}^*$$

Proof. Evidently, the elements of $\mathcal{L}\mathcal{M}_{mn}$ and $\mathcal{L}^*\mathcal{M}_{mn}^*$ satisfy the appropriate version of R_3 in (10-4). Also, if $\underline{A} \in \mathcal{M}_{mn}$, then Theorem 8-8 implies:

$$\underline{g}_{\underline{A}} \circ \underline{g}_{\underline{A}}^{*} \leq i_{E_m} \text{ and } \underline{g}_{\underline{A}}^{*} \circ \underline{g}_{\underline{A}} \geq i_{E_n}$$

so that $g_{\underline{A}}$ is a residuomorphism, with residual $g_{\underline{A}}^{*}$.

<u>Corollary 10-12. Let E_1 be a pre-residuated belt satisfying axiom X_{12}, and let $m,n \geq 1$ be given integers. Then $\mathcal{L}\mathcal{M}_{mn}$ and $\mathcal{L}^{*}\mathcal{M}^{*}_{mn}$ are conjugate.</u>

<u>Proof</u>. This follows from Theorem 10-11, and parts (iv) and (v) of Theorem 10-5.

Evidently, we may paraphrase Corollary 10-12 by: $(\mathcal{L}\mathcal{M})^{*} = \mathcal{L}^{*}\mathcal{M}^{*}$.

Now, suppose with the same notation that $\underline{A}, \underline{B} \in \mathcal{M}_{mn}$ are such that $g_{\underline{A}} = g_{\underline{B}}$. Since for any $\underline{X} \in \mathcal{M}_{np}$ the products $\underline{A} \otimes \underline{X}$, $\underline{B} \otimes \underline{X}$ are determined by the action of $g_{\underline{A}}$, $g_{\underline{B}}$ respectively on the columns of $\underline{X}$, we have $\underline{A} \otimes \underline{X} = \underline{B} \otimes \underline{X}$ for all $\underline{X} \in \mathcal{M}_{np}$, for all integers p. In particular, with p=m we have $\underline{A} \otimes \underline{A}^{*} = \underline{B} \otimes \underline{A}^{*}$, whence:

$$\begin{aligned} \underline{B} &\leq (\underline{B} \otimes \underline{A}^{*}) \otimes' A && \text{(by Theorem 8-8)} \\ &= (\underline{A} \otimes \underline{A}^{*}) \otimes' A \\ &= \underline{A} && \text{(by Theorem 8-5)} \end{aligned}$$

Hence $\underline{B} \leq \underline{A}$ and similarly $\underline{A} \leq \underline{B}$. Thus if $g_{\underline{A}} = g_{\underline{B}}$ we have $\underline{A} = \underline{B}$ and the mapping $\tau: \underline{A} \to g_{\underline{A}}$ is bijective. (The logic of the argument is essentially as in Lemma 10-7 and Theorem 10-8). Evidently, in the light of Theorem 10-11 we have established the following result.

<u>Proposition 10-13. If E_1 is a pre-residuated belt satisfying axiom X_{12} then $(\mathcal{M}_{mn}, \oplus)$ for given integers $m,n \geq 1$ has a faithful representation as a commutative band of residuomorphisms of E_n to E_m and hence of $\mathcal{M}_{np}$ to $\mathcal{M}_{mp}$ for each integer $p \geq 1$; and $(\mathcal{M}_{nn}, \oplus, \otimes)$ has a faithful representation as a belt of endo-residuomorphisms of E_n and hence of $\mathcal{M}_{np}$ for each integer $p \geq 1$.</u>

<u>10-5. Analogy with Hilbert Space.</u>

The asterisk notation, as in x^{*}, $\underline{A}^{*}$, etc., is used by us in a variety of ways which are in principle distinct. In Section 7-1 it denotes an abstract involution subject to certain axioms as in a *-ring. In Section 7-2 it denotes "transpose- and-conjugate", an operation strongly reminiscent of forming the adjoint of a Hermitian matrix. In Chapter 10, it denotes residuation. Our theory has aimed to show that, at least when E_1 is (say) a blog, these interpretations are all equivalent.

In classical linear operator theory the asterisk is used to denote the set $\mathrm{Hom}_{E_1}(E_n, E_1)$ of all linear functionals, and also to denote the adjoint of a linear operator. We show that in our theory too the asterisk has these significances.

<u>Theorem 10-14. Let E_1 be a blog. Then for each integer $n \geq 1$ we have:</u>

$$\underline{\mathrm{Hom}_{E_1}(E_n, E_1) = (E_n)^{*}}$$

<u>Proof</u>. Theorem 5-10 identifies $Hom_{E_1}(E_n, E_1)$ with $\mathcal{M}_{1n}$.
From Theorem 7-4 (since E_1 is self-conjugate):

$$\mathcal{M}_{1n} = (\mathcal{M}_{n1})^*$$

And since we may identify $\mathcal{M}_{n1}$ with E_n, the result follows. ●

For convenience, we may write E_n^* instead of $(E_n)^*$. This notation is slightly ambiguous, and we must note that E_n^* does <u>not</u> denote $(E^*)_n$.

Now let E_1 be self-conjugate and let $\underline{x}, \underline{y} \in E_n$. If we regard $\underline{x}$ as an element of $\mathcal{M}_{n1}$ then $\underline{x}^* \in \mathcal{M}_{1n}$ and so the multiplication $\underline{x}^* \otimes \underline{y}$ is defined, the product lying in E_1. Let us call this the <u>inner product</u> of $\underline{x}$ and $\underline{y}$, written $)\underline{x}, \underline{y}($. The following results are analogous to familiar properties of complex (pre-) Hilbert spaces.

<u>Theorem 10-15. Let E_1 be a blog and $n \geq 1$ a given integer. Then for each $x \in E_n$ there exists $g \in Hom_{E_1}(E_n, E_1)$, and for each $g \in Hom_{E_1}(E_n, E_1)$ there exists $x \in E_n$ such that</u>

$$g(\underline{y}) =)\underline{x}, \underline{y}(\quad , \forall \underline{y} \in E_n \tag{10-8}$$

<u>Proof</u>. For given $\underline{x} \in E_n$, let (10-8) define g. Then we easily confirm that $g \in Hom_{E_1}(E_n, E_1)$. Conversely, let $g \in Hom_{E_1}(E_n, E_1)$. Then by Theorem 5-10, there is a $u \in \mathcal{M}_{1n}$ such that:

$$\underline{u} \otimes \underline{y} = g(\underline{y}), \quad \forall \underline{y} \in E_n.$$

On taking $\underline{x} = \underline{u}^*$, (10-8) follows. ●

Now if $g \in Hom_{E_1}(E_n, E_m)$, let us define the <u>adjoint</u> $\tilde{g}$ of g to be a mapping from E_m to E_n such that:

$$)\underline{x}, g(\underline{y})(=)\tilde{g}(\underline{x}), \underline{y}(, \forall \underline{x} \in E_m, \quad \forall \underline{y} \in E_n \tag{10-9}$$

<u>Theorem 10-16. Let E_1 be a blog and let $m, n \geq 1$ be given integers. If $g \in Hom_{E_1}(E_n, E_m)$ then the adjoint $\tilde{g}$ of g exists uniquely and is its residual g^*.</u>

<u>Proof</u>. By Theorem 5-10, there is a matrix $\underline{A} \in \mathcal{M}_{mn}$ such that

$$g(\underline{y}) = \underline{A} \otimes \underline{y} \quad , \quad \forall \underline{y} \in E_n \tag{10-10}$$

Hence, if $\tilde{g}$ exists, then for all $\underline{x} \in E_m$ we have:

$$
\begin{aligned}
(\tilde{g}(\underline{x}))^* \otimes \underline{y} &= \,)\tilde{g}(\underline{x}),\ \underline{y}(&& \text{(by definition)} \\
&= \,)\underline{x},\ g(\underline{y})(&& \text{(by (10-9))} \\
&= \underline{x}^* \otimes g(\underline{y}) && \text{(by definition)} \\
&= \underline{x}^* \otimes \underline{A} \otimes \underline{y} && \text{(by (10-10))} \\
&= (\underline{A}^* \otimes' \underline{x})^* \otimes \underline{y}
\end{aligned}
$$

Since $\underline{y}$ is arbitrary, we therefore have:

$$\tau\ ((\tilde{g}(\underline{x}))^*) = \tau\ ((\underline{A}^* \otimes' \underline{x})^*)$$

where τ is the representation of $\mathcal{M}_{1n}$ as $\mathcal{L}\mathcal{M}_{1n}$. But by Theorem 5-10, τ is faithful, whence:

$$(\tilde{g}(\underline{x}))^* = (\underline{A}^* \otimes' \underline{x})^* \qquad \forall \underline{x} \in E_m.$$

$$\text{ie.} \qquad \tilde{g}(\underline{x}) = \underline{A}^* \otimes' \underline{x} \qquad \forall \underline{x} \in E_m.$$

Hence $\tilde{g} \in \mathcal{L}^*\mathcal{M}^*_{mn}$ and in fact $\tilde{g} = g^*$ by the proof of Theorem 10-11. Since g^* is unique it follows that $\tilde{g}$ is unique if it exists.

On the other hand, we have for all $\underline{x} \in E_m$, $\underline{y} \in E_n$:

$$
\begin{aligned}
)g^*(\underline{x}),\ \underline{y}(&= \,)\underline{A}^* \otimes' \underline{x},\ \underline{y}(\\
&= (\underline{A}^* \otimes' \underline{x})^* \otimes \underline{y} && \text{(by definition)} \\
&= x^* \otimes (\underline{A} \otimes \underline{y}) \\
&= \,)\underline{x},\ g(\underline{y})(&& \text{(by definition)}
\end{aligned}
$$

Hence g^* will always play the role of $\tilde{g}$, which accordingly always exists. ●

11. TRISECTIONS

11-1. The Demands of Reality.

Our study of minimax algebra was motivated by a number of practical examples discussed in Chapter 1. The extended real numbers, used in these examples, are taken up in our theory in the form of blogs, as defined in Section 4-1, and for algebraic convenience we ascribe values to the products $-\infty \otimes +\infty$ and $-\infty \otimes' +\infty$, following [37]. However, it is evident that these ascriptions do not correspond to any particular aspect of reality and it is difficult to say what interpretation we might put upon numerical results arrived at using manipulations which involve them.

Unfortunately, the obvious step of modifying the definition of a blog by reducing the domain of the multiplication operations, so that such products are simply not defined, leads to severe technical problems and interminable circumlocutions. We can however, see what happens if we allow these products to exist as defined, but then deliberately ostracise them in the algebra.

Defining a blog as in Section 4-1, therefore, let us say for two elements x,y in a given blog W that the products $x\otimes y$ and $x\otimes' y$ are /-undefined if one of x,y is $-\infty$ and the other is $+\infty$. Otherwise we say the products are /-defined, or that they /-exist.

Question 1. How wide a range of matrix operations can we carry out without forming /-undefined products in E_1?

Before answering this question, let us look back again at Section 7-3. The sequence $\{\underline{A}^r \otimes \underline{x}\}$ shows how the vector $\underline{x}(r)$ of starting-times of the rth cycles of activity evolves under the action of the operator $\underline{A}$. Now, we have seen that for formulation purposes we must sometimes admit infinite elements in the matrix $\underline{A}$. However, we shall generally regard our solutions as having no relevance to reality if the elements of $\underline{x}(r)$ do not remain finite. This leads us to:

Question 2. How can we characterise the set of left matrix multiplications which take the set of finite n-tuples into itself?

We shall see that both these questions lead us to the same set of matrices, which we call G-astic. However, in discussing the topics of G-astic matrices and /-existence, to which we devote several chapters, it is convenient to relate the ideas to a class of belts somewhat more general than the blogs. This enables us later to apply the terminology of the next two chapters in another context, without duplicating a whole set of ideas. We introduce next, therefore, the idea of a belt with a trisection.

11-2. Trisections

Let $(W, \oplus, \otimes, \oplus', \otimes')$ be a belt with a duality. We say that W has a trisection if W is the union of three non-empty, pairwise disjunct sets $\sigma_{-\infty}$, σ_{ϕ} and $\sigma_{+\infty}$ such that the operations in W satisfy the conditions set out in the table of Fig 11-1.

x	y	x ⊕ y	x ⊗ y	x ⊕' y	x ⊗' y
$\sigma_{-\infty}$	$\sigma_{-\infty}$	$\sigma_{-\infty}$	$\sigma_{-\infty}$	$\sigma_{-\infty}$	$\sigma_{-\infty}$
$\sigma_{-\infty}$	σ_{ϕ}	σ_{ϕ}	$\sigma_{-\infty}$	$\sigma_{-\infty}$	$\sigma_{-\infty}$
$\sigma_{-\infty}$	$\sigma_{+\infty}$	$\sigma_{+\infty}$	/	$\sigma_{-\infty}$	/
σ_{ϕ}	$\sigma_{-\infty}$	σ_{ϕ}	$\sigma_{-\infty}$	$\sigma_{-\infty}$	$\sigma_{-\infty}$
σ_{ϕ}	σ_{ϕ}	σ_{ϕ}	σ_{ϕ}	σ_{ϕ}	σ_{ϕ}
σ_{ϕ}	$\sigma_{+\infty}$	$\sigma_{+\infty}$	$\sigma_{+\infty}$	σ_{ϕ}	$\sigma_{+\infty}$
$\sigma_{+\infty}$	$\sigma_{-\infty}$	$\sigma_{+\infty}$	/	$\sigma_{-\infty}$	/
$\sigma_{+\infty}$	σ_{ϕ}	$\sigma_{+\infty}$	$\sigma_{+\infty}$	σ_{ϕ}	$\sigma_{+\infty}$
$\sigma_{+\infty}$	$\sigma_{+\infty}$	$\sigma_{+\infty}$	$\sigma_{+\infty}$	$\sigma_{+\infty}$	$\sigma_{+\infty}$

Fig 11-1. A trisection

For example, the first row of the table asserts: " if $x \in \sigma_{-\infty}$ and $y \in \sigma_{-\infty}$, then $x \oplus y$, $x \otimes y$, $x \oplus' y$, $x \otimes' y$ all belong to $\sigma_{-\infty}$". The symbol / indicates that no assertion at all is made regarding the corresponding products in rows 3 and 7.

In the obvious way, we shall speak of the trisection$(\sigma_{-\infty}, \sigma_{\phi}, \sigma_{+\infty})$. The set σ_{ϕ} will be called the middle of the trisection. Evidently the notion belt with a trisection generalises the notion blog; in fact a trivial verification of the table of Fig 11-1 yields the following result.

Proposition 11-1. If W is a blog with group G, then W has a trisection with middle G, namely the trisection $(\{-\infty\}, G, \{+\infty\})$. ●

The following result is analogous to Proposition 4-1.

Lemma 11-2. If the belt W with duality has a trisection $(\sigma_{-\infty}, \sigma_{\phi}, \sigma_{+\infty})$ then each of the sets $\sigma_{-\infty}$, σ_{ϕ}, $\sigma_{+\infty}$, $\sigma_{-\infty} \cup \sigma_{\phi}$, $\sigma_{\phi} \cup \sigma_{+\infty}$ is a belt with duality, under the appropriate restrictions of ⊕, ⊗, ⊕', ⊗'.

Proof. From the first line of the table in Fig 11-1 we see that all requisite sums and products exist within $\sigma_{-\infty}$, which is therefore a belt with duality. Similarly σ_{ϕ} and $\sigma_{+\infty}$ are belts with duality. Again, the first, second, fourth and fifth lines of the table show that all requisite sums and products exist with $\sigma_{-\infty} \cup \sigma_{\phi}$, which is therefore a belt with duality. Similarly $\sigma_{\phi} \cup \sigma_{+\infty}$ is a belt with duality. ●

For convenience in later manipulative work, we list the following facts which follow self-evidently from the table in Fig 11-1. (See also Proposition 4-3).

Proposition 11-3. Let the belt W with duality have a trisection $(\sigma_{-\infty}, \sigma_{\phi}, \sigma_{+\infty})$. Then for all $x, y, x_1, \ldots, y_n \in W$:

(i) If $x \otimes y \in \sigma_{+\infty}$ or $x \oplus y \in \sigma_{+\infty}$ then either $x \in \sigma_{+\infty}$ or $y \in \sigma_{+\infty}$ (or both)

(ii) <u>If</u> $x \otimes y \in \sigma_{-\infty}$ <u>then either</u> $x \in \sigma_{-\infty}$ <u>or</u> $y \in \sigma_{-\infty}$ (or <u>both</u>)

(iii) <u>If</u> $x \oplus y \in \sigma_{-\infty}$ <u>then both</u> $x \in \sigma_{-\infty}$ <u>and</u> $y \in \sigma_{-\infty}$

(iv) <u>If</u> $\sum_{i=1}^{n}{}_{\oplus} (x_i \otimes y_i) \in \sigma_{+\infty}$ <u>then at least one of</u>

$x_1,\ldots,x_n,\ y_1,\ldots,y_n \in \sigma_{+\infty}$

(v) <u>If</u> $\sum_{i=1}^{n}{}_{\oplus} (x_i \otimes y_i) \in \sigma_{-\infty}$ <u>then at least one of</u> $x_i,\ y_i \in \sigma_{-\infty}$ <u>for each</u> $i=1,\ldots,n$. ●

11-3 . Convex Subgroups

Evidently the existence of a trisection for a given belt W with duality imposes constraints on the structure of W. We investigate now what some of the consequences are.

<u>Lemma 11-4. If W is a blog and has a trisection $(\sigma_{-\infty},\sigma_{\phi},\sigma_{+\infty})$, then $-\infty \in \sigma_{-\infty}$ and $+\infty \in \sigma_{+\infty}$</u>

<u>Proof</u>. Suppose if possible that $-\infty \in \sigma_{\phi}$. Let q lie in the non-empty set $\sigma_{-\infty}$. We have, using the table of Fig. 11-1:

$$q = q \oplus -\infty \in \sigma_{-\infty} \oplus \sigma_{\phi} \subset \sigma_{\phi}$$

But this is impossible since $\sigma_{-\infty}$, σ_{ϕ} are disjunct by definition. Similarly, if $-\infty \in \sigma_{+\infty}$, let $h \in \sigma_{\phi}$. Then $h = h \oplus -\infty \in \sigma_{\phi} \oplus \sigma_{+\infty} \subset \sigma_{+\infty}$

This is similarly impossible. Since $W = \sigma_{-\infty} \cup \sigma_{\phi} \cup \sigma_{+\infty}$, we conclude that $-\infty \in \sigma_{-\infty}$ and similarly $+\infty \in \sigma_{+\infty}$. ●

Let us call a subset H of an ordered set W <u>convex</u> if the following holds for all a, b, $x \in W$:

If a, $b \in H$ and $a \leq x \leq b$ then $x \in H$.

<u>Theorem 11-5. The middle of a trisection of a belt W with duality is a convex sub-belt of W, with duality.</u>

Proof. If σ_{ϕ} is the middle of a trisection $(\sigma_{-\infty},\sigma_{\phi},\sigma_{+\infty})$ of a belt W with duality, then σ_{ϕ} is a sub-belt with duality by Lemma 11-2. Now let a, $b \in \sigma_{\phi}$ and $x \in W$ such that $a \leq x \leq b$. Suppose if possible that $x \in \sigma_{-\infty}$. Using the table of Fig. 11-1:

$$x = a \oplus x \in \sigma_{\phi} \oplus \sigma_{-\infty} \subset \sigma_{\phi}.$$

But this is impossible since $\sigma_{-\infty}$, σ_{ϕ} are disjunct by definition. Similarly, if $x \in \sigma_{+\infty}$:

$$b = x \oplus b \in \sigma_{+\infty} \oplus \sigma_{\phi} \subset \sigma_{+\infty}$$

This is similarly impossible. Since $W = \sigma_{-\infty} \cup \sigma_{\phi} \cup \sigma_{+\infty}$ we conclude that $x \in \sigma_{\phi}$, and that σ_{ϕ} is convex. ●

<u>Theorem 11-6. Let $H = \sigma_{\phi}$ be the middle of a trisection $(\sigma_{-\infty},\sigma_{\phi},\sigma_{+\infty})$ of a belt W with duality. Then $\sigma_{-\infty} = Q$ and $\sigma_{+\infty} = R$, where:</u>

$Q = \{x \mid x \varepsilon W, \text{ and no } h \varepsilon H \text{ satisfies } h \leq x\}$

$R = \{x \mid x \varepsilon W, \text{ and no } h \varepsilon H \text{ satisfies } h \geq x\}$ (11-1)

Proof. If $x \varepsilon \sigma_{-\infty}$ and $k \varepsilon H = \sigma_{\phi}$ then by the table of Fig 11-1:

$$x \oplus k \ \varepsilon \ \sigma_{\phi}$$

If $b = x \oplus k$, then $b \geq x$; and $b \varepsilon \sigma_{\phi}$ (just proved). Hence we cannot have $h \leq x$ for any $h \varepsilon \sigma_{\phi}$ otherwise by the convexity of σ_{ϕ} (Theorem 11-5), follows $x \varepsilon \sigma_{\phi}$ contradicting $x \varepsilon \sigma_{-\infty}$. Hence $x \varepsilon Q$, so $\sigma_{-\infty} \subseteq Q$.

Now if $y \varepsilon \sigma_{+\infty}$ and $k \varepsilon H = \sigma_{\phi}$ then by the table of Fig 11-1:

$$y \oplus' k \varepsilon \sigma_{\phi}$$

If $c = y \oplus' k$, then $c \leq y$; and $c \varepsilon \sigma_{\phi}$ (just proved). Hence $y \notin Q$. Hence $Q \cap \sigma_{+\infty}$ is empty, and evidently $Q \cap \sigma_{\phi} = Q \cap H$ is empty. But $W = \sigma_{-\infty} \cup \sigma_{\phi} \cup \sigma_{+\infty}$, so $Q \subset \sigma_{-\infty}$, i.e. $Q = \sigma_{-\infty}$. Similarly $R = \sigma_{+\infty}$. ●

Corollary 11-7. A given subset H of a belt W with duality can be the middle of at most one trisection of W.

Proof. As Theorem 11-6 shows, any trisection $(\sigma_{-\infty}, \sigma_{\phi}, \sigma_{+\infty})$ with $\sigma_{\phi} = H$ is uniquely identified as (Q, H, R). ●

If W is a division belt or a blog, we shall say that a subset H of W is a convex subgroup of W if H is a convex subset of W and:

If W is a division belt then $(H, \otimes)$ is a subgroup of $(W, \otimes)$.

If W is a blog then $(H, \otimes)$ is a subgroup of the group of W.

If a convex subgroup H is a proper subset of W, then we say H is a convex proper subgroup of W (as will, therefore, always be the case if W is a blog).

Theorem 11-8. The middle of a trisection of a division belt or blog W is a convex proper subgroup of W.

Proof. Let the trisection be $(\sigma_{-\infty}, \sigma_{\phi}, \sigma_{+\infty})$. Note that, if W is a blog, then by Lemma 11-4, σ_{ϕ} contains only finite elements. Suppose $a, b \varepsilon \sigma_{\phi}$. Then $a \otimes b^{-1} \varepsilon \sigma_{\phi}$. For suppose if possible that $a \otimes b^{-1} \varepsilon \sigma_{-\infty}$. Then $a = (a \otimes b^{-1}) \otimes b \varepsilon \sigma_{-\infty} \otimes \sigma_{\phi} \subset \sigma_{-\infty}$, a contradiction. Hence $a \otimes b^{-1} \notin \sigma_{-\infty}$ and similarly $a \otimes b^{-1} \notin \sigma_{+\infty}$. Hence $a \otimes b^{-1} \varepsilon \sigma_{\phi}$ so $(\sigma_{\phi}, \otimes)$ is a (convex) subgroup of W. It is necessarily a proper subset of W since $\sigma_{-\infty} \cup \sigma_{+\infty}$ is not empty. ●

Corollory 11-9. Let the belt W with duality have a trisection $(\sigma_{-\infty}, \sigma_{\phi}, \sigma_{+\infty})$. If W is a division belt, then $(\sigma_{\phi}, \oplus, \otimes, \oplus', \otimes')$ is a division sub-belt of W. If W is a blog then $(\{-\infty\} \cup \sigma_{\phi} \cup \{+\infty\}, \oplus, \otimes, \oplus', \otimes')$ is a sub-blog of W.

Proof. Follows directly from Theorem 11-8, and the algebraic structures of division belts and blogs. ●

11-4. The Linear Case

The question arises as to whether each proper convex subgroup of a division belt or blog W can be the middle of a trisection. We investigate this question in the present section, in relation to the specific case where W is linear.

Lemma 11-10. Let H be a non-empty subset of a given partially ordered set W. If W is linear, then the sets P and Q of (11-1) can equivalently be characterised by:

$$\left.\begin{aligned} Q &= \{x \mid x\in W \text{ and } h>x \text{ for all } h\in H\} \\ R &= \{x \mid x\in W \text{ and } h<x \text{ for all } h\in H\}, \end{aligned}\right\} \qquad (11\text{-}2)$$

If, additionally, H is a convex subset of W, then the sets P and Q can equivalently be characterised by:

$$\left.\begin{aligned} Q &= \{x \mid x\in W,\ x\notin H \text{ and } h>x \text{ for some } h\in H\} \\ R &= \{x \mid x\in W,\ x\notin H \text{ and } h<x \text{ for some } h\in H\} \end{aligned}\right\} \qquad (11\text{-}3)$$

Proof. If W is linear then for all $x\in W$ and $h\in H$, $x \not> h$ is true if and only if $x \geqslant h$ is false, so Q of (11-1) and Q of (11-2) are the same. Similarly for R.

Suppose additionally that H is convex, and that $x\in Q$ of (11-3). If $x\notin Q$ of (11-2) then for some $k\in H$, $k>x$ is false, i.e. (by linearity), $k\leqslant x$. Hence $k\leqslant x<h$, whence $x\in H$ by convexity, contradicting (11-3). Thus Q of (11-3) is contained in Q of (11-2), and the converse is trivial. Similarly for R. ●

Lemma 11-11. Let H be a non-empty convex subset of a given linearly ordered set W. Then H, together with Q and R (of (11-3)) form three pairwise disjunct sets whose union is W.

Proof. Let $h\in H$ and $x\in W$. If $x\in H$ then by (11-2), $x\notin Q\cup R$. Suppose $x\notin H$. By linearity, there holds exactly one of the relations.

$$x<h, \qquad x=h, \qquad x>h.$$

But $x=h$ is inconsistent with $x\notin H$, hence (by (11-3)) x belongs to one of Q,R but (by (11-2)) not to both. ●

Theorem 11-12. Let W be either a linear division belt other than $\{\emptyset\}$, or a linear blog. Let H be a proper convex subgroup of W and let Q,R be as in (11-3). Then W has the trisection (Q,H,R).

Proof. H is non-empty, since $\emptyset\in H$. Hence by Lemma 11-11, H, Q and R are pairwise disjunct with union W. And Q and R are non-empty. For if W is a blog then $-\infty\in Q$ whilst $+\infty\in R$. And if W is a division belt then we can find $x\in W$ such that $x\notin H$, since H is a proper subset of W. Then $x\in Q\cup R$ and it is easy to see that $x^{-1}\in Q$ if and only if $x\in R$ (using Lemma 4-6). So W is the union of the pairwise disjunct non-empty sets Q, H,R.

We must now verify all 36 entries in the table of Fig 11-1, with Q,H,R in the roles of $\sigma_{-\infty}$, $\sigma_{\emptyset}$, $\sigma_{+\infty}$ respectively. So let $x, y \in W$.

Now, if W is a blog, and one of $x,y\in W$ is either $-\infty$ or $+\infty$, then (since $-\infty\in Q$ and $+\infty\in R$), the table of Fig 4-1 guarantees the success of the verification in all these cases. We may restrict ourselves therefore to considering finite elements $x,y\in W$, in other words we may argue further as for the case that W is a division belt.

But in a division belt, multiplication is self-dual, so we need only consider the columns of the table for $\oplus$, $\otimes$ and $\oplus'$. Verification of the columns for $\oplus$ and $\oplus'$ is immediate because of the linearity of W and the fact that $q < h < r$ for all $q \varepsilon Q$, $h \varepsilon H$ and $r \varepsilon R$, by (11-2)

We are left with the column headed $x \otimes y$, for which we must verify the top five entries, the botton four being analogous to the top four. We take the cases separately, our notation being that $q_1, q_2 \varepsilon Q$ and $h_1, h_2 \varepsilon H$, but all are otherwise arbitrary.

1. By (11-2), $q_1 < h_1$ and $q_2 < \emptyset$. Hence by Lemma 4-11:

$$q_1 \otimes q_2 < h_1$$

   But h_1 is arbitrary, so $q_1 \otimes q_2 \varepsilon Q$ by (11-2), whence $Q \otimes Q \subset Q$

2. By (11-2), $q_1 < h_1 \otimes h_2^{-1} \varepsilon H$. Hence by Lemma 4-11:

$$q_1 \otimes h_2 < h_1$$

   Again, h_1 is arbitrary, so $q_1 \otimes h_2 \varepsilon Q$ by (11-2), whence $Q \otimes H \varepsilon Q$.

3. The table of Fig 11-1 requires no assertion to be made regarding $Q \otimes R$.

4. The case $H \varepsilon Q$ is similar to the case $Q \otimes H$ discussed under 2.

5. Since H is a group, $H \otimes H \subset H$. ●

   As a special case of the foregoing analysis, we may take H to be the trivial group $\{\emptyset\}$, as in the following result.

Theorem 11-13. Let W be either a division belt other than $\{\emptyset\}$, or a blog. Then $\{\emptyset\}$ is the middle of a trisection of W if and only if W is linear; and the trisection is then $(N, \{\emptyset\}, P)$ where N, P are respectively the negative and positive cone of W.

Proof. If W is linear, then since $\{\emptyset\}$ is a (trivial) convex subgroup of W, Theorem 11-12 implies that W has the trisection $(Q, \{\emptyset\}, R)$. But from (11-2) with $H = \{\emptyset\}$, we see that $Q = N$ and $R = P$.

Conversely, suppose W has a trisection $(Q, \{\emptyset\}, R)$. Let $x, y \varepsilon W$. If y is not finite then $x \leqslant y$ or $y \leqslant x$ according as y is $+\infty$ or $-\infty$. If y is finite then $x \otimes y^{-1} \varepsilon W = (Q \cup \{\emptyset\}) \cup (\{\emptyset\} \cup R)$. If $x \otimes y^{-1} \varepsilon Q \cup \{\emptyset\}$ then using the table of Fig 11-1: $(x \otimes y^{-1}) \oplus \emptyset = \emptyset$; whence $x \otimes y^{-1} \leqslant \emptyset$ so $x \leqslant y$. Similarly if $x \otimes y^{-1} \varepsilon \{\emptyset\} \cup R$ then $x \geqslant y$. Hence in all cases either $x \leqslant y$ or $x \geqslant y$, so W is linear. ●

11-5. Two Examples

We give now a couple of examples to illustrate the above results. Let $(J, \otimes)$ be an infinite linearly ordered group, and let W be the set of 3-tuples of elements of J. We make W into a group be defining multiplication componentwise:

$$(a_1, a_2, a_3) \otimes (b_1, b_2, b_3) = (a_1 \otimes b_1, a_2 \otimes b_2, a_3 \otimes b_3).$$

Now order the elements of W lexicographically, using the ordering in J, i.e. $(b_1,b_2,b_3) \geq (a_1,a_2,a_3)$ if and only if:

$$(b_3 > a_3) \text{ or } (b_3 = a_3 \text{ and } b_2 > a_2) \text{ or } (b_3 = a_3 \text{ and } b_2 = a_2 \text{ and } b_1 \geq a_1).$$

It is readily verified that $(W,\otimes)$ becomes a linearly ordered group, which we may construe as a linear division belt $(W,\oplus,\otimes,\oplus',\otimes')$ as in Section 2-8. Now define

$$H = \{(x,\phi,\phi) \mid x \varepsilon J\}.$$

Then H is a convex proper subgroup of W. Defining Q,R as in (11-1) we confirm that W has the trisection (Q,H,R).

Now let us take the same group $(W,\oplus)$ of 3-tuples of elements of J, but this time we take the ordering of $(W,\otimes)$ to be that of J^3:

$$(a_1,a_2,a_3) \geq (b_1,b_2,b_3) \text{ if and only if:}$$

$$a_1 \geq b_1 \text{ and } a_2 \geq b_2 \text{ and } a_3 \geq b_3.$$

$(W,\otimes)$ is now a lattice-ordered group, which we may construe as a (non-linear) division belt $(W,\oplus,\otimes,\oplus',\otimes')$ as in Section 2-8. The same subgroup H is again a convex proper subgroup of W, but now if $p \neq \phi$, we have:

$$(\phi,p,p^{-1}) \varepsilon Q \cap R.$$

Hence $Q \cap R$ is not empty and H cannot be the middle of a trisection.

12. σ_ϕ-ASTIC MATRICES

12-1 . σ_ϕ-asticity.

In Section 11-1, we posed two questions. In the present chapter, we answer the second of these questions, deferring the first to Chapter 13.

Let us recall from the theory of probability that a row-stochastic matrix is a (non-negative) matrix in which the sum of the elements in each row is unity. A column-stochastic matrix has the sum of the elements in each column equal to unity, and a doubly stochastic matrix is both row- and column-stochastic. It is easy to prove that the product of two (row-, column- or doubly) stochastic matrices, conformable for multiplication, is again (respectively row-, column- or doubly) stochastic.

An analogous result holds for matrices in minimax algebra; moreover the sum of "stochastic" matrices is then also "stochastic". To develop these ideas, we require the following definition.

Let $(E_1, \oplus, \otimes, \oplus', \otimes')$ be a belt with duality and $(\sigma_\phi, \oplus, \otimes, \oplus', \otimes')$ a sub-belt of E_1 with duality. We shall say that a finite subset $S \subset E_1$ is σ_ϕ-astic if there holds:

$$\sum_{x \varepsilon S}{}^{\oplus} x \;\varepsilon\; \sigma_\phi \tag{12-1}$$

In (12-1), we see the analogy to the stochastic case. Specifically, if σ_ϕ is just $\{\phi\}$, then a σ_ϕ-astic set satisfies:

$$\sum_{x \varepsilon S}{}^{\oplus} x = \text{the multiplicative identity } \phi$$

whereas a set of probabilities satisfies:

$$\sum_{x \varepsilon S} x = \text{the multiplicative identity } 1.$$

A matrix over E_1 will be called row-σ_ϕ-astic (respectively column-σ_ϕ-astic, or doubly σ_ϕ-astic) if the elements in each row (respectively each column, or each row and each column) form a σ_ϕ-astic set.
A convenient terminology is: α-σ_ϕ-astic where α may represent any one of the prefix words row-, column- or doubly.

To simplify later terminology, we shall admit two special variants of the above usage. Specifically, if σ_ϕ is the group G of a blog then we shall say G-astic as an alternative to σ_ϕ-astic; and if σ_ϕ is the trivial group $\{\phi\}$ then we shall say ϕ-astic as an alternative to σ_ϕ-astic.

Lemma 12-1. Let S be a subset of a belt E_1 with duality having a trisection $(\sigma_{-\infty}, \sigma_\phi, \sigma_{+\infty})$. If S contains a finite number of elements, then S is σ_ϕ-astic if and only if:

$$S \cap \sigma_{+\infty} \text{ is empty, but } S \cap \sigma_\phi \text{ is not empty.} \tag{12-2}$$

Proof. By Proposition 11-2:

$\sum_{x \in S}\oplus\, x \in \sigma_{+\infty}$ if and only if $S \cap \sigma_{+\infty}$ is non-empty.

$\sum_{x \in S}\oplus\, x \in \sigma_{-\infty}$ if and only if $S \cap (\sigma_\phi \cup \sigma_{+\infty})$ is empty.

So, since $\sum_{x \in S}\oplus\, x \in \sigma_\phi$ if and only if $\sum_{x \in S}\oplus\, x \notin \sigma_{-\infty} \cup \sigma_{+\infty}$,

$\sum_{x \in S}\oplus\, x \in \sigma_\phi$ if and only if $S \cap \sigma_{+\infty}$ is empty but $S \cap \sigma_\phi$ is not empty. ●

By virtue of Lemma 12-1, we may informally characterise G-astic matrices as having at least one finite element, and no element equalling $+\infty$; and ϕ-astic matrices as being non-positive and containing at least one element equalling ϕ.

We notice en passant that α-σ_ϕ-astic matrices are all particular cases of matrices over the belt $\sigma_{-\infty} \cup \sigma_\phi$, when E_1 has the trisection $(\sigma_{-\infty}, \sigma_\phi, \sigma_{+\infty})$.

Lemma 12-2. Suppose the belt E_1 with duality has two trisections $(\sigma_{-\infty}, \sigma_\phi, \sigma_{+\infty})$ and $(\dot\sigma_{-\infty}, \dot\sigma_\phi, \dot\sigma_{+\infty})$ with $\dot\sigma_\phi \subset \sigma_\phi$. If a matrix $\underline{A}$ with elements in E_1 is α-$\dot\sigma_\phi$-astic, then $\underline{A}$ is also α-σ_ϕ-astic. In particular, if E_1 is a linear blog with group G, and $\underline{A}$ is α-$\emptyset$-astic then $\underline{A}$ is α-G-astic.

Proof. Using Lemma 12-1, if $\underline{A} \in \mathcal{M}_{mn}$ is row-$\dot\sigma_\phi$-astic then:

$$\sum_{j=1}^{n}\oplus \{\underline{A}\}_{ij} \in \dot\sigma_\phi \qquad (i=1,\ldots, m)$$

Hence:
$$\sum_{j=1}^{n}\oplus \{\underline{A}\}_{ij} \in \sigma_\phi \qquad (i=1,\ldots, m)$$

Thus $\underline{A}$ is row-σ_ϕ-astic. Similar arguments apply if α is column- or doubly.

Finally, if E_1 is a linear blog then by Proposition 11-1 and Theorem 11-13, E_1 has the two trisections $(\{-\infty\}, G, \{+\infty\})$ and $(N,\{\emptyset\},P)$. And evidently, $\{\emptyset\} \subset G$. ●

The following result is that adumbrated in the introductory remarks of this chapter.

Theorem 12-3. Let $(\sigma_\emptyset, \oplus, \otimes, \oplus', \otimes')$ be a convex sub-belt with duality of a belt $(E_1, \oplus, \otimes, \oplus', \otimes')$ with duality, and for any integers $m, n \geq 1$ let $\mathcal{N}_{mn} \subset \mathcal{M}_{mn}$ denote the set of all $(m \times n)$ α-$\sigma_\emptyset$-astic matrices for a fixed meaning of α. Then the $(\mathcal{N}, \mathcal{N}, \sigma_\emptyset)$-hom-rep statement is true. In particular, this holds if $\sigma_\emptyset$ is the middle of a trisection of E_1, or if $\sigma_\emptyset$ is the trivial belt $\{\emptyset\}$.

Proof. First, let α denote the prefix "row", and let $\underline{A}, \underline{C} \in \mathcal{N}_{mn}$. Then:

$$\forall i=1,\ldots, m, \qquad \sum_{j=1}^{n}{}_{\oplus} \{\underline{A} \oplus \underline{C}\}_{ij} = \Big(\sum_{j=1}^{n}{}_{\oplus} \{\underline{A}\}_{ij}\Big) \oplus \Big(\sum_{j=1}^{n}{}_{\oplus} \{\underline{C}\}_{ij}\Big) \in \sigma_\emptyset$$

Hence: $\underline{A} \oplus \underline{C}$ is row-$\sigma_\emptyset$-astic. Moreover, for any $\lambda \in \sigma_\emptyset$:

$$\forall i=1,\ldots, m, \qquad \sum_{j=1}^{n}{}_{\oplus} \{\underline{A} \otimes \lambda\}_{ij} = \Big(\sum_{j=1}^{n}{}_{\oplus} \{A\}_{ij}\Big) \otimes \lambda \in \sigma_\emptyset$$

Hence: $\underline{A} \otimes \lambda$ is row-$\sigma_\emptyset$-astic. Moreover, for any $\underline{B} \in \mathcal{N}_{np}$:

$$\forall i=1,\ldots, m, \qquad \sum_{j=1}^{p}{}_{\oplus} \{\underline{A} \otimes \underline{B}\}_{ij} = \sum_{j=1}^{p}{}_{\oplus} \Big(\sum_{r=1}^{n}{}_{\oplus} \big(\{\underline{A}\}_{ir} \otimes \{\underline{B}\}_{rj}\big)\Big)$$

$$= \sum_{r=1}^{n}{}_{\oplus} \Big(\{\underline{A}\}_{ir} \otimes \sum_{j=1}^{p}{}_{\oplus} \{\underline{B}\}_{rj}\Big) \tag{12-3}$$

Now $\sum_{j=1}^{p}{}_{\oplus} \{\underline{B}\}_{rj} \in \sigma_\emptyset$ by hypothesis, for all $r=1,\ldots, n$. Hence $u, v \in \sigma_\emptyset$ where by definition:

$$u = \sum_{r=1}^{n}{}_{\oplus'} \Big(\sum_{j=1}^{p}{}_{\oplus} \{\underline{B}\}_{rj}\Big) \;; \qquad v = \sum_{r=1}^{n}{}_{\oplus} \Big(\sum_{j=1}^{p}{}_{\oplus} \{\underline{B}\}_{rj}\Big) \tag{12-4}$$

Evidently, by (3-18):

$$\forall r=1,\ldots, n, \qquad u \leq \sum_{j=1}^{p}{}_{\oplus} \{\underline{B}\}_{rj} \leq v$$

Hence for all $i=1,\ldots, m$; $r=1,\ldots, n$, we have:

$$\{\underline{A}\}_{ir} \otimes u \leq \{\underline{A}\}_{ir} \otimes \sum_{j=1}^{p}{}_{\oplus} \{\underline{B}\}_{rj} \leq \{\underline{A}\}_{ir} \otimes v$$

Closing w.r.t. index r and using (12-3):

$$\forall i=1,\ldots, m, \qquad \Big(\sum_{r=1}^{n}{}_{\oplus} \{\underline{A}\}_{ir}\Big) \otimes u \leq \sum_{j=1}^{p}{}_{\oplus} \{\underline{A} \otimes \underline{B}\}_{ij} \leq \Big(\sum_{r=1}^{n}{}_{\oplus} \{\underline{A}\}_{ir}\Big) \otimes v \tag{12-5}$$

Now, $\sum_{r=1}^{n}{}_{\oplus} \{\underline{A}\}_{ir} \in \sigma_\emptyset$ by hypothesis, so (12-5) exhibits $\sum_{j=1}^{p}{}_{\oplus} \{\underline{A} \otimes \underline{B}\}_{ij}$ lying between two elements of $\sigma_\emptyset$. Since $\sigma_\emptyset$ is convex, we infer:

$$\forall i=1,\ldots, m, \qquad \sum_{j=1}^{p}{}_{\oplus} \{\underline{A} \otimes \underline{B}\}_{ij} \in \sigma_\phi$$

Hence $\underline{A} \otimes \underline{B}$ is row-σ_ϕ-astic.

Summing up, we have shown that $\underline{A} \oplus \underline{C} \in \mathcal{N}_{mn}$, $\underline{A} \otimes \lambda \in \mathcal{N}_{mn}$ and $\underline{A} \otimes \underline{B} \in \mathcal{N}_{mp}$ when $\underline{A} \in \mathcal{N}_{mn}$, $\underline{B} \in \mathcal{N}_{np}$, $\underline{C} \in \mathcal{N}_{mn}$ and $\lambda \in \sigma_\phi$. Apart from some routine verifications, this proves the $(\mathcal{N}, \mathcal{N}, \sigma_\phi)$-hom-rep statement when α is "row". The proofs when α is "column" or "doubly" are similar. The particular case when σ_ϕ is the middle of a trisection follows now from Theorem 11-5. Moreover, the trivial belt $\{\phi\}$ is obviously convex, so the result holds for this case also. ●

12-2 . The Generalised Question 2

Question 2 from Section 11-1 asks: how can we characterise the set of left matrix multiplications which take the space of finite n-tuples into itself?

This question is addressed in the following results.

<u>Theorem 12-4. Let $(\sigma_\phi, \oplus, \otimes, \oplus', \otimes')$ be a convex sub-belt with duality of a belt $(E_1, \oplus, \otimes, \oplus', \otimes')$ with duality, and for any integers $m, n \geqslant 1$ let $\mathcal{N}_{mn} \subset \mathcal{M}_{mn}$ denote the set of all (mxn) row-σ_ϕ-astic matrices and let $\mathcal{O}_{mn} \subset \mathcal{M}_{mn}$ denote the set of all (mxn) matrices $\underline{A}$ for which $\{\underline{A}\}_{ij} \in \sigma_\phi$ for all $i=1,\ldots, m$; $j=1,\ldots, n$. Then the $(\mathcal{N}, \mathcal{O}, \sigma_\phi)$- and $(\mathcal{O}, \mathcal{O}, \sigma_\phi)$-hom-rep statements are true.</u>

<u>Proof</u>. The $(\mathcal{O}, \mathcal{O}, \sigma_\phi)$-hom-rep statement follows from Theorem 5-6 with σ_ϕ in the role of E_1 and $\mathcal{O}_{mn}$ in the role of $\mathcal{M}_{mn}$.

Now, since a sum of elements of σ_ϕ is again an element of σ_ϕ, we infer that $\mathcal{O}_{mn} \subset \mathcal{N}_{mn}$ for all integers $m, n \geqslant 1$. So, (using the $(\mathcal{O}, \mathcal{O}, \sigma_\phi)$-hom-rep statement), $(\mathcal{O}_{mn}, \oplus)$ is actually a subspace of $(\mathcal{N}_{mn}, \oplus)$ over σ_ϕ.

Let $\underline{A} \in \mathcal{N}_{mn}$, $\underline{B} \in \mathcal{O}_{np}$ for given integers $m, n, p \geqslant 1$. Then $u, v \in \sigma_\phi$ where by definition:

$$u = \sum_{r=1}^{n}{}_{\oplus'} \sum_{j=1}^{p}{}_{\oplus'} \{\underline{B}\}_{rj} \; ; \qquad v = \sum_{r=1}^{n}{}_{\oplus} \sum_{j=1}^{p}{}_{\oplus} \{\underline{B}\}_{rj} \tag{12-6}$$

Evidently, by (3-18):

$$\forall r=1,\ldots, n \; ; \; j=1,\ldots, p \, , \qquad u \leqslant \{\underline{B}\}_{rj} \leqslant v$$

Hence for all $i=1,\ldots, m$; $r=1,\ldots, n$; $j=1,\ldots, p$ we have:

$$\{\underline{A}\}_{ir} \otimes u \leqslant \{\underline{A}\}_{ir} \otimes \{\underline{B}\}_{rj} \leqslant \{\underline{A}\}_{ir} \otimes v$$

Closing w.r.t. index we have for all $i=1,\ldots, m$; $j=1,\ldots, p$:

$$\left(\sum_{r=1}^{n}{}_{\oplus} \{\underline{A}\}_{ir}\right) \otimes u \leq \sum_{r=1}^{n}{}_{\oplus} \left(\{\underline{A}\}_{ir} \otimes \{\underline{B}\}_{rj}\right) \leq \left(\sum_{r=1}^{n}{}_{\oplus} \{\underline{A}\}_{ir}\right) \otimes v \qquad (12\text{-}7)$$

Now $\sum_{r=1}{}_{\oplus} \{\underline{A}\}_{ir} \in \sigma_{\phi}$ by hypothesis, so (12-7) exhibits $\{\underline{A} \otimes \underline{B}\}_{ij}$ lying between two elements of σ_{ϕ}. Since σ_{ϕ} is convex, we infer: $\underline{A} \otimes \underline{B} \in \mathcal{O}_{mp}$.

In short, if $\underline{A} \in \mathcal{N}_{mn}$ and $\underline{B} \in \mathcal{O}_{np}$ then $\underline{A} \otimes \underline{B} \in \mathcal{O}_{mp}$. Apart from some routine verifications, this proves the $(\mathcal{N}, \mathcal{O}, \sigma_{\phi})$-hom-rep statement. ●

Theorem 12-4 shows that if $\underline{A} \in \mathcal{N}_{mn}$ is row-σ_{ϕ}-astic (with σ_{ϕ} convex), then:

$$g_{\underline{A}} \in \mathrm{Stab}_{E_1} \left((\sigma_{\phi})^n, (\sigma_{\phi})^m\right) \qquad (12\text{-}8)$$

where $g_{\underline{A}} : (\sigma_{\phi})^n \mapsto (\sigma_{\phi})^m$ is the left-multiplication defined by $g_{\underline{A}} : \underline{x} \mapsto \underline{A} \otimes \underline{x}$ for each $x \in (\sigma_{\phi})^n$; and $(\sigma_{\phi})^n$ denotes the set of n-tuples all of whose elements lie in σ_{ϕ}.

To answer Question 2, we must prove the converse also (under appropriate conditions). For this, we require the following definition. A subset T of a belt $(E_1, \oplus, \otimes)$ will be called <u>right-cancellative</u> or simply <u>cancellative</u> if there holds:

$$\forall \lambda \in E_1, \ \forall x \in T, \ \text{if } \lambda \otimes x \in T \ \text{ then } \ \lambda \in T \qquad (12\text{-}9)$$

Analogously we may define <u>left-cancellative</u>, and <u>right</u>- and <u>left-dually cancellative</u> (when E_1 has a duality). The following is then easily proved.

<u>Proposition 12-5. Let $(E_1, \oplus, \otimes, \oplus', \otimes')$ be a given belt with duality. If E_1 contains an element ϕ which acts as multiplicative identity element in $(E_1, \otimes)$ and in $(E_1, \otimes')$ then $\{\phi\}$ is right- and left-cancellative and dually cancellative. If E_1 has a trisection $(\sigma_{-\infty}, \sigma_{\phi}, \sigma_{+\infty})$ then σ_{ϕ} is right- and left-cancellative and dually cancellative. In particular, the group of a blog is right- and left-cancellative and dually cancellative.</u> ●

The following result now gives the desired converse to (12-8):

<u>Theorem 12-6. Let $(\sigma_{\phi}, \oplus, \otimes, \oplus', \otimes')$ be a cancellative sub-belt with duality of a belt $(E_1, \oplus, \otimes, \oplus' \otimes')$ with duality and let $\underline{A} \in \mathcal{M}_{mn}$ for given integers $m, n \geq 1$ be such that for each integer $p \geq 1$ the left-multiplication $g_A : \mathcal{M}_{np} \mapsto \mathcal{M}_{mp}$ defined by $g_{\underline{A}} : \underline{B} \mapsto \underline{A} \otimes \underline{B}$ satisfies:</u>

$$g_{\underline{A}} \in \mathrm{Stab}_{E_1} (\mathcal{O}_{np}, \mathcal{O}_{mp})$$

<u>where $\mathcal{O}_{np}, \mathcal{O}_{mp}$ are the sets of all (nxp) and (mxp) matrices respectively which have all their elements in σ_{ϕ}. Then $\underline{A}$ is row-σ_{ϕ}-astic.</u>

Proof. Let $u \in \sigma_\phi$ be arbitrary and let $\underline{B} \in \mathcal{O}_{np}$ have:

$$\{\underline{B}\}_{rj} = u \ (r=1,\dots, n; \ j=1,\dots, p). \tag{12-10}$$

By hypothesis, $\underline{A} \otimes \underline{B} \in \mathcal{O}_{mp}$, whence for all $i=1,\dots, m$:

$$\left(\sum_{r=1}^{n}{}_{\oplus} \{\underline{A}\}_{ir}\right) \otimes u = \sum_{r=1}^{n}{}_{\oplus} \left(\{\underline{A}\}_{ir} \otimes \{\underline{B}\}_{rj}\right) = \{\underline{A} \otimes \underline{B}\}_{ij} \in \sigma \tag{12-11}$$

Since $u \in \sigma_\phi$ and σ_ϕ is cancellative, (12-11) implies:

$$\sum_{r=1}^{n}{}_{\oplus} \{\underline{A}\}_{ir} \in \sigma_\phi$$

Hence $\underline{A}$ is row-σ_ϕ-astic. ●

Corollary 12-7. Let $(\sigma_\phi, \oplus, \otimes, \oplus', \otimes')$ be a convex cancellative sub-belt with duality of a belt $(E_1, \oplus, \otimes, \oplus', \otimes')$ with duality and let $\underline{A} \in \mathcal{M}_{mn}$ for given integers $m, n \geqslant 1$. Then for each integer $p \geqslant 1$, the left-multiplication $g_{\underline{A}} : \mathcal{M}_{np} \mapsto \mathcal{M}_{mp}$ defined by $g_{\underline{A}} : \underline{B} \mapsto \underline{A} \otimes \underline{B}$ satisfies:

$$g_{\underline{A}} \in \text{Stab}_{E_1} (\mathcal{O}_{np}, \mathcal{O}_{mp})$$

(where $\mathcal{O}_{np}, \mathcal{O}_{mp}$ are the sets of all (nxp) and (mxp) matrices respectively which have all their elements in σ_ϕ) if and only if $\underline{A}$ is row-σ_ϕ-astic. In particular, this holds when σ_ϕ is the middle of a trisection of E_1.

Proof. The main result follows immediately from Theorems 12-4 and 12-6. The particular case when σ_ϕ is the middle of a trisection follows because σ_ϕ is then convex and cancellative, by Theorem 11-5 and Proposition 12-5. ●

Our answer to Question 2, applicable for the principal interpretation, is therefore: "left matrix multiplications which preserve finiteness are exactly those induced by row-σ_ϕ-astic matrices". We can, in fact, prove a stronger result to the effect that "right-linear homomorphisms which preserve finiteness are exactly those induced by row-σ_ϕ-astic matrices". This we now do.

Corollary 12-8. Let E_1 be a blog with group G and for given integers $m, n \geqslant 1$ let $\mathcal{N}_{mn}$ denote the set of row-G-astic matrices over E_1, and G_n the set of finite n-tuples over E_1. Then:

$$\text{Stab}_{E_1} (G_n, G_m) = \{ g_{\underline{A}} \mid g_{\underline{A}} : E_n \to E_m \text{ and } \exists\, \underline{A} \in \mathcal{N}_{mn} \text{ such that:}$$

$$g_{\underline{A}}(\underline{x}) = \underline{A} \otimes \underline{x} \ (\forall\ \underline{x} \in E_n)\} \tag{12-12}$$

The representation $\tau : \underline{A} \mapsto g_{\underline{A}}$ is an isomorphism of $(\mathcal{N}_{mn}, \oplus)$ to $(\text{Stab}_{E_1} (G_n, G_m), \oplus)$; and if $m = n$ then τ is an isomorphism of the belt $(\mathcal{N}_{nn}, \oplus, \otimes)$ to the belt

$(\mathrm{Stab}_{E_1}(G_n, G_n), \oplus, \otimes)$.

Proof. Call the set defined on the right-hand side of (12-12): $\mathcal{L}\mathcal{N}_{mn}$. Since G is the middle of a trisection of E_1, Corollary 12-7 with $p = 1$ shows that:

$$\mathcal{L}\mathcal{N}_{mn} \subset \mathrm{Stab}_{E_1}(G_n, G_m) \tag{12-13}$$

Conversely, let $g \in \mathrm{Stab}_{E_1}(G_n, G_m)$. Then $g \in \mathrm{Hom}_{E_1}(E_n, E_n)$ by (5-6).

Now, since E_1 is a blog, it satisfies axioms X_8 and X_{10} and therefore g is necessarily a left-multiplication $g_{\underline{A}}$ for some $\underline{A} \in \mathcal{M}_{mn}$ by the proof of Theorem 5-10. But since $g \in \mathrm{Stab}_{E_1}(G_n, G_m)$, we infer $\underline{A} \in \mathcal{N}_{mn}$ by Corollary 12-7. Hence $g \in \mathcal{L}\mathcal{N}_{mn}$, and (12-12) is proved.

Thus τ is surjective of $\mathcal{N}_{mn}$ onto $\mathrm{Stab}_{E_1}(G_n, G_m)$, and is injective by Theorem 5-10. Hence τ is bijective, and by the usual arguments is a homomorphism. ●

13. /-EXISTENCE

13-1 . /-Existence Defined.

In Section 11-1, as part of a general discussion, we introduced the notion of /-existence as a formal device for avoiding the use of products in which $-\infty$ and $+\infty$ are combined. Generalising this notion, let E_1 be a belt with duality having a trisection $(\sigma_{-\infty}, \sigma_\phi, \sigma_{+\infty})$, and let $x,y \varepsilon E_1$. We shall say that the products $x \otimes y$ and $x \otimes' y$ are /-undefined if one of x,y lies in $\sigma_{-\infty}$ whilst the other lies in $\sigma_{+\infty}$. Similarly, we shall say that a matrix product, or other expression, is /-undefined if its evaluation requires the formation of a /-undefined product of elements of E_1. Otherwise, we shall say that such expressions or products are /-defined or that they /-exist.

In the following results we shall use the notation $\mathcal{O}_{mn}$ to denote the set of all $(m \times n)$ matrices whose elements all lie in σ_ϕ.

Proposition 13-1. Let E_1 be a belt with duality having a trisection $(\sigma_{-\infty}, \sigma_\phi, \sigma_{+\infty})$. Let $\underline{A} \varepsilon \mathcal{M}_{mn}$ and $\underline{B} \varepsilon \mathcal{M}_{np}$ for integers $m,n,p \geq 1$. Then $A \otimes B$ /-exists if either $\underline{A} \varepsilon \mathcal{O}_{mn}$ or $B \varepsilon \mathcal{O}_{np}$. Similarly $\underline{A} \otimes' \underline{B}$ /-exists if either $\underline{A} \varepsilon \mathcal{O}_{mn}$ or $\underline{B} \varepsilon \mathcal{O}_{np}$. ●

Lemma 13-2. Let $\underline{A}$, $\underline{X}$ be matrices over a belt E_1 with duality, having a trisection $(\sigma_{-\infty}, \sigma_\phi, \sigma_{+\infty})$.
If $\underline{A} \otimes \underline{X}$ /-exists and is row-σ_ϕ-astic then $\underline{A}$ is row-σ_ϕ-astic and the elements of $\underline{X}$ form a σ_ϕ-astic set

Proof. For element (i,j) of $\underline{A} \otimes \underline{X}$ we have:

$$\{\underline{A} \otimes \underline{X}\}_{ij} = \sum_{r}{}_{\oplus} (\{\underline{A}\}_{ir} \otimes \{\underline{X}\}_{rj})$$

Using the table of Fig 11-1 we conclude:

(i) If $\{\underline{A}\}_{ir} \varepsilon \sigma_{+\infty}$ for some i,r, then row r of $\underline{X}$ does not contain an element of $\sigma_{-\infty}$ (since $\underline{A} \otimes \underline{X}$/-exists), and then $\{\underline{A} \otimes \underline{X}\}_{ij} \varepsilon \sigma_{+\infty}$ for all j for that i.

(ii) If $\{\underline{A}\}_{ir} \varepsilon \sigma_{-\infty}$ for all r for some i, then no column of $\underline{X}$ contains an element of $\sigma_{+\infty}$, and then $\{\underline{A} \otimes \underline{X}\}_{ij} \varepsilon \sigma_{-\infty}$ for all j for that i.

Similarly:

(iii) If $\{\underline{X}\}_{rj} \varepsilon \sigma_{+\infty}$ for some r,j then $\{\underline{A} \otimes \underline{X}\}_{ij} \varepsilon \sigma_{+\infty}$ for all i for that j.

(iv) If $\{\underline{X}\}_{rj} \varepsilon \sigma_{-\infty}$ for all r,j then $\{\underline{A} \otimes \underline{X}\}_{ij} \varepsilon \sigma_{-\infty}$ for all i,j.

Since all these conclusions contradict the hypothesis that $\underline{A} \otimes \underline{X}$ is row-σ_ϕ-astic, we conclude (from (i) and (ii)) that $\underline{A}$ is row-σ_ϕ-astic and (from (iii) and (iv)) that the elements $\{\underline{X}\}_{ij}$ form a σ_ϕ-astic set. ●

Lemma 13-3. Consider the following four conditions for given matrices $\underline{A} \varepsilon \mathcal{M}_{mn}$,

$\underline{X} \varepsilon \mathcal{M}_{np}$ over a belt E_1 with duality, having a trisection $(\sigma_{-\infty}, \sigma_{\phi}, \sigma_{+\infty})$:

(i) $\underline{A}$ contains no element of $\sigma_{+\infty}$

(i)' $\underline{A}$ is row-σ_{ϕ}-astic

(ii) $\underline{A} \otimes \underline{X}$ /-exists and contains no element of $\sigma_{-\infty}$

(ii)' $\underline{A} \otimes \underline{X}$ /-exists and each column of $\underline{A} \otimes \underline{X}$ contains elements of $\sigma_{+\infty}$ only or elements of σ_{ϕ} only, according as the corresponding column of $\underline{X}$ does or does not contain an element of $\sigma_{+\infty}$.

Then the conditions: (i) with (ii); (i) with (ii)'; (i)' with (ii); (i)' with (ii)'; are all equivalent.

Proof. Since clearly (i)' implies (i) and (ii)' implies (ii), it suffices to show that (i) with (ii) implies (i)' with (ii)'. So suppose that $\underline{A}$ contains no element of $\sigma_{+\infty}$ and that $\underline{A} \otimes \underline{X}$ /-exists and contains no element of $\sigma_{-\infty}$.
Now if for some i,j we have:

$$\{\underline{A} \otimes \underline{X}\}_{ij} \varepsilon \sigma_{+\infty}, \text{ i.e. } \sum_{r}{}_{\oplus} (\{\underline{A}\}_{ir} \otimes \{\underline{X}\}_{rj}) \varepsilon \sigma_{+\infty},$$

then by (iv) of Proposition 11-3, some $\{\underline{X}\}_{rj} \varepsilon \sigma_{+\infty}$, since no $\{\underline{A}\}_{ir} \varepsilon \sigma_{+\infty}$. Hence by (iii) in the proof of Lemma 13-2 we infer that every element of column j of $\{\underline{A} \otimes \underline{X}\}$ lies in $\sigma_{+\infty}$, proving (ii)'.

Moreover, if $\underline{A}$ is not row-σ_{ϕ}-astic then some row i of $\underline{A}$ contains only elements of $\sigma_{-\infty}$, since $\underline{A}$ contains no element of $\sigma_{+\infty}$. But then by (ii) in the proof of Lemma 13-2, that is also true for $\underline{A} \otimes \underline{X}$, contradicting (ii). This proves (i)'. ●

We shall not bother to list the obvious duals and left-right variants of Lemmas 13-2 and 13-3, since these lemmas are very specifically intended as stepping stones for later theorems and have little interest in their own right.

13-2 . Compatible Trisections

Question 1 of Section 11-1 asked: How wide a range of matrix operations can we carry out without forming /-undefined products in E_1? Since our interest lies in the use of matrices and their duals as operators, we shall give this vague question a rather more precise formulation:

Can we characterise the matrices $\underline{A} \varepsilon \mathcal{M}_{mn}$ for which there exists subspaces S and T such that the left-multiplications $g_{\underline{A}}: S \to T$ and $g^*_{\underline{A}}: T \to S$ can be carried out without the use of /-undefined products? Evidently this leads us to the question of the /-existence of matrix triple products of the general form $\underline{A}^* \otimes' (\underline{A} \otimes \underline{X})$ or $\underline{A} \otimes (\underline{A}^* \otimes' \underline{Y})$, as used for example in Theorem 8-8.

Moreover, in Theorem 8-9 we considered some quadruple products which (as stated in that context), will be used in a later discussion, namely in connection with the solution of linear equations in Chapter 14 .

We address ourselves therefore, in Section 13-3 below, to investigating the

$\underline{P}$	α	$\underline{Q}$	ω_1	ω_2
$\underline{A}^* \otimes' (\underline{A} \otimes \underline{X})$	Row -	$\underline{A} \otimes \underline{X}$	$\sigma_{+\infty}$	$\sigma_{-\infty}$
$\underline{A} \otimes' (\underline{A}^* \otimes \underline{X})$	Column-dually -	$\underline{A}^* \otimes \underline{X}$	$\sigma_{-\infty}$	$\sigma_{-\infty}$
$\underline{A}^* \otimes (\underline{A} \otimes' \underline{X})$	Row-dually -	$\underline{A} \otimes' \underline{X}$	$\sigma_{-\infty}$	$\sigma_{+\infty}$
$\underline{A} \otimes (\underline{A}^* \otimes' \underline{X})$	Column -	$\underline{A}^* \otimes' \underline{X}$	$\sigma_{+\infty}$	$\sigma_{+\infty}$
$(\underline{X} \otimes \underline{A}) \otimes' \underline{A}^*$	Column -	$\underline{X} \otimes \underline{A}$	$\sigma_{+\infty}$	$\sigma_{-\infty}$
$(\underline{X} \otimes \underline{A}^*) \otimes' \underline{A}$	Row-dually -	$\underline{X} \otimes \underline{A}^*$	$\sigma_{-\infty}$	$\sigma_{-\infty}$
$(\underline{X} \otimes' \underline{A}) \otimes \underline{A}^*$	Column-dually	$\underline{X} \otimes' \underline{A}$	$\sigma_{-\infty}$	$\sigma_{+\infty}$
$(\underline{X} \otimes' \underline{A}^*) \otimes \underline{A}$	Row -	$\underline{X} \otimes' \underline{A}^*$	$\sigma_{+\infty}$	$\sigma_{+\infty}$

Fig 13-1 <u>Triple Products</u>

$\underline{P}$	α	$\underline{Q}$
$\underline{A} \otimes (\underline{A}^* \otimes' (\underline{A} \otimes \underline{X}))$	Doubly	$\underline{A} \otimes \underline{X}$
$\underline{A}^* \otimes (\underline{A} \otimes' (\underline{A}^* \otimes \underline{X}))$	Doubly dually -	$\underline{A}^* \otimes \underline{X}$
$\underline{A} \otimes' (\underline{A}^* \otimes (\underline{A} \otimes' \underline{X}))$	Doubly dually -	$\underline{A} \otimes' \underline{X}$
$\underline{A}^* \otimes' (\underline{A} \otimes (\underline{A}^* \otimes' \underline{X}))$	Doubly	$\underline{A}^* \otimes' \underline{X}$
$((\underline{X} \otimes \underline{A}) \otimes' \underline{A}^*) \otimes \underline{A}$	Doubly	$\underline{X} \otimes \underline{A}$
$((\underline{X} \otimes \underline{A}^*) \otimes' \underline{A}) \otimes \underline{A}^*$	Doubly dually -	$\underline{X} \otimes \underline{A}^*$
$((\underline{X} \otimes' \underline{A}) \otimes \underline{A}^*) \otimes' \underline{A}$	Doubly dually -	$\underline{X} \otimes' \underline{A}$
$((\underline{X} \otimes' \underline{A}^*) \otimes \underline{A}) \otimes' \underline{A}^*$	Doubly	$\underline{X} \otimes' \underline{A}^*$

Fig 13-2 <u>Quadruple Products</u>

/-existence of the various triple and quadruple matrix products used in Chapter 8. This question makes sense, of course, only for belts in which we can define both a conjugacy and a trisection, and we must therefore first develop the necessary ideas for this.

Accordingly, let (E_1, ⊕, ⊗, ⊕', ⊗') be a self-conjugate belt. We shall say that E_1 has a compatible trisection ($\sigma_{-\infty}$, $\sigma_{\emptyset}$, $\sigma_{+\infty}$) if:

$$\sigma^*_{-\infty} = \sigma_{+\infty}, \text{ i.e. if } x^* \varepsilon\ \sigma_{+\infty} \text{ if and only if } x \quad \sigma_{-\infty}. \tag{13-1}$$

Proposition 13-4. A blog with group G, under the self-conjugacy (7-2) has a compatible trisection ({$-\infty$}, G, {$+\infty$}). A linear blog, under the self-conjugacy (7-2), and a linear division belt, under the self-conjugacy (7-1), have the compatible trisection (N, {$\emptyset$},P), where N,P are respectively the negative and positive cone. ●

Evidently the condition $\sigma^*_{-\infty} = \sigma_{+\infty}$ implies $\sigma_{\emptyset}^* = \sigma_{\emptyset}$ so in the light of Lemma 11-2, we have the following

Proposition 13-5. If the self-conjugate belt E_1 has a compatible trisection ($\sigma_{-\infty}$, $\sigma_{\emptyset}$, $\sigma_{+\infty}$), then the belt ($\sigma_{\emptyset}$, ⊕, ⊗, ⊕' ⊗') is self-conjugate and the belts ($\sigma_{-\infty}$, ⊕, ⊗), ($\sigma_{-\infty} \cup \sigma_{\emptyset}$, ⊕, ⊗) are conjugate to the belts ($\sigma_{+\infty}$, ⊕', ⊗'), ($\sigma_{\emptyset} \cup \sigma_{+\infty}$,⊕',⊗') respectively. ●

Lemma 13-6. Let E_1 be a self-conjugate belt having a compatible trisection ($\sigma_{-\infty}$,$\sigma_{\emptyset}$, $\sigma_{+\infty}$). Let $\underline{A} \varepsilon \mathcal{M}_{mn}$, $\underline{B} \varepsilon \mathcal{M}_{mp}$. Then each product $\underline{A}^* \otimes \underline{B}$, $\underline{A}^* \otimes' \underline{B}$ /-exists if and only if there is no index i ($1 \leqslant i \leqslant m$) such that $\underline{A}$ and $\underline{B}$ both have an element of $\sigma_{+\infty}$ or both have an element of $\sigma_{-\infty}$, somewhere in row i.

Proof. Obviously $\underline{A}^*$ and $\underline{B}$ are conformable for both $\underline{A}^* \otimes \underline{B}$ and $\underline{A}^* \otimes' \underline{B}$ and so the products will both /-exist unless there are indices $i(1 \leqslant i \leqslant m)$, $r(1 \leqslant r \leqslant n)$, $s(1 \leqslant s \leqslant p)$ such that $\{\underline{A}^*\}_{ri} \otimes \{\underline{B}\}_{is}$ is not /-defined.

This will only happen if one of $\{\underline{A}^*\}_{ri}$, $\{\underline{B}\}_{is}$ is in $\sigma_{+\infty}$ and the other is in $\sigma_{-\infty}$. But from (13-1), this means that $\{\underline{A}\}_{ir}$ and $\{\underline{B}\}_{is}$ are either both in $\sigma_{+\infty}$, or in $\sigma_{-\infty}$. ●

Lemma 13-7 Let E_1 be a self-conjugate belt with a compatible trisection ($\sigma_{-\infty}$, $\sigma_{\emptyset}$, $\sigma_{+\infty}$) For $\underline{A} \varepsilon \mathcal{M}_{mn}$, each product $\underline{A} \otimes \underline{A}^*$, $\underline{A} \otimes' \underline{A}^*$, $\underline{A}^* \otimes \underline{A}$, $\underline{A}^* \otimes' \underline{A}$ /-exists if and only if all elements of $\underline{A}$ lie in $\sigma_{\emptyset}$, i.e. if and only if $\underline{A} \varepsilon \mathcal{O}_{mn}$.

Proof. Define $\underline{B} = \underline{A}$. Then $\underline{A}$ fails to have all its elements in $\sigma_{\emptyset}$ if and only if $\underline{A}$ and $\underline{B}$ both have an element of $\sigma_{-\infty}$, or both have an element of $\sigma_{+\infty}$, on some one row; and by Lemma 13-6 this is necessary and sufficient that the products $\underline{A}^* \otimes \underline{A}$, $\underline{A}^* \otimes' \underline{A}$ should not /-exist. Similarly for the other two products. ●

The following theorem gives the tie-up between /-existence and $\sigma_{\emptyset}$-asticity, already adumbrated in Section 11-1.

Theorem 13-8 Let E_1 be a self-conjugate belt having a compatible trisection

<u>$(\sigma_{-\infty}, \sigma_{\phi}, \sigma_{+\infty})$. Then $\underline{A} \in \mathcal{M}_{mn}$ is row-(respectively column- or doubly) σ_{ϕ}-astic if and only if there exists $\underline{X} \in \mathcal{M}_{np}$ (respectively $\underline{Y} \in \mathcal{M}_{qm}$ or both $\underline{X} \in \mathcal{M}_{np}$ and $\underline{Y} \in \mathcal{M}_{qm}$) such that $\underline{A}^* \otimes' (\underline{A} \otimes \underline{X})$ (respectively $(\underline{Y} \otimes \underline{A}) \otimes' \underline{A}^*$ or both $\underline{A}^* \otimes' (\underline{A} \otimes \underline{X})$ and $(\underline{Y} \otimes \underline{A}) \otimes' \underline{A}^*$) /-exist(s); and then such products are /-defined for every $\underline{X} \in \mathcal{O}_{np}$ (respectively every $\underline{Y} \in \mathcal{O}_{qm}$ or every $\underline{X} \in \mathcal{O}_{np}$ and every $\underline{Y} \in \mathcal{O}_{qm}$).</u>

<u>Proof</u>. Evidently the matrices are conformable for the relevant products, so /-existence turns upon the presence of elements from $\sigma_{-\infty}$ or $\sigma_{+\infty}$. Suppose first that $\underline{A} \in \mathcal{M}_{mn}$ is row-σ_{ϕ}-astic and that $\underline{X} \in \mathcal{O}_{np}$. Then $\underline{A} \otimes \underline{X}$ is /-defined by Proposition 13-1, and $\underline{A} \otimes \underline{X} \in \mathcal{O}_{mp}$ by Theorem 12-4. Hence by Proposition 13-1, $\underline{A}^* \otimes' (\underline{A} \otimes \underline{X})$ is /-defined.

Conversely, let $\underline{A} \in \mathcal{M}_{mn}$ and $\underline{X} \in \mathcal{M}_{np}$ be such that $\underline{A}^* \otimes' (\underline{A} \otimes \underline{X})$ is /-defined. Then $\underline{A} \otimes \underline{X}$ is necessarily /-defined, (being part of $\underline{A}^* \otimes' (\underline{A} \otimes \underline{X})$), and by Lemma 13-6 it cannot be that both $\underline{A}$ and $\underline{A} \otimes \underline{X}$ have an element of $\sigma_{+\infty}$ or both have an element of $\sigma_{-\infty}$ in any row i. Hence, making free use of the table of Fig 11-1, and the /-existence of $\underline{A} \otimes \underline{X}$, we argue as follows. $\underline{A}$ cannot have any element of $\sigma_{+\infty}$ in any row i, otherwise $\underline{A} \otimes \underline{X}$ would have exclusively elements of $\sigma_{+\infty}$ in row i. And $\underline{A}$ cannot have exclusively elements of $\sigma_{-\infty}$ in any row i otherwise $\underline{A} \otimes \underline{X}$ would have exclusively elements of $\sigma_{-\infty}$ in row i. But this all amounts to saying that $\underline{A}$ is row-σ_{ϕ}-astic.

Hence there exists $\underline{X} \in \mathcal{M}_{np}$ such that $\underline{A}^* \otimes' (\underline{A} \otimes \underline{X})$ is /-defined, if and only if $\underline{A}$ is row-σ_{ϕ}-astic; and $\underline{A}^* \otimes' (\underline{A} \otimes \underline{X})$ then /-exists for all $\underline{X} \in \mathcal{O}_{np}$. The discussion for $(\underline{Y} \otimes \underline{A}) \otimes' \underline{A}^*$ is analogous, and the doubly σ_{ϕ}-astic case is the intersection of the other two cases. ●

<u>13-3. Dually σ_{ϕ}-astic matrices</u>

If E_1 has a trisection $(\sigma_{-\infty}, \sigma_{\phi}, \sigma_{+\infty})$, we define a finite subset $S \subset E_1$ to be <u>dually σ_{ϕ}-astic</u> if:

$$\sum_{x \in S}{}_{\oplus'} x \in \sigma_{\phi}$$

The dual of Lemma 12-1 is the following.

<u>Proposition 13-9. Let S be a finite subset of a belt E_1 with duality having a trisection $(\sigma_{-\infty}, \sigma_{\phi}, \sigma_{+\infty})$. Then S is dually σ_{ϕ}-astic if and only if:</u>

<u>$S \cap \sigma_{-\infty}$ is empty, but $S \cap \sigma_{\phi}$ is not empty.</u> ●

We can now define row-dually, column-dually, doubly dually and α-dually σ_ϕ-astic matrices over E_1 by obvious analogy to the definitions in Section 12-1.

It will be convenient to extend the terminology α-σ_ϕ-astic matrix to allow α also to represent the prefixes column-dually, row-dually or doubly dually, and to introduce the terminology α*-σ_ϕ-astic matrix, where α* represents the prefix row, column, doubly, row-doubly, column-doubly or doubly dually exactly when α represents the prefix column-dually, row-dually, doubly dually, row, column or doubly respectively.

If $\mathcal{N}_{mn}$ denotes the commutative band of α-σ_ϕ-astic matrices in a given context, then $\mathcal{N}^*_{mn}$ will denote the commutative band of α*-σ_ϕ-astic matrices, a notation justified by the following self-evident proposition.

Proposition 13-10. Let E_1 be a self-conjugate belt with a compatible trisection $(\sigma_{-\infty},\sigma_\phi,\sigma_{+\infty})$, and $\underline{A}\varepsilon\mathcal{N}_{mn}$ for given integers $m,n>1$. Then $\underline{A}$ is α-σ_ϕ-astic if and only if $\underline{A}^*$ is α*-σ_ϕ-astic. ●

From Theorem 13-8 we can now infer a number of variants by substituting $\underline{A}^*$ for $\underline{A}$, and/or dualising. All these cases are conveniently summarised in tabular form, as in the following proposition.

Proposition 13-11. Let E_1 be a self-conjugate belt with a compatible trisection $(\sigma_{-\infty},\sigma_\phi,\sigma_{+\infty})$. Let $\underline{A}$ be a given matrix. Then the statement:

"Product $\underline{P}$ /-exists for some matrix $\underline{X}$ if and only if $\underline{A}$ is α-σ_ϕ-astic: and product $\underline{P}$ then /-exists for all matrices $\underline{X}$ with elements in σ_ϕ"

is valid for the combinations of $\underline{P}$ and α given in the table of Fig 13-1 provided only that the relevant matrices are conformable for the relevant products. ●

Corollary 13-12. Let E_1 be a self-conjugate belt with a compatible trisection $(\sigma_{-\infty},\sigma_\phi,\sigma_{+\infty})$. Let $\underline{A}\varepsilon\mathcal{M}_{mn}$. Then the /-existence of the quadruple product $\underline{A} \otimes (\underline{A}^* \otimes' (\underline{A} \otimes \underline{X}))$ for some $\underline{X}\varepsilon\mathcal{M}_{np}$, and the /-existence of the quadruple product $((\underline{Y} \otimes \underline{A}) \otimes' \underline{A}^*) \otimes \underline{A}$ for some $\underline{Y}\varepsilon\mathcal{M}_{qm}$, are each a necessary and sufficient condition that $\underline{A}$ be doubly σ_ϕ-astic. Similarly the /-existence of the quadruple product $\underline{A} \otimes' (\underline{A}^* \otimes (\underline{A} \otimes' \underline{X}))$ for some $\underline{X}\varepsilon\mathcal{M}_{np}$, and the existence of the quadruple product $((\underline{Y} \otimes' \underline{A}) \otimes \underline{A}^*) \otimes' \underline{A}$ for some $\underline{Y}\varepsilon\mathcal{M}_{qm}$, are each a necessary and sufficient condition that $\underline{A}$ be doubly dually σ_ϕ-astic. The given products are then all /-defined for all $\underline{X}\varepsilon\mathcal{O}_{np}$ and all $\underline{Y}\varepsilon\mathcal{O}_{qm}$.

Proof. The product $\underline{A} \otimes(\underline{A}^* \otimes'(\underline{A} \otimes \underline{X})$ /-exists if and only if the products $\underline{A}^* \otimes' (\underline{A} \otimes \underline{X})$ and $\underline{A} \otimes(\underline{A}^* \otimes' \underline{Z})$ both /-exist, where $\underline{Z} = \underline{A} \otimes \underline{X}$. By Theorem 13-8 and its dual, this happens if and only if both $\underline{A}$ is row-σ_ϕ-astic and $\underline{A}^*$ is row-dually σ_ϕ-astic, i.e. if and only if $\underline{A}$ is doubly σ_ϕ-astic. The other cases are handled similarly. ●

Now Theorem 13-8, and its extensions Proposition 13-11 and Corollary 13-12 state, for given matrix $\underline{A}$, that $\underline{A}$ is α-σ_ϕ-astic if and only if a particular form of product

/-exists for <u>some</u> matrix <u>X</u>. The following two results enable us to specify more closely for what range of matrices <u>X</u> such products <u>do</u> /-exist for given (α-σ_ϕ-astic) <u>A</u>.

<u>Theorem 13-13. Let E_1 be a self-conjugate belt with a compatible trisection $(\sigma_{-\infty}, \sigma_\phi, \sigma_{+\infty})$. Let $A \in \mathcal{M}_{mn}$ and $X \in \mathcal{M}_{np}$ be given. Then the product $A^* \otimes' (A \otimes X)$ /-exists if and only if:</u>

Either	$A \in \mathcal{O}_{mn}$
Or	<u>A</u> does not contain any element of $\sigma_{+\infty}$ and $A \otimes X$ /-exists and does not contain any element of $\sigma_{-\infty}$.

<u>Proof</u>. If $A^* \otimes' (A \otimes X)$ /-exists then <u>A</u> is row-σ_ϕ-astic by Theorem 13-8 and so does not contain any element of $\sigma_{+\infty}$. And clearly $A \otimes X$ /-exists if $A^* \otimes' (A \otimes X)$ /-exists. Moreover, if $\{A \otimes X\}_{ij} \in \sigma_{-\infty}$ for some i and j, then by Lemma 13-6 with $A \otimes X$ as <u>B</u>, we infer that row i of <u>A</u> contains no element of $\sigma_{-\infty}$ and so row i of <u>A</u> contains only elements of σ_ϕ. But then $\{A \otimes X\}_{ij} \in \sigma_{-\infty}$ implies (by (v) of Proposition 11-3) that column j of <u>X</u> consists entirely of elements of $\sigma_{-\infty}$. But this implies that $\{A \otimes X\}_{ij} \in \sigma_{-\infty}$ for <u>every</u> i and this j. Hence <u>every</u> row of <u>A</u> consists of elements of σ_ϕ only, i.e. $A \in \mathcal{O}_{mn}$.

Conversely, if <u>A</u> does not contain any element of $\sigma_{+\infty}$ then A^* does not contain any element of $\sigma_{-\infty}$, so if $A \otimes X$ /-exists and also does not contain any element of $\sigma_{-\infty}$ then product $A^* \otimes' (A \otimes X)$ is /-defined because the matrices are conformable, and have elements in the belt $\sigma_\phi \cup \sigma_{+\infty}$.

Lastly, if $A \in \mathcal{O}_{mn}$, then $A \otimes X$ and $A^* \otimes' (A \otimes X)$ /-exist by Proposition 13-1 because the matrices are conformable. ●

<u>Corollary 13-14. Let E_1 be a self-conjugate belt with a compatible trisection $(\sigma_{-\infty}, \sigma_\phi, \sigma_{+\infty})$. Let <u>A</u>, <u>X</u> be given matrices. Then the statement:</u>

<u>"Product <u>P</u> /-exists if and only if:</u>

<u>Either</u>	<u>A contains elements of σ_ϕ only</u>
<u>Or</u>	<u>A does not contain any element of ω_1 and product <u>Q</u> /-exists and does not contain any element of ω_2"</u>

<u>is valid for the combinations of <u>P</u>, <u>Q</u>, ω_1, ω_2 in the table of Fig 13-1, provided only that the relevant matrices are conformable for the relevant products.</u>

<u>Proof</u>. Similarly and dually to Theorem 13-13. ●

<u>Corollary 13-15. Let E_1 be a self-conjugate belt with a compatible trisection $(\sigma_{-\infty}, \sigma_\phi, \sigma_{+\infty})$. Let <u>A</u>, <u>X</u> be given matrices. Then the statement:</u>

<u>"Product <u>P</u> /-exists if and only if:</u>

<u>Either</u>	<u>A contains elements of σ_ϕ only</u>
<u>Or</u>	<u>A is α-σ_ϕ-astic and product <u>Q</u> /-exists and</u>

does not contain any element of ω_2"

is valid for the combinations of P, α, Q, ω_2 in the table of Fig 13-1 provided only that the relevant matrices are conformable for the relevant products.

Proof. According to Lemma 13-3, if A⊗X /-exists and does not contain any element of $\sigma_{-\infty}$, then the conditions "A does not contain any element of $\sigma_{+\infty}$", "A is row-σ_ϕ-astic" are equivalent. Hence the equivalence of Corollary 13-14 and Corollary 13-15 is proved for the first row of the table of Fig 13-1 and the other cases are proved similarly and dually. ●

Corollary 13-16. Let E_1 be a self-conjugate belt with a compatible trisection $(\sigma_{-\infty}, \sigma_\phi, \sigma_{+\infty})$. Let A, X be given matrices. Then the statement:

"Product P /-exists if and only if:

Either	A contains elements of σ_ϕ only
Or	A is α-σ_ϕ-astic and product Q /-exists and contains elements of σ_ϕ only"

is valid for the combinations of P,α, Q, in the table of Fig 13-2 provided only that the relevant matrices are conformable for the relevant products.

Proof. The product A ⊗ (A*⊗'(A⊗X)) /-exists if and only if the products A⊗(A*⊗'Y) and A*⊗'(A⊗X) both /-exist (where Y is A⊗X). By applying Corollary 13-15 to these two triple products, we see that we have just two possibilities, if the quadruple product A⊗(A*⊗'(A⊗X)) /-exists:

either (i)	A contains elements of σ_ϕ only
or (ii)	A is doubly σ_ϕ-astic, A* ⊗' Y /-exists and does not contain any element of $\sigma_{+\infty}$ and A ⊗ X /-exists and does not contain any element of $\sigma_{-\infty}$.

In case (ii), Lemma 13-3 applies and we infer that if A ⊗ X contains any elements of $\sigma_{+\infty}$ at all, then it contains a complete column of elements of $\sigma_{+\infty}$. But this would imply that A* ⊗' Y, which is A* ⊗' (A ⊗ X), would contain elements of $\sigma_{+\infty}$ also. Hence A ⊗ X contains no elements of either $\sigma_{-\infty}$ or $\sigma_{+\infty}$, i.e. in case (ii), A ⊗ X contains elements of σ_ϕ only.

Conversely, if A contains elements of σ_ϕ only, then by Proposition 13-1, the product A ⊗ (A* ⊗' (A ⊗ X))/-exists provided the matrices are conformable. And if Y = A⊗X /-exists and contains elements of σ_ϕ only and A is doubly σ_ϕ-astic, then A ⊗ (A* ⊗' Y) /-exists by Proposition 13-11.

This proves the corollary for the first line of the table of Fig 13-2 and the proof for the other lines proceeds similarly and dually. ●

13-4./-Defined Residuomorphisms.

As explained in Section 13-2, our aim in the present chapter is to investigate how much of our theory of residuomorphisms, as developed in Chapters 8 to 10, can survive a limitation to /-defined algebraic operations. If we are to combine the results of Chapters 8 to 10 with those of Chapters 11 to 13, however, we shall evidently need to require that E_1 be pre-residuated and satisfy axiom X_{12}, as well as have a compatible trisection. Such a belt E_1 will be called a generalised blog. Restrictive as this combination of requirements appears to be, it still leaves us with a class of belts considerably more general than the extended real numbers, including all blogs under the $(\{-\infty\}, G, \{+\infty\})$ trisection and all linear division belts or linear blogs under the $(N, \{\emptyset\}, P)$ trisection.

Now, if left-multiplication by a matrix $\underline{A}$ is to be a /-defined residuomorphism of some set S of matrices to some set T of matrices, we shall require:

$\underline{A}^* \otimes' (\underline{A} \otimes \underline{X})$ and $\underline{A} \otimes (\underline{A}^* \otimes' \underline{Y})$ to be /-defined for all $\underline{X} \in S$, $\underline{Y} \in T$.

According to Proposition 13-11, this at least requires $\underline{A}$ to be doubly $\sigma_\emptyset$-astic; but if the sets S and T are taken sufficiently general, the total set of /-defined residuomorphisms from S to T may obviously be forced to be smaller than the total set of doubly $\sigma_\emptyset$-astic matrices.

It is evidently relevant to ask two questions:

1. What is the largest subset of $\mathcal{M}_{np}$ on which every doubly $\sigma_\emptyset$-astic (mxn) matrix induces a /-defined residuated left-multiplication?
2. What is the largest subset of the set $\mathcal{N}_{mn}$ of doubly $\sigma_\emptyset$-astic (mxn) matrices, every member of which reduces a /-defined residuated left-multiplication on the whole of $\mathcal{M}_{np}$?

We use the notations of Section 10-4 and restrict ourselves as in Section 10-4 to the case $p = 1$, i.e. to operands which are columns. The notation $(\sigma_\emptyset)_n$ denotes the subset of E_n of all n-tuples with all elements in $\sigma_\emptyset$.

Theorem 13-17. Let E_1 be a generalised blog, having the trisection $(\sigma_{-\infty}, \sigma_\emptyset, \sigma_{+\infty})$. For given integers $m,n \geq 1$ let $\mathcal{N}_{mn}$ be the commutative band of doubly $\sigma_\emptyset$-astic (mxn) matrices.

Then (i) $\mathcal{L}\mathcal{N}_{mn} : (\sigma_\emptyset)_n \Leftrightarrow (\sigma_\emptyset)_m : \mathcal{L}^* \mathcal{N}^*_{mn}$ and all multiplications are /-defined.

(ii) Let $\mathcal{L}\mathcal{N}_{mn} : S \Leftrightarrow T : \mathcal{L}^* \mathcal{N}^*_{mn}$, where $S \subset E_n$ and $T \subset E_m$, and $m,n > 1$. If all multiplications are /-defined, then $S \subset (\sigma_\emptyset)_n$ and $T \subset (\sigma_\emptyset)_m$.

Proof (i) From Theorem 10-11:

$$\mathcal{L}\mathcal{M}_{mn} : E_n \Leftrightarrow E_m : \mathcal{L}^* \mathcal{M}^*_{mn}$$

Hence from Theorem 12-4 and its dual (and Proposition 13-10)

$$\mathcal{L}\mathcal{N}_{mn} : (\sigma_\emptyset)_n \Leftrightarrow (\sigma_\emptyset)_m : \mathcal{L}^* \mathcal{N}^*_{mn}$$

And the multiplications are all /-defined by Proposition 13-1.

(ii) If, under the given conditions, we could find $\underline{u} \in S$ but $\underline{u} \notin (\sigma_\emptyset)_n$, we should have $\{\underline{u}\}_i \in \sigma_{-\infty} \cup \sigma_{+\infty}$ for some i $(1 \leq i \leq n)$. If in fact $\{\underline{u}\}_i \in \sigma_{+\infty}$ for some i $(1 \leq i \leq n)$, let $\underline{A} \in \mathcal{N}_{mn}$ be such that $\{\underline{A}\}_{1i} \in \sigma_{-\infty}$ but $\{\underline{A}\}_{rs} \in \sigma_\phi$ otherwise. Then $\underline{A} \in \mathcal{N}_{mn}$ but $\underline{A} \otimes \underline{u}$ is not /-defined. And if $\{\underline{u}\}_i \in \sigma_{-\infty}$ for some i $(1 \leq i \leq n)$, let $\underline{A} \in \mathcal{N}_{mn}$ be such that $\{\underline{A}\}_{1i} \in \sigma_\phi$, $\{\underline{A}\}_{1s} \in \sigma_{-\infty}$ otherwise and $\{\underline{A}\}_{rs} \in \sigma_\phi$ $(r \neq 1)$ otherwise. We easily verify that $\underline{A}^* \otimes' \underline{w}$ is not /-defined, where $\underline{w} = \underline{A} \otimes \underline{u} \in T$. Hence if $\underline{u} \in S$ then $\underline{u} \in (\sigma_\phi)_n$, and similarly if $\underline{v} \in T$ then $\underline{v} \in (\sigma_\phi)_m$. ●

Let us now recall the notations of Section 8-4. Then from Theorem 8-9, Proposition 8-10 and Theorem 13-17 we immediately derive the following.

Proposition 13-18. Let E_1 be a generalised blog, having the trisection $(\sigma_{-\infty}, \sigma_\phi, \sigma_{+\infty})$. If the notation $\mathcal{W}_{np}$ (etc) is taken to be synonymous with the notation $\mathcal{O}_{np}$ (etc) then the diagram of Fig. 8-3 is valid when $\underline{A}$ is doubly dually σ_ϕ-astic and the diagram of Fig. 8-4 is valid when $\underline{A}$ is doubly σ_ϕ-astic, all multiplications being /-defined. ●

13-5. $\mathcal{O}_{mn}$ As Operators

Theorem 13-17 contains an asymmetry, in that part (ii) requires $m, n > 1$ whereas part (i) only requires $m, n \geq 1$. The reason for this is that $\mathcal{N}_{mn} = \mathcal{O}_{mn}$ if either m or n is 1, and then all multiplications are /-defined by Proposition 13-1. If $m, n > 1$, then $\mathcal{O}_{mn}$ is a proper subset of $\mathcal{N}_{mn}$. Part (ii) of Theorem 13-17 tells us that $(\sigma_\phi)_n$ is the greatest subset of E_n on which all multiplications are /-defined; if we wish to enlarge the set of operands from $(\sigma_\phi)_n$ to E_n we must therefore reduce the set of operators, and the following result shows in fact that we must reduce it from $\mathcal{N}_{mn}$ to $\mathcal{O}_{mn}$. This result thus answers question 2 of the previous section, and also takes care of the case $m = 1$ or $n = 1$.

Theorem 13-19. Let E_1 be a generalised blog having the trisection $(\sigma_{-\infty}, \sigma_\phi, \sigma_{+\infty})$. For given integers $m, n \geq 1$, we have:

(i) $\mathcal{L}\mathcal{O}_{mn} : E_n \Leftrightarrow E_m : \mathcal{L}^*\mathcal{O}^*_{mn}$

and all multiplications are /-defined.

(ii) Let $\mathcal{L}\mathcal{S} : E_n \Leftrightarrow E_m : \mathcal{L}^*\mathcal{S}^*$

where $\mathcal{S} \subset \mathcal{M}_{mn}$. If all multiplications are /-defined, then $\mathcal{S} \subset \mathcal{O}_{mn}$.

Proof. Since evidently $\mathcal{O}^*_{mn} = \mathcal{O}_{nm}$, part (i) follows from Theorem 10-11 and Proposition 13-1. Conversely, since E_n contains $(\sigma_{-\infty})_n \cup (\sigma_{+\infty})_n$, it is clear that $\mathcal{O}_{mn}$ is the largest subset of $\mathcal{M}_{mn}$ which induces left-multiplications which are /-defined for all operands in E_n. The proves part (ii). ●

Finally, from Theorem 8-11 and Proposition 13-1 (and its dual) we have the following result.

Proposition 13-20. Let E_1 be a generalised blog having the trisection $(\sigma_{-\infty}, \sigma_{\phi}, \sigma_{+\infty})$. Then the diagrams of Figs. 8-3 and 8-4 are valid for arbitrary $\underline{A} \in \mathcal{G}_{mn}$, when the notation $\mathcal{W}_{np}$ (etc) is taken to be synonymous with the notation $\mathcal{M}_{np}$ (etc). Moreover, all the relevant transformations are /-defined. ●

13-6. Some Questions Answered

As foreshadowed in Section 11-1, we have now related σ_{ϕ}-asticity to two apparently separate questions. From Theorem 12-6, we know that if the left-multiplication $g_{\underline{A}}$ takes $(\sigma_{\phi})_n$ into $(\sigma_{\phi})_m$ then $\underline{A}$ must be row-σ_{ϕ}-astic. By the dual hereof, if the residual $g^*_{\underline{A}}$ is to take $(\sigma_{\phi})_m$ into $(\sigma_{\phi})_n$, then $\underline{A}^*$ must be row-duallyσ_{ϕ}-astic. And this is exactly the conclusion we arrive at if we demand that $g_{\underline{A}}$ and $g^*_{\underline{A}}$ can be applied in combination without producing expressions where are /-undefined.

For the principal interpretation, therefore, we shall regard as particularly important the cases in which we describe the state of a system by means of finite n-tuples upon which we operate with doubly G-astic matrices and their duals. By so doing, we do not exclude any residuomorphisms which, together with their residuals, preserve finiteness since all such residuomorphisms can be induced by matrix multiplications as the following result shows.

Theorem 13-21. Let E_1 be a blog with group G, and for given integers, m, n $\geq$ 1, let $\mathcal{N}_{mn}$ denote the set of (mxn) doubly G-astic matrices. If $\mathcal{H} \subset \mathrm{Hom}_{E_1}(E_n, E_m)$, then $\mathcal{H}$ satisfies:

$$\mathcal{H} : G_n \Leftrightarrow G_m : \mathcal{H}^* \tag{13-1}$$

if and only if $\mathcal{H} \subset \mathcal{L}\mathcal{N}_{mn}$.

Proof. The "if" follows from Theorem 13-17. On the other hand, by Theorem 5-10:

$$\mathcal{H} \subset \mathcal{L}\mathcal{M}_{mn}$$

Hence, if (13-1) holds, we have:

$$\mathcal{H} \subset \mathcal{L}\mathcal{M}_{mn} \cap \mathrm{Stab}_{E_1}(G_n, G_m)$$

So, by Theorem 12-6, $\mathcal{H}$ consists of certain left-multiplications $g_{\underline{A}}$ where each $\underline{A}$ is row-G-astic. Hence, by Corollary 10-12 and the uniqueness of the conjugate, $\mathcal{H}^*$ consists of the corresponding dual left-multiplications $g^*_{\underline{A}}$:

$$g^*_{\underline{A}} : \underline{B} \mapsto \underline{A}^* \otimes' \underline{B} \tag{13-2}$$

Hence by (13-1) and the dual of Theorem 12-6, $\underline{A}^*$ is row-dually G-astic, so by

Proposition 13-10, $\underline{A}$ is also column-G-astic and so doubly G-astic. Hence:

$$\mathcal{H} \subset \mathcal{L}\mathcal{W}_{mn}$$

●

14. THE EQUATION $\underline{A} \otimes \underline{x} = \underline{b}$ OVER A BLOG

14-1. Some Preliminaries

We turn now to the analysis of a practical problem - the solution of a set of simultaneous linear equations in minimax-algebra.

We have seen in Section 1-2.1 how such a problem can arise. In what follows, we use the notations established in earlier chapters. We shall discuss the following problem:

Given $\underline{A} \in \mathcal{M}_{mn}$ and $\underline{b} \in E_m$ to find $\underline{x} \in E_n$ such that $\underline{A} \otimes \underline{x} = \underline{b}$ (14-1)

Now, if $\underline{x}, \underline{y} \in E_n$ are solutions to (14-1), we have:

$$\underline{A} \otimes (\underline{x} \oplus \underline{y}) = (\underline{A} \otimes \underline{x}) \oplus (\underline{A} \otimes \underline{y}) = \underline{b} \oplus \underline{b} = \underline{b}$$

Evidently the following proposition then follows.

Proposition 14-1. The set of solutions of (14-1) is either empty or is a commutative band.

In order to develop a non-trivial theory of solutions of (14-1), we must set extra conditions on E_1 - taking care always to include the principal interpretation as a possible case. In fact, we shall base our analysis on the identities of Chapter 8, and so shall usually take E_1 to be pre-residuated and satisfying axiom X_{12}. Since we wish our results to have some practical relevance, we shall also investigate the question of /-existence of the expressions which we use, in the case that E_1 is a generalised blog - that is to say that:

(i) E_1 is pre-residuated

(ii) E_1 satisfies axiom X_{12}

(iii) E_1 has a compatible trisection $(\sigma_{-\infty}, \sigma_{\phi}, \sigma_{+\infty})$.

Lemma 14-2. Let E_1 be a belt with duality having a trisection $(\sigma_{-\infty}, \sigma_{\phi}, \sigma_{+\infty})$. Let m, n, r be given integers ≥ 1, and $\underline{A} \in \mathcal{M}_{mn}$ a given matrix. Then the set $\mathcal{R}_{nr}(\underline{A})$ of all matrices $\underline{X} \in \mathcal{M}_{nr}$ such that $\underline{A} \otimes \underline{X}$ is /-defined, is a commutative sub-band of $\mathcal{M}_{nr}$. Hence the set of solutions $\underline{x}$ of (14-1) such that $\underline{A} \otimes \underline{x}$ /-exists is either empty or is a commutative sub-band of E_n.

Proof. Evidently, $\mathcal{R}_{nr}(\underline{A})$ is not empty since $\mathcal{O}_{nr} \subset \mathcal{R}_{nr}(\underline{A})$ by Proposition 13-1. Let $\underline{X}, \underline{Y} \in \mathcal{R}_{nr}(\underline{A})$. Then the product $\underline{A} \otimes (\underline{X} \oplus \underline{Y})$ is /-defined unless we can find i, j, k, $(1 \leq i \leq m;\ 1 \leq j \leq n;\ 1 \leq k \leq r)$ such that one of $\{\underline{A}\}_{ij}$, $\{\underline{X} \oplus \underline{Y}\}_{jk}$ lies in $\sigma_{+\infty}$ and the other in $\sigma_{-\infty}$. But then (using Proposition 11-3) we should have that $\{\underline{A}\}_{ij}$ lies in $\sigma_{+\infty}$ or $\sigma_{-\infty}$ whilst one of $\{\underline{X}\}_{jk}$, $\{\underline{Y}\}_{jk}$ lies in $\sigma_{-\infty}$ or $\sigma_{+\infty}$ respectively, so that one of $\underline{A} \otimes \underline{X}$, $\underline{A} \otimes \underline{Y}$ is not /-defined - a contradiction. Hence $\underline{X} \oplus \underline{Y} \in \mathcal{R}_{nr}(\underline{A})$. Finally, the set of solutions $\underline{x}$ of (14-1) such that $\underline{A} \otimes \underline{x}$ /-exists is evidently the intersection of the commutative band $\mathcal{R}_{n1}(\underline{A})$

with the set of all solutions mentioned in Proposition 14-1.

We shall call any solution of (14-1), which is such that $\underline{A} \otimes \underline{x}$ /-exists, a /-solution of (14-1).

14-2 . The Principal Solution

Theorem 14-3. Let E_1 be a pre-residuated belt satisfying axiom X_{12}. Then (14-1) has at least one solution if and only if $\underline{x} = \underline{A}^* \otimes' \underline{b}$ is a solution; and $\underline{x} = \underline{A}^* \otimes' \underline{b}$ is then the greatest solution.

Proof. The 'if' is trivial. So suppose $\underline{u} \varepsilon E_n$ is a solution of (14-1). From Theorem 8-9:

$$\underline{A} \otimes (\underline{A}^* \otimes' \underline{b}) = \underline{A} \otimes (\underline{A}^* \otimes' (\underline{A} \otimes \underline{u})) = \underline{A} \otimes \underline{u} = \underline{b}$$

Hence $\underline{x} = \underline{A}^* \otimes' \underline{b}$ is a solution of (14-1). Moreover, using Theorem 8-8:

$$\underline{u} \leqslant \underline{A}^* \otimes' (\underline{A} \otimes \underline{u}) = \underline{A}^* \otimes' \underline{b} = \underline{x}$$

Hence $\underline{x}$ is the greatest solution of (14-1).

Theorem 14-4. Let E_1 be a generalised blog, and $\underline{A} \varepsilon \mathcal{M}_{mn}$, $\underline{b} \varepsilon E_m$ be such that (14-1) is soluble. Then $\underline{x} = \underline{A}^* \otimes' \underline{b}$ /-exists and is a /-solution of (14-1), if and only if one of the following cases arises:

(i) $\underline{A} \varepsilon \mathcal{O}_{mn}$, and $\underline{b} \varepsilon (\sigma_{+\infty})_m$.

(ii) $\underline{A} \varepsilon \mathcal{O}_{mn}$, and $\underline{b} \varepsilon (\sigma_{-\infty})_m$.

(iii) $\underline{A}$ is doubly σ_ϕ-astic, and $\underline{b} \varepsilon (\sigma_\phi)_m$.

Moreover, every solution of (14-1) is then a /-solution, and $\underline{A}^* \otimes' \underline{b}$ lies in $(\sigma_{+\infty})_n$, $(\sigma_{-\infty})_n$ or $(\sigma_\phi)_n$ respectively according as case (i), (ii) or (iii) arises.

Proof. The 'if' follows from Propositions 13-1 and 13-11, with Theorem 14-3. Conversely, if $\underline{A}^* \otimes' \underline{b}$ /-exists and is a /-solution of (14-1) then $\underline{A}^* \otimes' \underline{b} \varepsilon \mathcal{R}_{n1}(\underline{A})$ and $\underline{A} \otimes (\underline{A}^* \otimes' \underline{b}) = \underline{b}$. Hence,

$$\underline{A}^* \otimes' (\underline{A} \otimes (\underline{A}^* \otimes' \underline{b})) = \underline{A}^* \otimes' \underline{b}$$

/-exists, so $\underline{A}$ is in any case doubly σ_ϕ-astic (and perhaps $\underline{A} \varepsilon \mathcal{O}_{mn}$) by Corollary 13-16. In the ensuing argument we make free use of Fig.11-1. First, suppose $\underline{b} = \underline{A} \otimes \underline{x}$ does not contain any element of $\sigma_{-\infty}$. Then we have the case (i) with (ii) of Lemma 13-3, so (ii)' of Lemma 13-3 applies. Hence, either $\underline{b} = \underline{A} \otimes \underline{x} \varepsilon (\sigma_\phi)_m$ (and then $\underline{A}^* \otimes' \underline{b} \varepsilon (\sigma_\phi)_n$ by Theorem 13-17, and we have case (iii) of the present theorem) or $\underline{b} \varepsilon (\sigma_{+\infty})_m$.

In the latter case, all elements of $\underline{x} = \underline{A}^* \otimes' \underline{b}$ must lie in $\sigma_{+\infty}$ also and hence $\underline{A}$ cannot contain any element of $\sigma_{-\infty}$ because the product $\underline{A} \otimes \underline{x}$ /-exists. Hence $\underline{A} \varepsilon \mathcal{O}_{mn}$, and we have case (i) of the present theorem.

Alternatively, if $\underline{b}$ does contain an element of $\sigma_{-\infty}$, then $\underline{x} = \underline{A}^* \otimes' \underline{b}$ must have all elements in $\sigma_{-\infty}$, and then so must $\underline{b} = \underline{A} \otimes \underline{x}$. Hence $\underline{A}^*$ cannot contain any element of $\sigma_{+\infty}$ because the product $\underline{A}^* \otimes' \underline{b}$ /-exists. Hence $\underline{A}$ does not contain any element of $\sigma_{-\infty}$, so $\underline{A} \in \mathcal{G}_{mn}$, and we have case (ii) of the present theorem.

Finally, every solution is a /-solution in cases (i) and (ii) by Proposition 13-1. In case (iii): every solution $\underline{u}$ satisfies $\underline{u} \leqslant \underline{v}$ where $\underline{v} = \underline{A}^* \otimes' \underline{b}$ (Theorem 14-3), so $\underline{u} \oplus \underline{v} = \underline{v}$ with $\underline{v} \in (\sigma_\phi)_n$, whence $\underline{u} \in (\sigma_{-\infty} \cup \sigma_\phi)_n$ by Fig. 11-1. But $\underline{A}$ (being σ_ϕ-astic) also has all elements in $\sigma_{-\infty} \cup \sigma_\phi$, so $\underline{A} \otimes \underline{u}$ is /-defined in case (iii) also. ●

For future reference, we record in the following result a number of dual and left-right generalisations of Theorems 14-3 and 14-4.

<u>Corollary 14-5. Let E_1 be a pre-residuated belt satisfying axiom X_{12} and let $\underline{A}$, $\underline{B}$ be given matrices over E_1. Then for all combinations of α, P, $\underline{Q}$ and β given in Fig. 14-1, the following statement is valid:</u>

<u>"The matrix equation P has at least one solution if and only if the product $\underline{Q}$ is a solution; and the product $\underline{Q}$ is then the β solution. Furthermore, if E_1 is a generalised blog, and product Q is /-defined, and equation P is /-defined when $\underline{X} = \underline{Q}$, then equation P is /-defined when $\underline{X}$ is any solution of equation P."</u>

<u>Proof</u>. We consider the equation $\underline{A} \otimes \underline{X} = \underline{B}$. Since each column of $\underline{B}$ is produced by the action of matrix $\underline{A}$ on the corresponding column of $\underline{X}$, the validity of the statement for this case follows directly from Theorems 14-3 and 14-4. (The conformability of the matrices for the relevant operations is implicit in the assumption of the existence of solutions.) The other cases are proved similarly and dually. ●

P	α	$\underline{Q}$	β
$\underline{A} \otimes \underline{X} = \underline{B}$	Doubly	$\underline{A}^* \otimes' \underline{B}$	Greatest
$\underline{A}^* \otimes \underline{X} = \underline{B}$	Doubly dually	$\underline{A} \otimes' \underline{B}$	Greatest
$\underline{A} \otimes' \underline{X} = \underline{B}$	Doubly dually	$\underline{A}^* \otimes \underline{B}$	Least
$\underline{A}^* \otimes' \underline{X} = \underline{B}$	Doubly	$\underline{A} \otimes \underline{B}$	Least
$\underline{X} \otimes \underline{A} = \underline{B}$	Doubly	$\underline{B} \otimes' \underline{A}^*$	Greatest
$\underline{X} \otimes \underline{A}^* = \underline{B}$	Doubly dually	$\underline{B} \otimes' \underline{A}$	Greatest
$\underline{X} \otimes' \underline{A} = \underline{B}$	Doubly dually	$\underline{B} \otimes \underline{A}^*$	Least
$\underline{X} \otimes' \underline{A}^* = \underline{B}$	Doubly	$\underline{B} \otimes \underline{A}$	Least

<u>Fig. 14-1</u>
<u>Principal Solutions</u>

For each combination of P and $\underline{Q}$ in Fig. 14-1, we shall refer to $\underline{Q}$ as the <u>principal solution</u> to equation P, when $\underline{Q}$ actually is a solution to equation P.

Finally, we may paraphrase Theorems 14-3 and 14-4 (and, analogously, Corollary 14-5) as a <u>solubility criterion</u>:

Problem (14-1) is soluble if and only if $\underline{A} \otimes (\underline{A}^* \otimes' \underline{b}) = \underline{b}$; and every solution is a /-solution if $\underline{A} \otimes (\underline{A}^* \otimes' \underline{b})$ /-exists. (14-2)

14-3 · The Boolean Case

Theorem 14-4 identifies the cases in which (14-1) has a /-defined /-solution. All solutions are then /-solutions and the question arises: can we find all solutions? Obviously, we require further assumptions on E_1 before this is possible, and our analysis will assume that E_1 is a blog.

Accordingly, we now address the following problem:

To find all solutions of (14-1), given that E_1 is a blog and that $\underline{A} \otimes (\underline{A}^* \otimes' \underline{b})$ /-exists and equals $\underline{b}$. (14-3)

Theorem 14-4 shows that three cases are relevant. Cases (i) and (ii) are disposed of in the following result, which is an immediate application of Proposition 4-3 and Fig. 4-1 (bearing in mind that $\underline{A}$ must be finite in cases (i) and (ii)).

Proposition 14-6. Let E_1 be a blog. If all elements of $\underline{b}$ are $-\infty$, then (14-3) has $\underline{b}$ as its unique solution. If all elements of $\underline{b}$ are $+\infty$, then solutions to (14-3) are exactly those elements of E_n which have at least one element equal to $+\infty$. ●

Case (iii), which now remains, is the important case where $\underline{A}$ is doubly G-astic and $\underline{b}$ is finite. In Chapter 15, we obtain all solutions of (14-1) for this case, assuming that E_1 is a linear blog. As a preliminary step, we consider now the particular case that E_1 is the 3-element blog ③. Then $\underline{b}$, being finite, has all elements equal to ϕ.

Lemma 14-7. Let E_1 be the 3-element blog ③, let $\underline{A}$ be doubly G-astic and $\underline{b}$ be finite. Then (14-1) is soluble, having as principal /-solution $\underline{x} = \underline{e}$ where $\underline{e}$ has all components equal to ϕ. Hence no solution to (14-1) contains $+\infty$ as element, and all solutions are /-solutions.

Proof. From Theorem 13-17, $\underline{A}^* \otimes' \underline{b} \in (\sigma_\phi)_n$ and $\underline{A} \otimes (\underline{A}^* \otimes' \underline{b}) \in (\sigma_\phi)_m$.

But $(\sigma_\phi)_n$ contains only $\underline{e}$ and $(\sigma_\phi)_m$ contains only $\underline{b}$. Hence:

$$\underline{A}^* \otimes' \underline{b} = \underline{e} \text{ and } \underline{A} \otimes \underline{e} = \underline{b},$$

and both products are /-defined by Theorem 13-17.

Hence (14-1) has $\underline{e}$ as principal /-solution. By Theorem 14-3, each solution $\underline{x}$ satisfies $\underline{x} \leqslant \underline{e}$ and so does not contain $+\infty$. Hence $\underline{A} \otimes \underline{x}$ is /-defined and $\underline{x}$ is a /-solution to (14-1). ●

Now let us suppose that we wish to solve the following example of (14-1) with $E_1 =$ ③, $\underline{A}$ doubly G-astic and $\underline{b}$ finite:

$$\begin{bmatrix} \emptyset & \emptyset & \emptyset \\ -\infty & \emptyset & \emptyset \\ \emptyset & -\infty & \emptyset \\ \emptyset & \emptyset & \emptyset \end{bmatrix} \otimes \begin{bmatrix} y_1 \\ y_2 \\ y_3 \end{bmatrix} = \begin{bmatrix} \emptyset \\ \emptyset \\ \emptyset \\ \emptyset \end{bmatrix} \tag{14-4}$$

The procedure in effect uses standard Boolean analysis, as discussed, for example in [34]. Since we are not using standard Boolean notation, however, we now show how the solution is completed, so that the present treatment is self-contained.

A set of simultaneous linear equations such as (14-4):

$$\sum_{j=1}^{n}{}_{\oplus}\ \alpha_{ij} \otimes y_j = \emptyset \quad (i = 1, \ldots, m)$$

where each α_{ij} and each y_j is $-\infty$ or $\emptyset$, is equivalent to:

$$\prod_{i=1}^{m}{}_{\otimes}\ \left(\sum_{j=1}^{n}{}_{\oplus}\ \alpha_{ij} \otimes y_j \right) = \emptyset$$

(for under the given conditions such a product equals $\emptyset$ if and only if all its factors equal $\emptyset$ (Proposition 4-3 with E_1 = ③)).

Multiplying out, and using the fact that $x \otimes x = x = x \oplus x$ when $x \in$ ③, we have:

$$\sum_{r=1}^{k}{}_{\oplus}\ p_r = \emptyset \tag{14-5}$$

where each p_r is a product of some of the variables y_j, each product containing at least one y_j, no y_j occurring to power greater than one in a given product p_r, and no two products p_r containing exactly the same selection of y_j's.

(Terms containing a coefficient $-\infty$ may be dropped, so that all coefficients in (14-5) are equal to the multiplicative identity $\emptyset$, which need not to be written. Not <u>all</u> terms get dropped since a coefficient $\neq -\infty$ is present at least once in each row of the G-astic matrix. Identical p_r's are amalgamated, since $p_r \oplus p_r = p_r$.) We now introduce the n functions $\bar{y}_j$ $(j = 1, \ldots, n)$ where:

$$\bar{y}_j = \emptyset \text{ if } y_j = -\infty$$
$$= -\infty \text{ otherwise.}$$

Now, according to Lemma 14-7, no solution contains $+\infty$. But:

$$\left.\begin{aligned} y_j \otimes \bar{y}_j &= -\infty \\ y_j \oplus \bar{y}_j &= \emptyset \end{aligned}\right\} \quad (j = 1, \ldots, n) \text{ if each } y_j \text{ is either } -\infty \text{ or } \emptyset. \tag{14-6}$$

So, for given product p_r in (14-5) let $y_{n_1}, \ldots, y_{n_t}$ be the variables y_j which do <u>not</u> occur in p_r.

From (14-6) we have:

$$p_r = p_r \otimes (y_{n_1} \oplus \bar{y}_{n_1}) \otimes \ldots \otimes (y_{n_t} \oplus \bar{y}_{n_t}) \text{ if each } y_j \text{ is either } -\infty \text{ or } \emptyset.$$

On multiplying out the right-hand side of this expression, recalling that p_r is a product of y_j's, we obtain an expression q_r, such that $q_r = p_r$ when each y_j is either $-\infty$ or $\emptyset$, and q_r consists of a sum of 2^t distinct products, each product containing, for every $j = 1, \ldots, n$ either y_j or $\bar{y}_j$, but not both, exactly once. A product of this kind is called a minterm.

Clearly, there are exactly 2^n different possible minterms, and each given minterm is capable of assuming the value $\emptyset$ in one and only one way, namely by giving each variable y_j, for $j = 1 \ldots, n$, the value $\emptyset$ if y_j itself occurs in the minterm, and the value $-\infty$ if $\bar{y}_j$ occurs in the minterm. For each of the 2^n different ways of assigning one of the values $-\infty$, $\emptyset$ to each y_j, exactly one of the 2^n possible minterms takes the value $\emptyset$, and the others take the value $-\infty$.

Since p_r has been expressed as a sum of minterms, we now can express $\sum_{r=1}^{k}{}_{\oplus}\, p_r$ as a sum of minterms, i.e. we arrive at a disjunctive normal form [27]. In view of the law $x \oplus x = x$, we may assume that any given minterm occurs at most once in the sum. Such a sum takes the value $\emptyset$ if and only if (exactly) one of the minterms which occur in it takes the value $\emptyset$, and this happens for a unique combination of assignments of $-\infty$ or $\emptyset$ to the variables y_j. The solutions to (14-5) are therefore precisely those such assignments which make some minterm in the disjunctive normal form of $\sum_{r=1}^{k}{}_{\oplus}\, p_r$ equal to $\emptyset$.

Applying this procedure to equations (14-4), we first consider:

$$(y_1 \oplus y_2 \oplus y_3) \otimes (y_2 \oplus y_3) \otimes (y_1 \oplus y_3) \otimes (y_1 \oplus y_2 \oplus y_3) = \emptyset.$$

Multiplying out and using $y \otimes y = y = y \oplus y$, etc., we obtain:

$$(y_1 \otimes y_2 \otimes y_3) \oplus (y_1 \otimes y_2) \oplus (y_1 \otimes y_3) \oplus (y_2 \otimes y_3) \oplus y_3 = \emptyset.$$

Introducing $\bar{y}_1$, $\bar{y}_2$ and $\bar{y}_3$:

$$(y_1 \otimes y_2 \otimes y_3) \oplus (y_1 \otimes y_2 \otimes (y_3 \oplus \bar{y}_3)) \oplus (y_1 \otimes (y_2 \oplus \bar{y}_2) \otimes y_3)$$
$$\oplus ((y_1 \oplus \bar{y}_1) \otimes y_2 \otimes y_3) \oplus ((y_1 \oplus \bar{y}_1) \otimes (y_2 \oplus \bar{y}_2) \otimes y_3) = \emptyset$$

Multiplying out, and condensing equal minterms into one:

$$(y_1 \otimes y_2 \otimes y_3) \oplus (y_1 \otimes y_2 \otimes \bar{y}_3) \oplus (y_1 \otimes \bar{y}_2 \otimes y_3) \oplus (\bar{y}_1 \otimes y_2 \otimes y_3)$$
$$\oplus (\bar{y}_1 \otimes \bar{y}_2 \otimes y_3) = \emptyset$$

The left-hand side is in disjunctive normal form, and the solutions to (14-4) are thus:

y_1	y_2	y_3
$\emptyset$	$\emptyset$	$\emptyset$
$\emptyset$	$\emptyset$	$-\infty$
$\emptyset$	$-\infty$	$\emptyset$
$-\infty$	$\emptyset$	$\emptyset$
$-\infty$	$-\infty$	$\emptyset$

15. LINEAR EQUATIONS OVER A LINEAR BLOG

15-1. All Solutions of (14-3)

We describe now a procedure for deriving all solutions to problem (14-1) when E_1 is any linear blog, $\underline{A}$ is doubly G-astic and $\underline{b}$ is finite. An important consequence of the assumption of linearity is that in any given finite set of elements there is a greatest (the attained maximum). As usual, G is the group of E_1.

We first transform the equations $\underline{A} \otimes \underline{x} = \underline{b}$ so that all the right-hand sides equal $\emptyset$. We do this by multiplying each equation

$\sum_{j=1}^{n}{}_{\oplus} a_{ij} \otimes x_j = b_i$ through from the left by b_i^{-1}, obtaining as an equivalent set of equations:

$$\sum_{j=1}^{n}{}_{\oplus} \alpha_{ij} \otimes x_j = \emptyset \quad (i = 1, \ldots, m) \tag{15-1}$$

where $\alpha_{ij} = b_i^{-1} \otimes a_{ij}$.

In the matrix $[\alpha_{ij}]$, we inspect each column, and we mark every element α_{ij} which is greatest in its column. At least one element per column will thus be marked.

For example, if E_1 has the principal interpretation, let the given equations be:

$$\begin{bmatrix} 7 & -5 \\ 2 & -\infty \end{bmatrix} \otimes \begin{bmatrix} x_1 \\ x_2 \end{bmatrix} = \begin{bmatrix} 4 \\ -1 \end{bmatrix}$$

Applying the above procedure, we obtain

$$\begin{bmatrix} \textcircled{3} & \textcircled{-9} \\ \textcircled{3} & -\infty \end{bmatrix} \otimes \begin{bmatrix} x_1 \\ x_2 \end{bmatrix} = \begin{bmatrix} \emptyset \\ \emptyset \end{bmatrix} \tag{15-2}$$

The encircled elements are greatest in their particular columns.

If β_j is the value of the greatest element(s) in column j then clearly β_j is finite, since $\underline{A}$ was doubly G-astic.

Opening (15-1) w.r.t. index i (see Section 3-3) we have:

$$\forall\, i = 1, \ldots, m; \forall j = 1, \ldots, n, \quad \alpha_{ij} \otimes x_j \leq \emptyset \tag{15-3}$$

Restricting ourselves to such α_{ij} as are finite, we infer:

$$\forall\, i = 1, \ldots, m; \forall j = 1, \ldots, n \text{ with } \alpha_{ij} \in G, \quad x_j \leq \alpha_{ij}^{-1}$$

Closing w.r.t. i,
$$x_j \leq \sum_{\substack{i \\ \alpha_{ij} \in G}}{}_{\oplus'} \alpha_{ij}^{-1} = \Big(\sum_{\substack{i \\ \alpha_{ij} \in G}}{}_{\oplus} \alpha_{ij}\Big)^{-1} \text{ (by (3-1))} = \beta_j^{-1} \tag{15-4}$$

Relation (15-4) is thus a necessary condition to be satisfied by any solution to (14-1) when E_1 is a linear blog, $\underline{A}$ is doubly G-astic and $\underline{b}$ is finite.

<u>Lemma 15-1. Let E_1 be a linear blog. Suppose there is some row i of the matrix $[\alpha_{ij}]$ in (15-1) in which no element is marked. That is to say, for this i:</u>

$$\forall\ j = 1, \ldots, n, \quad \alpha_{ij} < \beta_j \tag{15-5}$$

<u>Then the equations (15-1) have no solution.</u>

<u>Proof</u>. Suppose that (15-5) holds for some row i, but that nevertheless we can find $x_1, \ldots, x_n \in E_1$ satisfying (15-1). Then:

$$\emptyset = \sum_{j=1}^{n}{}_{\oplus}\ \alpha_{ij} \otimes x_j = \alpha_{iJ} \otimes x_J \quad \text{say, where} \quad 1 \leqslant J \leqslant n$$

(the attained maximum).

Then x_J and α_{iJ} must be finite, since their product is $\emptyset$, and:

$$x_J^{-1} = \alpha_{iJ} < \beta_J \text{ (from (15-5))}$$

whereas (15-4) implies $x_J^{-1} \geqslant \beta_J$, a contradiction. ●

We are left now with the case that in each row and each column of $[\alpha_{ij}]$, at least one element is marked. We now transpose into a Boolean problem, as follows. Introduce the Boolean variables y_j ($j = 1, \ldots, n$) and the Boolean matrix $[\gamma_{ij}]$ where $\gamma_{ij} = \emptyset$ if α_{ij} is marked in $[\alpha_{ij}]$, and $\gamma_{ij} = -\infty$ otherwise.

Now solve the Boolean equations:

$$\sum_{j=1}^{n}{}_{\oplus}\ \gamma_{ij} \otimes y_j = \emptyset \quad (i = 1, \ldots, m) \tag{15-6}$$

as described in Section 14-3. (From the way in which it was derived, $[\gamma_{ij}]$ is clearly doubly G-astic.) We assert:

<u>Each solution to (15-6) consists of an assignment of one of the values $-\infty$, $\emptyset$ to each y_j. For each solution to (15-6) we obtain a set of solutions to (15-1) as follows:</u> (15-7)

<u>For each $j = 1, \ldots, n$: if $y_j = \emptyset$ then x_j is given the value β_j^{-1}; if $y_j = -\infty$ then x_j is given an arbitrary value such that $x_j < \beta_j^{-1}$.</u>

To illustrate the procedure, we derive for equations (15-2) the following Boolean problem:

$$\begin{bmatrix} \emptyset & \emptyset \\ \emptyset & -\infty \end{bmatrix} \otimes \begin{bmatrix} y_1 \\ y_2 \end{bmatrix} = \begin{bmatrix} \emptyset \\ \emptyset \end{bmatrix}$$

Proceeding as in Section 14-3 we obtain:

$$(y_1 \oplus y_2) \otimes y_1 = \emptyset$$

i.e. $\quad (y_1 \otimes y_2) \oplus y_1 = \emptyset$

Introducing $\bar{y}_2$:

$$(y_1 \otimes y_2) \oplus (y_1 \otimes (y_2 \oplus \bar{y}_2)) = \emptyset$$

i.e. $$(y_1 \otimes y_2) \oplus (y_1 \otimes \bar{y}_2) = \emptyset$$

with the solutions:

y_1	y_2
$\emptyset$	$\emptyset$
$\emptyset$	$-\infty$

Applying procedure (15-7) to these solutions, we derive for (14-8) the solutions:

x_1	x_2
β_1^{-1}	β_2^{-1}
β_1^{-1}	$<\beta_2^{-1}$

From (15-2) we have $\beta_1 = 3$ and $\beta_2 = -9$, so we have as the final complete set of solutions:

1	$x_1 = -3;\quad x_2 = 9$
2	$x_1 = -3;\quad -\infty \leqslant x_2 < 9$

(At this point, it is perhaps of interest to note that for a negligible extra effort we can solve equations of the form: $(\underline{A} \otimes \underline{x}) \oplus \underline{p} = \underline{b}$ with $\underline{A}$, $\underline{b}$ as before and $\underline{p} \in E_m$. Clearly, unless $\underline{p} \leqslant \underline{b}$, the equations are inconsistent. Now determine the column-maxima β_j as before, but in transposing into the Boolean problem, drop any equation for which $p_i = b_i$ since this equation is automatically satisfied by all solutions in which $\beta_j \otimes x_j \leqslant \emptyset$ $(j = 1, \ldots, n)$.
At the same time, drop the $\underline{p}$-column altogether, since if $p_i < b_i$ then this p_i does not constrain the variables.)

15-2. Proving the Procedure

Before proving that (15-7) actually yields solutions to (15-1) it is convenient to define an <u>admissible assignment</u> of values to the variables x_j as one for which each x_j receives some value such that $x_j \leqslant \beta_j^{-1}$. In view of (15-4) we know that all solutions to (15-1) must be admissible assignments.

Lemma 15-2. <u>Let E_1 be a linear blog. Then the solutions of (15-1) are exactly the admissible assignments such that for each $i = 1, \ldots, m$, there holds $x_j = \beta_j^{-1}$ for at least one j such that α_{ij} is marked.</u>

<u>Proof</u>. Since $\alpha_{ij} \leqslant \beta_j$ for all $i = 1, \ldots, m$ and $j = 1, \ldots, n$, we have:

$$\alpha_{ij} \otimes x_j \leqslant \beta_j \otimes x_j \leqslant \emptyset \qquad (15\text{-}8)$$

for all $i = 1, \ldots, m$ and $j = 1, \ldots, n$ and any admissible assignment.

Now suppose α_{ij} is an unmarked element in the matrix $[\alpha_{ij}]$, and that x_j receives some value under an admissible assignment. Then either $x_j = -\infty$, in which case $\alpha_{ij} \otimes x_j < \emptyset$, or x_j is finite in which case $\alpha_{ij} \otimes x_j < \emptyset$ also, for if $\alpha_{ij} \otimes x_j = \emptyset$ could occur in (15-8) we should have:

$$\alpha_{ij} = x_j^{-1} \geqslant \beta_j, \text{ (from (15-4))}$$

whereas α_{ij} was unmarked, and therefore $\alpha_{ij} < \beta_j$.

Hence for every admissible assignment, we have:

$$\left.\begin{array}{l} \alpha_{ij} \otimes x_j < \emptyset \text{ if } \alpha_{ij} \text{ is unmarked} \\ \alpha_{ij} \otimes x_j \leqslant \emptyset \text{ if } \alpha_{ij} \text{ is marked} \end{array}\right\}$$

Hence, if in a given row i there is at least one unmarked element, we have, by the linearity of E_1:

$$\sum_{\alpha_{ij} \text{ unmarked}}^{\oplus} \alpha_{ij} \otimes x_j < \emptyset$$

for every admissible assignment.

So a necessary and sufficient condition for $x_1, \ldots, x_n$ to be a solution to (15-1) is that it be an admissible assignment such that:

$$\sum_{\alpha_{ij} \text{ marked}}^{\oplus} \alpha_{ij} \otimes x_j = \emptyset \quad (i = 1, \ldots, m). \qquad (15\text{-}9)$$

Hence, since each $\alpha_{ij} \otimes x_j \leqslant \emptyset$, we infer that the solutions of (15-1) are exactly the admissible assignments such that for every $i = 1, \ldots, m$ there holds $\alpha_{ij} \otimes x_j = \emptyset$ for at least one j such that α_{ij} is marked. But α_{ij} is marked, if and only if $\alpha_{ij} = \beta_j$, and the result follows. ●

Now since ③ is a possible case for E_1, exactly the same procedures and reasoning apply to the Boolean equations (15-6), for which an admissible assignment is just any of the 2^n different assignments of the values $\emptyset$, $-\infty$, to y_j $(j=1,\ldots, n)$.

We begin by marking the column-maxima in (15-6), i.e. the $\emptyset$'s. But by the way the matrix $[\gamma_{ij}]$ was created, we see that element γ_{ij} is $\emptyset$ if and only if α_{ij} is marked. Using this fact in the version of Lemma 15-2 appropriate to the case $E_1 =$ ③, we infer for the Boolean problem (15-6):

<u>Proposition 15-3. The solutions of (15-6) are exactly the assignments of the values $\emptyset$ or $-\infty$ to the variables y_j such that for every $i = 1, \ldots, m$, there holds $y_j = \emptyset$ for at least one j such that α_{ij} is marked in (15-1).</u> ●

We can now prove our main result.

<u>Theorem 15-4. Let E_1 be a linear blog. Then procedure (15-7) yields all solutions to equations (15-1) without repetition.</u>

<u>Proof.</u> Examining procedure (15-7), we see that the assignments it generates are

admissible assignments, and that by Proposition 15-3 they ensure that for every $i = 1, \ldots, m$, there holds $x_j = \beta_j^{-1}$ for at least one j such that α_{ij} is marked. Hence by Lemma 15-2 every assignment produced by the procedure is a solution to (15-1).

Conversely, if a certain admissible assignment is a solution to (15-1), let us for $j = 1, \ldots, m$ give y_j the value $\emptyset$ if $x_j = \beta_j^{-1}$ in this assignment, and y_j the value $-\infty$ if $x_j < \beta_j^{-1}$ in this assignment.

From Lemma 15-2 and Proposition 15-3 we see that the resulting assignment of values to the y_j's is a solution to (15-6) and it is easy to confirm that the given solution to (15-1) belongs to exactly the class of solutions of (15-1) which arises by applying the solution procedure (15-7) to the solution of (15-6) just created.

Hence the class of solutions generated by the solution procedure (15-7) is exactly the class of solutions to (15-1).

Finally, it is clear that the solution procedure involves no duplication of solutions, since two distinct solutions to (15-6) give rise under the solution procedure to different combinations of variables x_j receiving the value β_j^{-1}. ●

15-3. Existence and Uniqueness

In the previous sections we have been concerned with an operational procedure for finding solutions to linear equations over a linear blog. From this procedure, we can now deduce some existence and uniqueness theorems, which are important for the later theory.

First we record the following lemma, which is virtually self-evident, but which will be useful later.

Lemma 15-5. Let E_1 be a linear blog and let $\underline{A} \in \mathcal{M}_{mn}$ be doubly G-astic and $\underline{b} \in E_m$ be finite. Let $\underline{\rho} \in E_m$ have all m coordinates equal to $\emptyset$, and let $\underline{C} \in \mathcal{M}_{mn}$ be defined by $\{\underline{C}\}_{ij} = \{\underline{b}\}_i^{-1} \otimes \{\underline{A}\}_{ij}$ $(i = 1, \ldots, m;\ j = 1, \ldots, n)$. Then the equations $\underline{A} \otimes \underline{x} = \underline{b}$ and $\underline{C} \otimes \underline{x} = \underline{\rho}$ have identical solution-sets, and all solutions are /-solutions.

Proof. Evidently, $\{\underline{C}\}_{ij} = -\infty$ if and only if $\{\underline{A}\}_{ij} = -\infty$, and $\{\underline{C}\}_{ij}$ is otherwise finite. So $\underline{C}$ is doubly G-astic, and clearly ρ is finite. Hence all solutions are /-solutions, by Theorem 14-4.

Also, $\underline{A} \otimes \underline{x} = \underline{b}$ if and only if $\sum_{j=1}^{n}{}_{\oplus} \{\underline{A}\}_{ij} \otimes x_j = \{\underline{b}\}_i$ $(i = 1, \ldots, m)$

and this clearly holds if and only if for $i = 1, \ldots, m$ we have

$$\sum_{j=1}^{n}{}_{\oplus} ((\{\underline{b}\}_i)^{-1} \otimes \{\underline{A}\}_{ij} \otimes x_j) = (\{\underline{b}\}_i)^{-1} \otimes \sum_{j=1}^{n}{}_{\oplus} (\{\underline{A}\}_{ij} \otimes x_j) = \emptyset = \{\underline{\rho}\}_i$$

i.e. if and only if $C \otimes \underline{x} = \underline{\rho}$. ●

Theorem 15-6. Let E_1 be a linear blog, and let $\underline{A} \in \mathcal{M}_{mn}$ be doubly G-astic and $\underline{b} \in E_m$ be finite. Then a necessary and sufficient condition that the equation

$\underline{A} \otimes \underline{x} = \underline{b}$ shall have at least one solution is that each row of the matrix $\underline{C} = [(\{\underline{b}\}_i)^{-1} \otimes \{\underline{A}\}_{ij}]$ contain at least one column-maximum.

Proof. The necessity of the condition was proved in Lemma 15-1. Conversely, if each row of $\underline{C} = [\alpha_{ij}]$ does indeed contain an element which is greatest in its column, then the procedure beginning with (15-1) goes through as far as (15-6), where it defines a Boolean problem $\underline{B} \otimes \underline{y} = \underline{c}$ with $\underline{B}$ doubly G-astic and $\underline{c}$ finite. But such an equation always has at least one solution, namely all $y_i = \phi$ (Lemma 14-7). Correspondingly, the original equations then have at least one solution, namely all $x_j = \beta_j^{-1}$.

We remark that the solution $x_j = \beta_j^{-1}$ $(j = 1, \ldots, m)$ discussed in the foregoing theorem, is exactly the principal solution in this case.

Theorem 15-7. Let E_1 be a linear blog and let $\underline{A} \in \mathcal{M}_{mn}$ be doubly G-astic and $\underline{b} \in E_m$ be finite. Then a necessary and sufficient condition that the equation $\underline{A} \otimes \underline{x} = \underline{b}$ shall have exactly one solution is, that each row of the matrix $\underline{C} = [(\{\underline{b}\}_i)^{-1} \otimes \{\underline{A}\}_{ij}]$ contain at least one column-maximum and for each $j = 1, \ldots, n$ there is an i $(1 \leq i \leq m)$ such that $\{\underline{C}\}_{ij}$ is the only column-maximum on row i.

Proof. In view of Theorem 15-6, we may assume that $\underline{C}$ has at least one column-maximum on each row.

Now in terms of Lemma 15-2, the solutions of $\underline{A} \otimes \underline{x} = \underline{b}$ consist of certain admissible assignments, whereby certain variables receive the value β_j^{-1} and others receive an arbitrary value $<\beta_j^{-1}$.

So if, for each column j in $\underline{C}$ we can find at least one row i in which only $\{\underline{C}\}_{ij}$ is marked, then Lemma 15-2 tells us that every solution has $x_j = \beta_j^{-1}$, so there is only one solution.

Conversely, if for a particular column k in $\underline{C}$ there is no row i in which only $\{\underline{C}\}_{ik}$ is marked then Lemma 15-2 tells us that all the assignments:

$$x_j = \beta_j^{-1} \qquad (j \neq k)$$
$$x_k \leq \beta_k^{-1}$$

are solutions, for arbitrary $x_k \leq \beta_k^{-1}$, so there is no unique solution. (This is true even if E_1 is ③ since the β_j's are always finite, and so equal ϕ if E_1 is ③. Hence "arbitrary $x_k \leq \beta_k^{-1}$" still implies a non-unique solution.)

We shall now give Theorems 15-6 and 15-7 another form, which will have applications in the immediately following chapters and also later when we discuss canonical forms of matrices.

Let ϕ as usual be the multiplicative identity element. From Section 12-1 we recall the use of the term *ϕ-astic*.

When E_1 is linear, a doubly $\emptyset$-astic matrix clearly has $\emptyset$ as the maximum element in each column, and each row contains at least one of these column-maxima.

Theorem 15-8. Let E_1 be a blog and let $A \in \mathcal{M}_{mn}$ be doubly G-astic and $\underline{b} \in E_m$ be finite. Then a necessary and sufficient condition that the equation $\underline{A} \otimes \underline{x} = \underline{b}$ shall have at least one solution is that we can find n finite elements $x_1, \ldots, x_n$ such that the matrix $\underline{B} = [(\{\underline{b}\}_i)^{-1} \otimes \{\underline{A}\}_{ij} \otimes x_j]$ is doubly $\emptyset$-astic.

Proof. $\underline{A} \otimes \underline{x} = \underline{b}$ is soluble, according to Theorems 14-3 and 14-4 if and only if it has a principal solution $\underline{x} = \underline{A}^* \otimes' \underline{b}$ which is finite, i.e. if and only if we can find finite elements $x_1, \ldots, x_n$ such that:

$$x_j = \sum_{i=1}^{m}{}_{\oplus'} (\{\underline{A}^*\}_{ji} \otimes' \{\underline{b}\}_i) \qquad (j=1,\ldots, n)$$

$$\{\underline{b}\}_i = \sum_{j=1}^{n}{}_{\oplus} (\{\underline{A}\}_{ij} \otimes x_j) \qquad (i=1,\ldots, m)$$

These relations are evidently equivalent to:

$$\left.\begin{aligned} &\Big(\sum_{i=1}^{m}{}_{\oplus'} (\{\underline{A}^*\}_{ji} \otimes' \{\underline{b}\}_i)\Big)^{-1} \otimes x_j = \emptyset \qquad (j=1,\ldots, n) \\ &(\{b\}_i)^{-1} \otimes \sum_{j=1}^{n}{}_{\oplus} (\{\underline{A}\}_{ij} \otimes x_j) = \emptyset \qquad (i=1,\ldots, m) \end{aligned}\right\} \qquad (15\text{-}10)$$

Now, making free use of the results of Section 7-1:

$$\Big(\sum_{i=1}^{m}{}_{\oplus'} (\{\underline{A}^*\}_{ji} \otimes' \{\underline{b}\}_i)\Big)^{-1}$$

$$= \Big(\sum_{i=1}^{m}{}_{\oplus'} (\{\underline{A}^*\}_{ji} \otimes' \{\underline{b}\}_i)\Big)^*$$

$$= \sum_{i=1}^{m}{}_{\oplus} (\{\underline{A}^*\}_{ji} \otimes' \{\underline{b}\}_i)^*$$

$$= \sum_{i=1}^{m}{}_{\oplus} ((\{\underline{b}\}_i)^* \otimes (\{\underline{A}^*\}_{ji})^*)$$

$$= \sum_{i=1}^{m}{}_{\oplus} \; ((\{b\}_i)^{-1} \otimes \{A\}_{ij})$$

Hence the first relation in (15-9) gives:

$$\sum_{i=1}^{m}{}_{\oplus} \; ((\{\underline{b}\}_i)^{-1} \otimes \{\underline{A}\}_{ij} \otimes x_j) = \phi \qquad (j=1, \ldots, n)$$

And evidently the second relation in (15-9) gives:

$$\sum_{j=1}^{n}{}_{\oplus} \; ((\{\underline{b}\}_i)^{-1} \otimes \{\underline{A}\}_{ij} \otimes x_j) = \phi \qquad (i=1, \ldots, m)$$

But these relations are just:

$$\sum_{i=1}^{m}{}_{\oplus} \; \{\underline{B}\}_{ij} = \phi \qquad (j=1, \ldots, n)$$

$$\sum_{j=1}^{n}{}_{\oplus} \; \{\underline{B}\}_{ij} = \phi \qquad (i=1, \ldots, m)$$

And clearly these state that $\underline{B}$ is doubly ϕ-astic.

Let us call a square matrix $\underline{A} \in \mathcal{M}_{nn}$ strictly doubly $\emptyset$-astic if it satisfies the following two requirements:

(i) $\{\underline{A}\}_{ij} \leqslant \emptyset \qquad (i=1, \ldots, n; \;\; j=1, \ldots, n)$

(ii) On each row, and on each column of $\underline{A}$, we can find one and only one element equal to $\emptyset$.

Evidently a strictly doubly $\emptyset$-astic matrix is doubly $\emptyset$-astic. Each column has a unique element as column-maximum, and each row contains one and only one such column-maximum.

Theorem 15-9. Let E_1 be a linear blog and let $\underline{A} \in \mathcal{M}_{mn}$ be doubly G-astic and $\underline{b} \in E_m$ be finite. Then a necessary and sufficient condition that the equation $\underline{A} \otimes \underline{x} = \underline{b}$ shall have exactly one solution is that we can find n finite elements $x_1, \ldots, x_n$ such that the matrix $\underline{B} = [(\{\underline{b}\}_i)^{-1} \otimes \{A\}_{ij} \otimes x_j]$ is doubly $\emptyset$-astic and contains a strictly doubly $\emptyset$-astic $(n \times n)$ submatrix.

Proof. The positions of the column-maxima in a matrix are clearly unchanged if all the elements in a given column are multiplied by the same finite constant x, since (when x is finite) there holds:

$$\alpha > \beta \quad \text{if and only if} \quad \alpha \otimes x > \beta \otimes x$$

Hence for any finite $x_1, \ldots, x_n$ the matrix $\underline{B}$ has its column-maxima in precisely the same positions as the matrix $\underline{C} = [(\{\underline{b}\}_i)^{-1} \otimes \{\underline{A}\}_{ij}]$.

So if finite $x_1, \ldots, x_n$ exist such that $\underline{B}$ is doubly $\emptyset$-astic with a strictly doubly $\emptyset$-astic $(n \times n)$ submatrix, then the column-maxima lie in $\underline{C}$ in a pattern which, according to Theorem 15-7 implies that the equation $\underline{A} \otimes \underline{x} = \underline{b}$ has precisely one solution.

Now let the column-maxima of the matrix $\underline{C} = [(\{\underline{b}\}_i)^{-1} \otimes \{\underline{A}\}_{ij}]$ have the values $\beta_1, \ldots, \beta_n$ respectively (necessarily finite), and let $x_j = \beta_j^{-1}$ $(j=1, \ldots, n)$. Evidently the matrix:

$$\underline{B} = [(\{\underline{b}\}_i)^{-1} \otimes \{\underline{A}\}_{ij} \otimes x_j]$$

then has all column-maxima equal to $\emptyset$, i.e. $\underline{B}$ is column $\emptyset$-astic.

If the equation $\underline{A} \otimes \underline{x} = \underline{b}$ is uniquely soluble then, according to Theorem 15-7, the column-maxima lie in $\underline{C}$ in a pattern which implies that $\underline{B}$ is doubly $\emptyset$-astic with a strictly doubly $\emptyset$-astic $(n \times n)$ submatrix. ●

15-4. A Linear Programming Criterion

The existence and uniqueness criteria so far developed apply, of course, to the principal interpretation in particular and we shall make use of this fact later. We can, however, obtain a quite different solubility criterion in this case, which it is convenient to record at the present point. In fact, for the principal interpretation, it is evident that problem (14-1) is equivalent to:

$$\left.\begin{array}{l}\text{Given } \underline{A} \in \mathcal{M}_{mn} \text{ and } \underline{b} \in E_m, \text{ to find } x \in E_n \text{ such that} \\ \max\limits_{j=1,\ldots,n} (a_{ij} + x_j) = b_i \quad (i = 1, \ldots, m)\end{array}\right\} \qquad (15\text{-}11)$$

i.e. such that:

$$\forall i = 1, \ldots, m;\ j = 1, \ldots, n:\ a_{ij} + x_j \leq b_i \qquad (15\text{-}12)$$

$$\text{and } \forall i, \ldots, m,\ a_{ij} + x_j = b_i \ \text{ for some } j \ (1 \leq j \leq n) \qquad (15\text{-}13)$$

(i.e. equality occurs at least once for each index i)

<u>Theorem 15-10. Let E_1 receive the principal interpretation, and let $\underline{A} \in \mathcal{M}_{mn}$ be doubly G-astic and $\underline{b} \in E_m$ be finite. Let σ be the set of pairs (i, j) such that a_{ij} is finite $(1 \leq i \leq m;\ 1 \leq j \leq n)$. Then a sufficient condition that the equation $\underline{A} \otimes \underline{x} = \underline{b}$ be soluble is that some solution $\{\xi_{ij} | (i, j) \in \sigma\}$ of the following optimisation problem in the variables y_{ij} $((i, j) \in \sigma)$:</u>

<u>Minimise</u>: $$\sum_{(i,j) \in \sigma} (b_i - a_{ij}) y_{ij}$$

<u>Subject to</u> $$\sum_{\substack{i=1 \\ (i,j) \in \sigma}}^{m} y_{ij} = 1 \ (j = 1, \ldots, n)$$

<u>and</u> $y_{ij} \geq 0$ $((i, j) \in \sigma)$

<u>shall satisfy</u>: $$\sum_{\substack{j=1 \\ (i,j) \in \sigma}}^{n} \xi_{ij} > 0 \ (i = 1, \ldots, m).$$

<u>Proof</u>. The given problem is one of linear programming [33]. It has a finite solution because it calls for the minimisation of a continuous function on a compact set. Let $\{\xi_{ij} | (i, j) \in \sigma\}$ be a solution.

From the theory of linear programming we know that the dual of the given problem is:

$$\left.\begin{array}{l} \text{Maximise } \sum_{j=1}^{n} x_j \\ \text{Subject to } x_j \leq b_i - a_{ij} \quad (\forall (i,j) \in \sigma) \end{array}\right\} \qquad (15\text{-}14)$$

This problem has a certain solution $\eta_1, \ldots, \eta_n$ which obviously satisfies (15-12), and by the theorem of complementary slacks [33] we have:

$$\xi_{ij}\,(b_i - a_{ij} - \eta_j) = 0 \quad (\forall (i,j) \in \sigma) \qquad (15\text{-}15)$$

Now, if we have $\sum_{\substack{j=1 \\ (i,j)\in\sigma}}^{n} \xi_{ij} > 0 \quad (i = 1, \ldots, m)$ then:

$\forall i = 1, \ldots, m, \quad \xi_{ij} > 0$ and $(i, j) \in \sigma$ for some $j(1 \leq j \leq n)$

(because $\xi_{ij} \geq 0$). Hence from (15-15):

$\forall i = 1, \ldots, m, \quad (b_i - a_{ij} - \eta_j) = 0$ and $(i, j) \in \sigma$ for some $j(1 \leq j \leq n)$

In other words, $(\eta_1, \ldots, \eta_n)$ satisfy (15-13) as well as (15-12), i.e. $(\eta_1, \ldots, \eta_n)$ are a solution to (15-11) and thus to (14-1). ●

We remark that it is evident by direct inspection that (15-14) always has the unique solution $m_j = \min (b_i - a_{ij})$. Theorem 15-10 therefore gives sufficient conditions that (14-1) shall have its principal solution $\underline{A}^* \otimes' \underline{b}$ since for the principal interpretation, component j of $\underline{A}^* \otimes \underline{b}$ is just m_j.

15-5. Left-right variants.

We remarked earlier that, in addition to the duality introduced through axioms $X_1', \ldots, X_6'$ et sequ., two other "involutions" may be operating in a given context, namely a "left-right" correspondence arising from non-commutativity of (matrix) multiplication, and a "row-column" correspondence for matrices. As a result, a given theorem concerning matrices may be susceptible of many variants, as illustrated by Figs. 8-3 and 8-4, or Table 14-1.

In particular, we may consider, in place of the problem (14-1) the following "row-variant":

Given $\underline{A} \in \mathcal{M}_{mn}$ and $\underline{c} \in \mathcal{M}_{1n}$, to find $\underline{y} \in \mathcal{M}_{1m}$ such that $\underline{y} \otimes \underline{A} = \underline{c}$ (15-16)

Evidently we may develop an appropriate version of the theory presented in Chapter 14 and the present chapter, covering principal solutions, routines for finding all solutions, and criteria for solubility and unique solubility. However, the "row theory" and "column theory" not only proceed along parallel lines, they positively interact, as we shall now show.

First we require the following lemma, which formally records what is already implicit in the proof of Theorem 15-8.

Lemma 15-11. Let E_1 be a blog and let $\underline{A} \varepsilon \mathcal{M}_{mn}$ be doubly G-astic and $\underline{b} \varepsilon E_n$ be finite. Suppose we can find finite elements $x_1, \ldots, x_n$ such that the matrix $\underline{B} = [(\{\underline{b}\}_i)^{-1} \otimes \{\underline{A}\}_{ij} \otimes x_j]$ is doubly ϕ-astic. Then $x_1, \ldots, x_n$ constitute the principal solution to the equation $\underline{A} \otimes \underline{x} = \underline{b}$. If moreover E_1 is linear and $\underline{B}$ contains a strictly doubly ϕ-astic (nxn) submatrix, then $x_1, \ldots, x_n$ constitute the unique solution to the equation $\underline{A} \otimes \underline{x} = \underline{b}$.

Proof. The first assertion follows from the fact that, in the proof of Theorem 15-8, there is established a logical equivalence (under the hypothesis of the theorem) between:

(i) $x_1, \ldots, x_n$ constitute the principal solution to the equation $\underline{A} \otimes \underline{x} = \underline{b}$

(ii) $x_1, \ldots, x_n$ make $\underline{B}$ doubly ϕ-astic.

The second assertion now follows in the light of Theorem 15-9. ●

Our "row-and-column" results now follow.

Theorem 15-12. Let E_1 be a linear blog, let $\underline{A} \varepsilon \mathcal{M}_{nn}$ be doubly G-astic and let $\underline{b} \varepsilon \mathcal{M}_{n1}$ and $\underline{c} \varepsilon \mathcal{M}_{1n}$ both be finite. Then the equation $\underline{A} \otimes \underline{x} = \underline{b}$ has the unique solution $\underline{x} = \underline{c}^*$ if and only if the equation $\underline{y} \otimes \underline{A} = \underline{c}$ has the unique solution $\underline{y} = \underline{b}^*$.

Proof. From Theorem 15-9, Lemma 15-11 and their "row-variants", it follows that the following condition is necessary and sufficient both for problem (14-1) to have the unique solution $\underline{x} = \underline{c}^*$ and for problem (15-16) to have the unique solution $\underline{y} = \underline{b}^*$.

The matrix $\underline{B} \varepsilon \mathcal{M}_{nn}$ is strictly doubly ϕ-astic, where

$$\{\underline{B}\}_{ij} = \{\underline{b}^*\}_i \otimes \{\underline{A}\}_{ij} \otimes \{c^*\}_j \qquad (i, j=1,\ldots, n)$$ ●

Corollary 15-13. Let E_1 be a linear blog, let $\underline{A} \varepsilon \mathcal{M}_{nn}$ be doubly G-astic and let $\underline{b} \varepsilon \mathcal{M}_{n1}$ and $\underline{c} \varepsilon \mathcal{M}_{1n}$ both be finite. Then the following conditions are all equivalent:

(i) Equation $\underline{A} \otimes \underline{x} = \underline{b}$ has unique solution $\underline{x} = \underline{c}^*$

(ii) Equation $\underline{y} \otimes \underline{A} = \underline{c}$ has unique solution $\underline{y} = \underline{b}^*$

(iii) Equation $\underline{A}^* \otimes' \underline{w} = \underline{c}^*$ has unique solution $\underline{w} = \underline{b}$

(iv) Equation $\underline{z} \otimes' \underline{A}^* = \underline{b}^*$ has unique solution $\underline{z} = \underline{c}$

Proof. Equivalence of (i) and (ii) was established in Theorem 15-12. Condition (iii) is obtained from condition (ii) by dualising and condition (iv) from condition (i) similarly. ●

16. LINEAR DEPENDENCE

16-1. Linear Dependency Over E_1

Let E_1 be a given belt, and let $\underline{A} \in \mathcal{M}_{mn}$ and $\underline{b} \in E_n$ be given. In Chapter 14, we looked upon the equation $\underline{A} \otimes \underline{x} = \underline{b}$ as the matrix embodiment of a set of simultaneous equations. There are, however, other ways of interpreting $\underline{A} \otimes \underline{x} = \underline{b}$. If $\underline{A} \in \mathcal{M}_{mn}$, then $\underline{A}$ has n columns, $\underline{a}(j)$, j=1,..., n each of which is m-tuple, i.e. $\underline{a}(j) \in E_m$.

Then the equation $\underline{A} \otimes \underline{x} = \underline{b}$ may be written:

$$\sum_{j=1}^{n}{}_{\oplus} \underline{a}(j) \otimes x_j = \underline{b} \qquad (16\text{-}1)$$

Relation (16-1) expresses the <u>(right) linear dependence (over E_1)</u> of $\underline{b} \in E_m$ on $\underline{a}(j) \in E_m$ (j=1,..., n), the modifier "right" indicating that the scalar multipliers x_j multiply from the right. By obvious analogy, we can define <u>left linear dependence</u> (over E_1). If dual addition ($\Sigma_{\oplus'}$) is used in place of addition ($\Sigma_{\oplus}$), we may define <u>dual left and dual right linear dependence (over E_1)</u>, in the obvious way. By <u>linear dependence</u>, without further modifiers, we shall always mean right linear dependence over whichever belt figures in the given context. We shall also say that $\underline{b}$ in (16-1) is a <u>linear combination</u> of $\underline{a}(1),\ldots, \underline{a}(n)$, (even when n=1).

Notice that for m > 1, left linear dependence and right linear dependence over E_1 do not imply one another, unless E_1 is commutative.

For let E_1 be a non-commutative division belt, and let a_1, c, x $\in E_1$ be chosen so that $c \otimes x \neq x \otimes c$, and define:

$$\left.\begin{aligned} a_2 &= a_1 \otimes c \\ b_1 &= a_1 \otimes x \\ b_2 &= a_2 \otimes x \end{aligned}\right\} ; \quad \underline{b} = \begin{bmatrix} b_1 \\ b_2 \end{bmatrix}, \quad \underline{a} = \begin{bmatrix} a_1 \\ a_2 \end{bmatrix} \qquad (16\text{-}2)$$

We have
$$\begin{bmatrix} b_1 \\ b_2 \end{bmatrix} = \begin{bmatrix} a_1 \\ a_2 \end{bmatrix} \otimes x \qquad (16\text{-}3)$$

hence $\underline{b} \in E_2$ is right linearly dependent on $\underline{a} \in E_2$. Suppose $\underline{b}$ were also left linearly dependent on $\underline{a}$, i.e. for some $y \in E_1$ we had:

$$\begin{bmatrix} b_1 \\ b_2 \end{bmatrix} = y \otimes \begin{bmatrix} a_1 \\ a_2 \end{bmatrix} \qquad (16\text{-}4)$$

Then $a_2 \otimes x = y \otimes a_2$

Substituting for a_2 from (16-2):

$$a_1 \otimes c \otimes x = y \otimes a_1 \otimes c$$

But from (16-3) and (16-4):

$$y \otimes a_1 = b_1 = a_1 \otimes x$$

Hence $a_1 \otimes c \otimes x = a_1 \otimes x \otimes c$

So $\quad c \otimes x = x \otimes c$, (since E_1 is a division belt) contradicting the hypothesis. Clearly we can extend this counterexample to E_m for $m > 2$ by extending the vectors $\underline{b}$ and $\underline{a}$ in any way.

Again, linear dependence and dual linear dependence do not imply one another. Consider for example in the principal interpretation the following $\underline{A} \in \mathcal{M}_{33}$ and $\underline{b} \in E_3$:

$$\underline{A} = \begin{bmatrix} 0 & 10 & 100 \\ 10 & 2 & 10 \\ 1 & 9 & 10 \end{bmatrix}, \quad \underline{b} = \begin{bmatrix} 0 \\ 2 \\ 1 \end{bmatrix}$$

The 3-tuple $\underline{b}$ is dually linearly dependent on the columns of $\underline{A}$, the multipliers x_j all being equal to zero ($\emptyset$). However, $\underline{b}$ is not linearly dependent on the columns of $\underline{A}$ as we see if we begin to solve $\underline{A} \otimes \underline{x} = \underline{b}$ following the procedure of Section 15-1. We bring all elements of $\underline{b}$ to zero, and mark the greatest elements in the columns of $\underline{A}$, to get:

$$\begin{bmatrix} 0 & \textcircled{10} & \textcircled{100} \\ \textcircled{8} & 0 & 8 \\ 0 & 8 & 9 \end{bmatrix} \otimes \begin{bmatrix} x_1 \\ x_2 \\ x_3 \end{bmatrix} = \begin{bmatrix} 0 \\ 0 \\ 0 \end{bmatrix}$$

Since no element in the third row is marked, the equations are inconsistent, by Lemma 15-1.

It is evident that we can extend this example to E_m with $m > 3$ by "filling " all the m-tuples with (m-3) components equal to zero.

Perhaps surprisingly, the case m=2 does not always follow the general rule, as the following result shows:

<u>Theorem 16-1. Suppose E_1 is a linear blog. Let $\underline{b} \in E_2$, and $\underline{a}(j) \in E_2$ $(j=1,\ldots, n)$, all be finite.</u>

<u>Then $\underline{b}$ is linearly dependent on $\underline{a}(1),\ldots, \underline{a}(n)$ if and only if $\underline{b}$ is dually linearly dependent on $\underline{a}(1),\ldots, \underline{a}(n)$.</u>

<u>Proof</u>. Let $\underline{A} \in \mathcal{M}_{2n}$ have $\underline{a}(1),\ldots, \underline{a}(n)$ as its columns, and let us begin to apply the procedure of Section 15-1 to the solution of $\underline{A} \otimes \underline{x} = \underline{b}$. We bring all the right-hand sides to $\emptyset$, and mark the resulting greatest elements in the columns of the matrix. By Theorem 15-6, the equation $\underline{A} \otimes \underline{x} = \underline{b}$ is soluble if and only if in each row, some element is marked as greatest in its column. But this happens if and only if in each row, some element is equal to the <u>least</u> in its column. And by the dual of Theorem 15-6, this is a necessary and sufficient condition that $\underline{A} \otimes' \underline{x} = \underline{b}$ have at least one solution.

16-2. The $\mathcal{A}$ test. Let E_1 be a given blog. Suppose we are given m-tuples $\underline{a}(j) \in E_m$ (j=1,..., n) and we wish to determine, for each of them, whether or not it is linearly dependent on the other (n-1) m-tuples. Obviously we can do this by n applications of the solubility criterion stated at the end of Section 14-2, with each $\underline{a}(j)$ in turn taking the role of $\underline{b}$.

The following theorem gives a more convenient mechanical procedure. Let $\underline{A} \in \mathcal{M}_{nn}$ be the matrix having $\underline{a}(j)$ as its jth column (j=1,..., n). Define a matrix $\underline{\mathcal{A}} \in \mathcal{M}_{nn}$ as follows:

$$\left.\begin{array}{ll} \{\underline{\mathcal{A}}\}_{ii} = -\infty & (i=1,\ldots, n) \\ \{\underline{\mathcal{A}}\}_{ij} = \{\underline{A}^* \otimes' \underline{A}\}_{ij} & (i=1,\ldots, n;\ j=1,\ldots, n;\ i \neq j) \end{array}\right\} \qquad (16\text{-}5)$$

In other words, $\underline{\mathcal{A}}$ is the matrix $\underline{A}^* \otimes' \underline{A}$ with its diagonal elements overwritten by $-\infty$'s. We now compare each column of $\underline{A}$ with the corresponding column of $\underline{A} \otimes \underline{\mathcal{A}}$, and make use of the following theorem:

Theorem 16-2. Let E_1 be a blog. Let the matrix $\underline{A} \in \mathcal{M}_{mn}$ have columns $\underline{a}(j) \in E_m$ (j=1,..., n≥2), not necessarily all different. Then for each j=1,.., n, the jth column of $\underline{A} \otimes \underline{\mathcal{A}}$ is identical with $\underline{a}(j)$ if and only if $\underline{a}(j)$ is linearly dependent on the other columns of $\underline{A}$. The elements of the jth column of $\underline{\mathcal{A}}$ then give suitable coefficients to express the linear dependence.

Proof. Suppose first that for a particular j, the jth column of $\underline{A} \otimes \underline{\mathcal{A}}$ is identical with $\underline{a}(j)$, i.e.:

$$\{\underline{A} \otimes \underline{\mathcal{A}}\}_{ij} = \{\underline{a}(j)\}_i \qquad (i=1,\ldots, m)$$

$$\text{i.e.} \quad \sum_{k=1}^{n}{}_{\oplus} (\{\underline{A}\}_{ik} \otimes \{\underline{\mathcal{A}}\}_{kj}) = \{\underline{a}(j)\}_i \qquad (i=1,\ldots, m) \qquad (16\text{-}6)$$

But this is:

$$\sum_{\substack{k=1\\k\neq j}}^{n}{}_{\oplus} (\underline{a}(k) \otimes \{\underline{\mathcal{A}}\}_{kj}) = \underline{a}(j) \qquad \text{(using the fact that } \{\mathcal{A}\}_{jj} = -\infty)$$

In other words, $\underline{a}(j)$ is linearly dependent on the other columns of $\underline{A}$, and the coefficients of the relation are just the elements $\{\underline{\mathcal{A}}\}_{kj}$ of the jth column of $\underline{\mathcal{A}}$. Conversely, suppose some column of $\underline{A}$ is linearly dependent on the others. For simplicity of notation and without loss of generality, suppose $\underline{a}(n)$ is linearly dependent on the other columns.

Then $\underline{B} \otimes \underline{x} = \underline{a}(n)$ is soluble, where matrix $\underline{B}$ has $\underline{a}(1),\ldots, \underline{a}(n-1)$ as its columns. Hence, by Theorem 14-3:

$$\underline{B} \otimes (\underline{B}^* \otimes' \underline{a}(n)) = \underline{a}(n). \qquad (16\text{-}7)$$

Now define $\underline{c} \in E_n$ as follows:

$$\left.\begin{array}{ll} c_j = \{\underline{B}^* \otimes' \underline{a}(n)\}_j & (j=1,\ldots,(n-1)) \\ c_n = -\infty & \end{array}\right\} \qquad (16\text{-}8)$$

Then $\qquad \underline{A} \otimes \underline{c} = \sum_{j=1}^{n}{}_{\oplus}\ \underline{a}(j) \otimes c_j$

$$= \sum_{j=1}^{n-1}{}_{\oplus}\ \underline{a}(j) \otimes c_j \qquad \text{(since } c_n = -\infty)$$

$$= \underline{B} \otimes (\underline{B}^* \otimes' \underline{a}(n)) \qquad \text{(by (16-8) and the definition of } \underline{B})$$

$$= \underline{a}(n) \qquad \text{(by (16-7))} \qquad (16\text{-}9)$$

Now, for $j=1,\ldots,(n-1)$, $c_j = \{\underline{B}^* \otimes' \underline{a}(n)\}_j$ (by (16-8))

$$= \underline{a}(j)^* \otimes' \underline{a}(n) \qquad \text{(by the definition of } \underline{B})$$

$$= \{\underline{\mathcal{A}}\}_{jn} \qquad \text{(by the definition of } \underline{\mathcal{A}})$$

Moreover, $c_n = -\infty = \{\underline{\mathcal{A}}\}_{nn}$. Hence $\underline{c}$ is precisely column n of $\underline{\mathcal{A}}$.

Hence (16-9) says:

$$\underline{a}(n) = \underline{A} \otimes (n^{\text{th}} \text{ column of } \underline{\mathcal{A}})$$

$$= n^{\text{th}} \text{ column of } (\underline{A} \otimes \underline{\mathcal{A}}).$$

●

We illustrate the technique with the following example in the principal interpretation. Given the 3-tuples:

$$\underline{a}(1) = \begin{bmatrix} 1 \\ 3 \\ 2 \end{bmatrix}, \quad \underline{a}(2) = \begin{bmatrix} 3 \\ 4 \\ 5 \end{bmatrix}, \quad \underline{a}(3) = \begin{bmatrix} 2 \\ 2 \\ 5 \end{bmatrix}, \quad \underline{a}(4) = \begin{bmatrix} 3 \\ 1 \\ 3 \end{bmatrix} \qquad (16\text{-}10)$$

Which of these is linearly dependent on the others? We form:

$$\underline{A}^* \otimes' \underline{A} = \begin{bmatrix} -1 & -3 & -2 \\ -3 & -4 & -5 \\ -2 & -2 & -5 \\ -3 & -1 & -3 \end{bmatrix} \otimes' \begin{bmatrix} 1 & 3 & 2 & 3 \\ 3 & 4 & 2 & 1 \\ 2 & 5 & 5 & 3 \end{bmatrix}$$

$$= \begin{bmatrix} 0 & 1 & -1 & -2 \\ -3 & 0 & -2 & -3 \\ -3 & 0 & 0 & -2 \\ -2 & 0 & -1 & 0 \end{bmatrix}$$

Whence:

$$\mathcal{A} = \begin{bmatrix} -\infty & 1 & -1 & -2 \\ -3 & -\infty & -2 & -3 \\ -3 & 0 & -\infty & -2 \\ -2 & 0 & -1 & -\infty \end{bmatrix} \qquad (16\text{-}11)$$

$$\underline{A} \otimes \mathcal{A} = \begin{bmatrix} 1 & 3 & 2 & 0 \\ 1 & 4 & 2 & 1 \\ 2 & 5 & 3 & 3 \end{bmatrix} \qquad (16\text{-}12)$$

Comparing (16-12) with (16-10), we see that $\underline{a}(2)$ is linearly dependent on $\underline{a}(1)$, $\underline{a}(3)$, $\underline{a}(4)$. From (16-11) we see that the coefficients are 1, 0, 0 respectively. Checking:

$$\begin{bmatrix} 1 & 2 & 3 \\ 3 & 2 & 1 \\ 2 & 5 & 3 \end{bmatrix} \otimes \begin{bmatrix} 1 \\ 0 \\ 0 \end{bmatrix} = \begin{bmatrix} \max(2, 2, 3) \\ \max(4, 2, 1) \\ \max(3, 5, 3) \end{bmatrix} = \begin{bmatrix} 3 \\ 4 \\ 5 \end{bmatrix} = \underline{a}(2).$$

Obviously, the jth column of $\mathcal{A}$ does not give the only possible expression of the linear dependence of $\underline{a}(j)$ on the other columns, since in general the equation $\underline{B} \otimes \underline{x} = \underline{a}(j)$ does not have a unique solution. The $\mathcal{A}$-procedure in fact provides the principal solution in each case (Section 14-2).

The following corollary defines the dual form of the $\mathcal{A}$-test.

Corollary 16-3. Let E_1 be a blog. Let the matrix $\underline{A} \in \mathcal{M}_{mn}$ have colums $\underline{a}(j)$ ($j=1,\ldots, n \geqslant 2$), not necessarily all different. Then for each $j=1,\ldots, n$, $\underline{a}(j)$ is dually (right) linearly dependent upon the other columns of $\underline{A}$ if and only if $\underline{a}(j)$ is identical with the jth column of $\underline{A} \otimes' \mathcal{A}$, where $\mathcal{A}$ is formed by overwriting the diagonal elements of $\underline{A}^* \otimes \underline{A}$ with $+\infty$; and the elements of the jth column of $\mathcal{A}$ then give suitable coefficients to express the dual linear dependence.

Proof. Dual to that of Theorem 16-2. ●

The following extension of the $\mathcal{A}$-test has some interest. Suppose $\underline{a}(1),\ldots, \underline{a}(r)$ and $\underline{a}(r+1),\ldots, \underline{a}(n)$ are two given sets of m-tuples. For each m-tuple, we wish to determine whether it is linearly dependent on the m-tuples in the other set. We define $\mathcal{A} \in \mathcal{M}_{nn}$ as follows:

$$\{\underline{\mathcal{A}}\}_{ij} = -\infty \text{ if } \underline{a}(i) \text{ and } \underline{a}(j) \text{ are in the same set}$$
$$= \underline{a}(i)^* \otimes' \underline{a}(j) \text{ otherwise.}$$

It is then readily seen that a comparison of each $\underline{a}(j)$ with the corresponding column of $\underline{A} \otimes \underline{\mathcal{A}}$ furnishes the required criterion.

<u>16-3. Some Dimensional Anomalies</u>. In conventional linear algebra, the concept of linear dependence gives a basis for a theory of <u>rank</u> and <u>dimension</u>. In the present section we shall point out some anomalies which make it clear that the situation in minimax algebra is more complicated. These considerations will lead us, in the following section, to introduce the idea of <u>strong linear independence</u>.

<u>Theorem 16-4. Suppose E_1 is a blog other than ③. Let $m > 2$ and $k > 1$ be arbitrary integers. Then we can always find k finite m-tuples, no one of which is linearly dependent on the others.</u>

<u>Proof</u>. Assume first of all that $m = 3$, and let $x > \emptyset$ be an arbitrary finite element. Then:

$$x^r < x^s < \emptyset \text{ for integers } r < s < 0 \tag{16-13}$$

Consider the k finite 3-tuples:

$$\begin{bmatrix} \emptyset \\ x^r \\ x^{k-r} \end{bmatrix} \quad (r=1,\ldots, k) \quad (\text{We define } x^0 = \emptyset) \tag{16-14}$$

By applying the $\mathcal{A}$-test we shall show that each is linearly independent of the others. Defining $\underline{A}$ and $\underline{\mathcal{A}}$ as usual, we find by direct computation:

$$\begin{aligned}
\{\underline{\mathcal{A}}\}_{ij} &= \underline{a}(i)^* \otimes' \underline{a}(j) && (i \neq j) \\
&= \emptyset \oplus' x^{(j-1)} \oplus' x^{(i-j)} && (i \neq j) \\
&= \begin{cases} x^{(i-j)} & \text{if } i < j \\ x^{(j-i)} & \text{if } i > j \end{cases}
\end{aligned} \tag{16-15}$$

Hence $\{\underline{\mathcal{A}}\}_{ij} < \emptyset$ for $i \neq j$ and $\{\mathcal{A}\}_{ii} = -\infty$ by definition.

For given j ($1 \leq j \leq k$):

$$\{\underline{A} \otimes \underline{\mathcal{A}}\}_{1j} = \sum_{i=1}^{k}{}_{\oplus} (\{\underline{A}\}_{1i} \otimes \{\underline{\mathcal{A}}\}_{ij})$$

$$= \sum_{i=1}^{k}{}_{\oplus} \{\underline{\mathcal{A}}\}_{ij} \text{ (since } \{\underline{A}\}_{1i} = \emptyset \text{ (i=1,..., k))}$$

$$= -\infty \oplus \sum_{\substack{i=1 \\ i \neq j}}^{k}{}_{\oplus} \{\underline{\mathcal{A}}\}_{ij}$$

$$= \sum_{i<j}{}_{\oplus} \{\underline{\mathcal{A}}\}_{ij} \oplus \sum_{i>j}{}_{\oplus} \{\underline{\mathcal{A}}\}_{ij} \qquad (16\text{-}16)$$

(where one of the summations may be vacuous).

Using (16-13), (16-15) and (16-16):

$\{\underline{A} \otimes \underline{\mathcal{A}}\}_{1j} = x^{-1} < \emptyset$ And this holds for arbitrary j.

Hence the first element (x^{-1}) in each column of $\underline{A} \otimes \underline{\mathcal{A}}$ is different from the first element $(\emptyset)$ in the corresponding 3-tuple in (16-14). Hence by Theorem 16-2, no one of these 3-tuples is linearly dependent on the others.

For m > 3, the result follows by extending the 3-tuples in any arbitrary way by (m-3) extra rows. ●

The assumption that $E_1 \neq ③$ is necessary in the above theorem in order that we can find $\emptyset < x < +\infty$, (so that $x^{-1}, x^{-2}, \ldots$ are all different). But even if $E_1 = ③$ we can produce a dimensional anomaly by using infinite elements, as the following theorem shows.

Theorem 16-5. Suppose $E_1 = ③$. Let m > 2. Then we can always find (at least) (m^2-m) m-tuples, no one of which is linearly dependent on the others.

Proof. Consider the set S of (m^2-m) m-tuples each of which has exactly one component equal to $+\infty$, exactly one component equal to $-\infty$, and all other components equal to $\emptyset$.

Suppose $\underline{b}, \underline{a}(1), \ldots, \underline{a}(k) \in S$ are all different and that, if possible:

$$\underline{b} = \sum_{r=1}^{k}{}_{\oplus} \underline{a}(r) \otimes \lambda_r \text{ with } \lambda_r \in E_1 \quad (r=1,\ldots, k) \qquad (16\text{-}17)$$

In the following we make free use of Proposition 4-3.

Since $\underline{b}$ has at least one finite component, not all λ_r (r=1,..., k) can be $-\infty$. Evidently we may suppose that from (16-17) we have dropped all terms in which λ_r is $-\infty$, so $\lambda_r = \emptyset$ or $\lambda_r = +\infty$ in (16-17). But if some $\lambda_r = +\infty$ then that $\underline{a}(r) \otimes \lambda_r$ has (since $\underline{a}(r) \in S$), (m-1) components $+\infty$; but $\underline{b} \geqslant \underline{a}(r) \otimes \lambda_r$ from (16-17), whence $\underline{b}$ must have at least (m-1) components $+\infty$, contradicting $\underline{b} \in S$. Hence every $\lambda_r = \emptyset$ in (16-17), whence:

$$\underline{b} = \sum_{r=1}^{k}{}_{\oplus} \underline{a}(r)$$

Using Proposition 4-3 it is now clear that $\underline{a}(1), \ldots, \underline{a}(k)$ all have their $-\infty$ in the same coordinate position as $\underline{b}$ does, and that one of $\underline{a}(1), \ldots, \underline{a}(k)$, (say $\underline{a}(j)$) has its $+\infty$ in the same position as $\underline{b}$ does. But this means that $\underline{a}(j) = \underline{b}$ contradicting the assumption that $\underline{b}$ was different from $\underline{a}(1), \ldots, \underline{a}(k)$. ●

Hence no linear dependence can exist.

Since every blog E_1 contains ③ isomorphically, it is evident that the dimensional anomaly in Theorem 16-5 can be extended to every blog.

Theorems 16-4 and 16-5 both assume m > 2. The following result shows that the situation when m = 2 is much more intuitive, at least when E_1 is linear.

Theorem 16-6. Suppose E_1 is a linear blog. Let there be given a set S containing $k > 2$ finite 2-tuples. Then we can find $\underline{a}, \underline{b} \in S$ such that each 2-tuple in S is linearly dependent on $\underline{a}, \underline{b}$.

Proof. Let the 2-tuples in S be:

$$\begin{bmatrix} x_{1r} \\ x_{2r} \end{bmatrix} \quad (r = 1, \ldots, k)$$

Since E_1 is linear we can choose to number these 2-tuples in such a way that:

$$x_{11} \otimes x_{21}^{-1} \leq x_{12} \otimes x_{22}^{-1} \leq \ldots\ldots \leq x_{1k} \otimes x_{2k}^{-1} \qquad (16\text{-}18)$$

Define $\underline{A} = \begin{bmatrix} x_{11} & x_{1k} \\ x_{21} & x_{2k} \end{bmatrix}$ and consider for r=1,..., k:

$$\underline{A} \otimes (\underline{A}^* \otimes' \begin{bmatrix} x_{1r} \\ x_{2r} \end{bmatrix}) = \underline{A} \otimes \begin{bmatrix} (x_{11}^{-1} \otimes x_{1r}) \oplus' (x_{21}^{-1} \otimes x_{2r}) \\ (x_{1k}^{-1} \otimes x_{1r}) \oplus' (x_{2k}^{-1} \otimes x_{2r}) \end{bmatrix}$$

$$= A \otimes \begin{bmatrix} x_{21}^{-1} \otimes x_{2r} \\ x_{1k}^{-1} \otimes x_{1r} \end{bmatrix} \quad \text{(using (16-18) and (2-9))}$$

$$= \begin{bmatrix} (x_{11} \otimes x_{21}^{-1} \otimes x_{2r}) \oplus x_{1r}) \\ x_{2r} \oplus (x_{2k} \otimes x_{1k}^{-1} \otimes x_{1r}) \end{bmatrix}$$

$$= \begin{bmatrix} x_{1r} \\ x_{2r} \end{bmatrix} \quad \text{(using (16-18) and (2-9))}$$

Hence, the equation:

$$\underline{A} \otimes \underline{z} = \begin{bmatrix} x_{1r} \\ x_{2r} \end{bmatrix}$$

is soluble for each r=1,..., k, or in other words, each $\begin{bmatrix} x_{1r} \\ x_{2r} \end{bmatrix}$ is linearly dependent on

$$\begin{bmatrix} x_{11} \\ x_{21} \end{bmatrix}, \quad \begin{bmatrix} x_{1k} \\ x_{2k} \end{bmatrix}$$

Corollary 16-7. Suppose E_1 is a linear blog. Let there be given a set S containing $k > 2$ finite 2-tuples. Then we can find $\underline{a}, \underline{b} \in S$ such that each 2-tuple in S is both linearly dependent and dually (right) linearly dependent on $\underline{a}, \underline{b}$. Moreover, if E_1 is commutative, and in particular if E_1 receives the principal interpretation, then each 2-tuple in S is also left linearly dependent and dually left linearly dependent on $\underline{a}, \underline{b}$.

Proof. Follows directly from Theorems 16-1 and 16-6, and the property of commutativity. ●

16-4. Strong Linear Independence. In conventional linear algebra, a number of different, but logically equivalent, definitions are possible of the notion of linear independence of a set of elements of a vector space. However, in [47] , we have formulated for minimax algebra analogous definitions of various alternative forms of linear independence of elements of a band-space, and shown that they are not logically equivalent, although certain logical implications may be demonstrated among them.

It is clear, then, that we must take care what definition we employ if we hope to develop a theory of rank and dimension applicable to the principal interpretation. For example, the results of the previous section show that the apparently logical step of defining linear independence as the mere negation of linear dependence does not lead to a satisfactory theory of dimension. These considerations motivate the following definition.

Let E_1 be a blog and let $\underline{a}(1),\ldots, \underline{a}(k) \in E_n$ $(k \geq 1)$. We shall say that $\underline{a}(1),\ldots, \underline{a}(k)$ are strongly linearly independent if there is at least one finite n-tuple $\underline{b} \in E_n$ which has a unique expression in the form:

$$\underline{b} = (\sum_{r=1}^{t}{}_{\oplus}\, \underline{a}(j_r) \otimes \lambda_{j_r}) \tag{16-19}$$

with $\lambda_{j_r} \in E_1$ $(r=1,\ldots, t)$, $1 \leq j_r \leq k$ $(r=1,\ldots, t)$

and $j_r < j_s$ if $r < s$. $(r=1,\ldots, t;\ s=1,\ldots, t)$

We shall abbreviate "strongly linearly independent" to SLI.

Lemma 16-8. Let E_1 be a blog with group G. Let $\underline{a}(1),\ldots, a(k), \underline{b} \in E_n$ $(k \geq 1)$ be such that $\underline{b}$ is finite and has a unique expression of the form (16-19). Then $t=k$; $j_1=1,\ldots, j_t=k$; $\lambda_{j_r} \in G$ $(r=1,\ldots, t)$; and $\underline{A}$ is doubly G-astic, where $\underline{A} \in \mathcal{M}_{nk}$ is the matrix whose columns are $\underline{a}(1),\ldots, \underline{a}(t)$ in that order.

Proof. If there is an index u $(1 \leq u \leq k)$ such that u does not occur in the list $j_1,\ldots, j_t$ then since

$$\underline{b} = (\sum_{r=1}^{t}{}_{\oplus}\, \underline{a}(j_r) \otimes \lambda_{j_r}) \oplus (\underline{a}(u) \otimes -\infty), \tag{16-20}$$

we may rearrange (16-20) to give another expression for $\underline{b}$ in the form (16-19), contradicting the uniqueness of the given form. Hence t=k, and $j_1=1,\ldots, j_t=k$. Moreover no λ_{j_r} can be $-\infty$ in (16-19), for if k=1 this would contradict the finiteness of $\underline{b}$, and for k>1 we could drop the particular term $\underline{a}(j_r) \otimes \lambda_{j_r}$ from (16-19), to give another such expression for $\underline{b}$, contradicting uniqueness. Hence no component of any of $\underline{a}(1),\ldots, \underline{a}(k)$ is $+\infty$, otherwise the same component of $\underline{b}$ would be $+\infty$.

Again, none of $\underline{a}(1),\ldots, \underline{a}(k)$ has all its components $-\infty$, for if k=1 this would contradict the finiteness of b, and for k>1 we could drop that term from (16-19) to give another expression for $\underline{b}$, contradicting uniqueness. And no one component can be $-\infty$ in all of $\underline{a}(1),\ldots, \underline{a}(k)$ else the same component would be $-\infty$ in $\underline{b}$, contradicting the finiteness of $\underline{b}$.

Hence $\underline{A}$ is doubly G-astic, and so $\underline{A}^* \otimes' \underline{b}$ is finite. Hence by Theorem 14-3, each $\lambda_{j_r} < +\infty$ in (16-19) and so each λ_{j_r} is finite.

The following is a simple, but useful, reformulation of some of the information in Lemma 16-8 :

Corollary 16-9. Let E_1 be a blog and let $\underline{a}(1),\ldots, \underline{a}(n) \in E_m$ for given integers $m,n\geq 1$. Then $\underline{a}(1),\ldots, \underline{a}(n)$ are SLI if and only if there exists a finite m-tuple $\underline{b} \in E_m$ such that the equation $\underline{A} \otimes \underline{x} = \underline{b}$ is uniquely soluble, where $\underline{A} \in \mathcal{M}_{mn}$ is the matrix whose columns are $\underline{a}(1),\ldots, \underline{a}(n)$ in that order.

Proof. Obviously, each solution of the equation $\underline{A}\otimes\underline{x}=\underline{b}$ furnishes an expression of the form (16-19). Conversely, each expression of the form (16-19) furnishes, as in (16-20), a similar expression with t=n, which in turn may be regarded as furnishing a solution to the equation $\underline{A}\otimes\underline{x}=\underline{b}$. Hence the uniqueness of the one implies the uniqueness of the other.

For a given belt E_1, define linear independence as the mere negation of linear dependence: $\underline{a}(1),\ldots, \underline{a}(n) \in E_m$ are linearly independent exactly when no one of them is linearly dependent on the others. It is natural to enquire how linear independence relates to strong linear independence.

Theorem 16-10. Let E_1 be a blog and $\underline{a}(1),\ldots, \underline{a}(k) \in E_n$. For $\underline{a}(1),\ldots, \underline{a}(k)$ to be linearly independent it is sufficient, but not necessary, that $\underline{a}(1),\ldots, \underline{a}(k)$ be SLI.

Proof. If $\underline{a}(1),\ldots, \underline{a}(k)$ are SLI then there is a finite $\underline{b} \in E_n$ with a unique expression (16-19) in which, according to Lemma 16-8, all of $\underline{a}(1),\ldots, \underline{a}(k)$ occur. So if one of $\underline{a}(1),\ldots, \underline{a}(k)$ were linearly dependent on the others it could be substituted out of (16-19) to give another such expression for $\underline{b}$, contradicting uniqueness.

However, for any linear blog E_1 consider $\underline{a}(1), \underline{a}(2), \underline{a}(3) \in E_3$ where

$$\underline{a}(1) = \begin{bmatrix} -\infty \\ \emptyset \\ \emptyset \end{bmatrix} ; \quad \underline{a}(2) = \begin{bmatrix} \emptyset \\ -\infty \\ \emptyset \end{bmatrix} ; \quad \underline{a}(3) = \begin{bmatrix} \emptyset \\ \emptyset \\ -\infty \end{bmatrix}$$

On the one hand, the $\mathcal{A}$ test shows (Theorem 16-2) that $\underline{a}(1)$, $\underline{a}(2)$, $\underline{a}(3)$ are linearly independent. On the other hand, for any finite $\underline{b} \in E_3$ define the matrix $\underline{C} = (\{\underline{b}\}_i)^{-1} \{\underline{a}(j)\}_i$, i.e.:

$$\underline{C} = \begin{bmatrix} -\infty & (\{\underline{b}\}_1)^{-1} & (\{\underline{b}\}_1)^{-1} \\ (\{\underline{b}\}_2)^{-1} & -\infty & (\{\underline{b}\}_2)^{-1} \\ (\{\underline{b}\}_3)^{-1} & (\{\underline{b}\}_3)^{-1} & -\infty \end{bmatrix}$$

Obviously, $\underline{C}$ has the property that each column-maximum occurs twice on some row showing by Theorem 15-7 that the equation $\underline{A} \otimes \underline{x} = b$ is not uniquely soluble, where $\underline{A}$ is the matrix whose columns are $\underline{a}(1)$, $\underline{a}(2)$, $\underline{a}(3)$. Hence by Corollary 16- 9, $\underline{a}(1)$, $\underline{a}(2)$, $\underline{a}(3)$ are not SLI.

We conclude this chapter with some definitions giving duals and analogues of the concept SLI:

When we wish to emphasize that the coefficients λ multiply from the right in (16-19) we shall say <u>right SLI</u> rather than merely SLI; and by obvious analogy we may define the concept <u>left SLI</u>.

If we replace the displayed formula in (16-19) by:

$$\underline{b} = \left(\sum_{r=1}^{t} {}_{\oplus'} a(j_r) \otimes' \lambda_{j_r} \right)$$

then the definition, so modified, will be taken as that of the concept <u>right dual SLI</u>, with an obvious analogous definition for <u>left dual SLI</u>.

17. RANK OF MATRICES

17-1. Regular Matrices: Let E_1 be a blog and $\underline{A} \varepsilon \mathcal{M}_{nn}$ a square matrix. We say that $\underline{A}$ is right or left column-regular if the columns of $\underline{A}$ are right or left SLI respectively. And we make corresponding dual definitions in the obvious way.

In conventional linear algebra, we are interested in dependencies among the rows of matrices as well as among the columns. Accordingly, let us say that $\underline{A} \varepsilon \mathcal{M}_{nn}$ is right or left row-regular if the transposed matrix $\underline{A}'$ is right or left column-regular respectively. And we make corresponding dual definitions in the obvious way.

By analogy with Corollary 16-9 , we derive the following:

Proposition 17-1. Let E_1 be a blog and let $\underline{A} \varepsilon \mathcal{M}_{nn}$ for given integer $n \geq 1$. Then A is left row-regular if and only if there exists a finite n-tuple $\underline{c} \varepsilon \mathcal{M}_{1n}$ such that the equation $\underline{y} \otimes \underline{A} = \underline{c}$ is uniquely soluble. ●

Theorem 17-2. Let E_1 be a linear blog and let $\underline{A} \varepsilon \mathcal{M}_{nn}$ for given integer $n \geq 1$. Then the following conditions are equivalent:

(i) $\underline{A}$ is right column-regular

(ii) $\underline{A}$ is left row-regular

Similarly, the following conditions are equivalent:

(iii) $\underline{A}$ is right row-regular

(iv) $\underline{A}$ is left column-regular

If any one of these conditions holds then $\underline{A}$ is doubly G-astic (where G is the group of E_1). And if E_1 is commutative then all four conditions are equivalent.

Proof. Suppose (i) holds. Then by Corollary 16-9 there exists a finite n-tuple $\underline{b} \varepsilon E_n$ such that the equation $\underline{A} \otimes \underline{x} = \underline{b}$ is uniquely soluble. And by Lemma 16-10, $\underline{A}$ is doubly G-astic. Let $\underline{c}^* \varepsilon \mathcal{M}_{n1}$ be the unique solution of the equation $\underline{A} \otimes \underline{x} = \underline{b}$. So $\underline{c} \varepsilon \mathcal{M}_{1n}$ is finite, and according to Theorem 15-12 the equation $\underline{y} \otimes \underline{A} = \underline{c}$ has the unique solution $\underline{y} = \underline{b}^*$. Hence by Proposition 17-1, $\underline{A}$ is left row-regular, so (ii) holds.

The converse follows similarly. The equivalent of (iii) and (iv) follows from the equivalence of (i) and (ii) for the transposed $\underline{A}'$ of $\underline{A}$. The equivalence of all conditions when E_1 is commutative is then trivial. ●

By dual arguments, we can prove:

Proposition 17-3. Let E_1 be a linear blog and let $\underline{A} \varepsilon \mathcal{M}_{nn}$. The the following conditions are equivalent:

(i) $\underline{A}$ is dually right column-regular

(ii) $\underline{A}$ is dually left row-regular

Similarly, the following conditions are equivalent:

(iii) $\underline{A}$ is dually right row-regular

(iv) $\underline{A}$ is dually left column-regular

If E_1 is commutative then all four conditions are equivalent.

When E_1 is a linear blog, we shall use the single terms regular to cover (i) and (ii) of Theorem 17-2 and dually regular to cover (i) and (ii) of Proposition 17-3.

Theorem 17-4. Let E_1 be a linear blog and let $\underline{A} \varepsilon \mathcal{M}_{nn}$ for given integer $n \geq 1$. Then $\underline{A}$ is regular if and only if $\underline{A}^*$ is dually regular.

Proof. The result follows directly from the definitions together with Corollary 15-12. ●

17-2. Matrix Rank Over A Linear Blog

Let E_1 be any blog and let $\underline{A} \varepsilon \mathcal{M}_{mn}$. Suppose that we can find r columns $(1 \leq r \leq n)$ of $\underline{A}$, but no more, which are SLI. Then we shall say that $\underline{A}$ has right column-rank equal to r. The epithet right will be dropped when it is not needed for emphasis. Similarly we may define left column-rank. We define the right row-rank and left row-rank of $\underline{A}$ as the right column-rank and left column-rank respectively of $\underline{A}'$, the transposed of $\underline{A}$. Finally, we make dual definitions of all these ranks in the obvious way.

Before proving relationships among these ranks, we need one more definition. Let us say that a given matrix $\underline{A} \varepsilon \mathcal{M}_{mn}$ has ϕ-astic rank equal to r (integral) if the following is true for k=r but not for k>r:

There are $\underline{x} \varepsilon E_n$ and $\underline{y} \varepsilon E_m$, both finite, such that $\underline{B} \varepsilon \mathcal{M}_{mn}$ is doubly ϕ-astic and contains a $(k \times k)$ strictly doubly ϕ-astic submatrix, where

$$\{\underline{B}\}_{ij} = \{\underline{y}\}_i \otimes \{\underline{A}\}_{ij} \otimes \{\underline{x}\}_j \qquad (i=1,\ldots, m;\ j=1,\ldots, n) \tag{17-1}$$

Lemma 17-5. Let E_1 be a linear blog with group G, and suppose that $\underline{A} \varepsilon \mathcal{M}_{mn}$ has ϕ-astic rank equal to r. Then $\underline{A}$ is doubly G-astic and contains a set of (at least) r columns which are SLI.

Proof. If $\underline{A}$ had any row or column consisting only of $-\infty$'s, or containing $+\infty$, then, from (17-1), so would $\underline{B}$. But $\underline{B}$ is doubly ϕ-astic. So $\underline{A}$ is doubly G-astic (by Lemma 12-1).

Now suppose without loss of generality that $\underline{B}$ contains an $(r \times r)$ strictly doubly ϕ-astic submatrix in its first r columns. Let $\underline{D} \varepsilon \mathcal{M}_{mr}$ consist of the first r columns of $\underline{A}$. Arguing as above, it is clear that $\underline{D}$ is doubly G-astic. Applying Theorem 15-9 and Corollary 16-9 (with $\underline{D}$ in the role of $\underline{A}$) we infer that the columns of $\underline{D}$ are SLI. ●

Lemma 17-6. Let E_1 be a linear blog with group G, and suppose that $\underline{A} \varepsilon \mathcal{M}_{mn}$ is doubly G-astic and contains a set of r columns which are SLI. Then $\underline{A}$ has ϕ-astic rank equal to (at least) r.

Proof. Let $\underline{D} \varepsilon \mathcal{M}_{mr}$ have as its columns the first r columns of $\underline{A}$, assumed (without loss of generality) to be SLI. By Lemma 16-8, $\underline{D}$ is doubly G-astic. Applying

Theorem 15-9 and Corollary 16-9 (with $\underline{D}$ in the role of $\underline{A}$) we infer that there exist finite elements $x_1,\ldots, x_r$ and $y_1,\ldots, y_m$ such that $\underline{C} \in \mathcal{M}_{mr}$ is doubly $\emptyset$-astic and contains an $(r\times r)$ strictly doubly $\emptyset$-astic submatrix, where:

$$\{\underline{C}\}_{ij} = y_i \otimes \{\underline{A}\}_{ij} \otimes x_j \quad (i=1,\ldots, m;\ j=1,\ldots, r) \tag{17-2}$$

If $r=n$ then we have satisfied (17-1) with $\underline{C}$ in the role of $\underline{B}$; hence $\underline{A}$ has $\emptyset$-astic rank at least equal to r in this case.

Suppose $r < n$. Since $\underline{A}$ is doubly G-astic, the sum

$$\sum_{i=1}^{m}{}_{\oplus}(y_i \otimes \{\underline{A}\}_{ij})$$

is finite for each $j=1,\ldots, n$, so we may define:

$$x_j = \Big(\sum_{i=1}^{m}{}_{\oplus} (y_i \otimes \{\underline{A}\}_{ij}\Big)^{-1} \quad (j=r+1,\ldots, n) \tag{17-3}$$

$$\text{i.e. } x_j^{-1} = \sum_{i=1}^{m}{}_{\oplus}(y_i \otimes \{\underline{A}\}_{ij}) \quad (j=r+1,\ldots, n) \tag{17-4}$$

Since E_1 is linear, the maximum is attained for some i_j $(1\leq i_j\leq m)$ for each $j=r+1,\ldots, n$ on the right-hand of (17-4) and we may write:

$$x_j^{-1} = y_{i_j} \otimes \{\underline{A}\}_{i_j j} \geq y_i \otimes \{\underline{A}\}_{ij} \quad (i=1,\ldots, m;\ j=r+1,\ldots, n) \text{ or equivalently:}$$

$$\emptyset = y_{i_j} \otimes \{\underline{A}\}_{i_j j} \otimes x_j \geq y_i \otimes \{\underline{A}\}_{ij} \otimes x_j \quad (i=1,\ldots, m,\ j=r+1,\ldots, n) \tag{17-5}$$

Now define $\underline{B}$ as in (17-1) using $x_1,\ldots, x_n$ and $y_1,\ldots, y_m$ as in (17-2) and (17-3). Then the first r columns of $\underline{B}$ (which constitute the matrix $\underline{C}$) are $\emptyset$-astic by (17-2) and the remaining columns of $\underline{B}$ are $\emptyset$-astic by (17-5). Moreover the rows of $\underline{B}$ are $\emptyset$-astic because they are obtained by adjoining $\emptyset$-astic columns to the row-$\emptyset$-astic matrix $\underline{C}$. Hence $\underline{B}$ is doubly $\emptyset$-astic and contains by (17-2) a strictly doubly $\emptyset$-astic $(r\times r)$ submatrix. Hence $\underline{A}$ has $\emptyset$-astic rank at least equal to r. ●

<u>Theorem 17-7. Let E_1 be a linear blog with group G and let $\underline{A} \in \mathcal{M}_{mn}$ be doubly G-astic. Then the following statements are all equivalent:</u>

(i) <u>$\underline{A}$ has $\emptyset$-astic rank equal to r</u>

(ii) <u>$\underline{A}$ has right column-rank equal to r</u>

(iii) <u>$\underline{A}$ has left row-rank equal to r</u>

(iv) <u>$\underline{A}^*$ has dual right column-rank equal to r</u>

(v) <u>$\underline{A}^*$ has dual left row-rank equal to r</u>

<u>Proof</u>. The equivalence of (i) and (ii) evidently follows from Lemmas 17-5 and 17-6. The equivalence of (i) and (iii) follows by analogous reasoning, noting the row-column symmetry of condition (17-1). And (iv) and (v) are duals of (iii) and (ii) respectively. ●

In view of Theorem 17-7, we may (for doubly G-astic $\underline{A}$) simply use the expression rank of $\underline{A}$ for any of the ranks (i) to (iii) and dual rank of $\underline{A}$ for the corresponding dual quantities. Thus Theorem 17-7 asserts the equality of the rank of $\underline{A}$ with the dual rank of $\underline{A}^*$.

Corollary 17-8. Let E_1 be a linear blog with group G and let $\underline{A} \in \mathcal{M}_{mn}$ be doubly G-astic. Then the following statements are all equivalent:

(i) $\underline{A}$ has left column-rank equal to r

(ii) $\underline{A}$ has right row-rank equal to r

(iii) $\underline{A}^*$ has dual left column-rank equal to r

(iv) $\underline{A}^*$ has dual right row-rank equal to r

Moreover, if E_1 is commutative then statements (i) to (iv) are all true if and only if $\underline{A}$ has rank r.

Proof. The equivalence of (i) to (iv) follows from Theorem 17-7 applied to $\underline{A}'$, the transposed of $\underline{A}$. Moreover, if E_1 is commutative then for given $x_1, \ldots, x_n$ and $y_1, \ldots, y_m$ we have:

$$y_i \otimes \{\underline{A}\}_{ij} \otimes x_j = x_j \otimes \{\underline{A}\}_{ij} \otimes y_i$$

$$= x_j \otimes \{\underline{A}'\}_{ji} \otimes y_i \qquad (i=1,\ldots, m;\ j=1,\ldots, n)$$

It then follows from (17-1) that $\underline{A}$ has $\emptyset$-astic rank r if and only if $\underline{A}'$ has $\emptyset$-astic rank r, and all ranks become equal. ●

17-3. Existence of Rank. In the foregoing results, we have demonstrated the equality of various ranks of a matrix, if they exist. We have not yet discussed whether a matrix necessarily has such ranks. This we now do.

Theorem 17-9. Let E_1 be a linear blog with group G and let $\underline{A} \in \mathcal{M}_{mn}$. Then there is an integer r such that $\underline{A}$ has $\emptyset$-astic rank r, if and only if $\underline{A}$ is doubly G-astic. And r is then an integer satisfying $1 \leq r \leq \min(m,n)$.

Proof. If $\underline{A}$ has $\emptyset$-astic rank r, then $\underline{A}$ is doubly G-astic by Lemma 17-5.

Conversely, let $\underline{A}$ be doubly G-astic. Define an integer s $(1 \leq s \leq n)$ as follows: s is the least integer such that some set of s columns of $\underline{A}$ constitute a doubly G-astic $(m \times s)$ submatrix of $\underline{A}$.

If s=1 then some column of $\underline{A}$, say $\underline{a}(1)$, is finite. The equation $\underline{a}(1) \otimes x = \underline{a}(1)$ then clearly has the unique solution $x=\emptyset$. Hence $\underline{A}$ has column-rank at least equal to one, and so has $\emptyset$-astic rank at least equal to one. If s>1 then some s columns of $\underline{A}$, say $\underline{a}(1), \ldots, \underline{a}(s)$, have the property that they, but no proper subset, constitute a doubly G-astic (mxs) matrix. Let $\underline{C}$ be the matrix $[\underline{a}(1), \ldots, \underline{a}(s)]$.

On removal of any one column, $\underline{C}$ obviously remains column-G-astic and hence must cease to be row-G-astic. Evidently this implies that for each column-number j ($j=1,\ldots,s$) there is a row i_j such that $\{\underline{C}\}_{i_j j}$ is finite but all other elements of row i_j are $-\infty$. Without loss of generality we can assume $i_j=j$ ($j=1,\ldots,s$). Then $\underline{C}$ may be partitioned as:

$$\underline{C} = \left[\begin{array}{c} g_1 \quad\quad -\infty \\ \ddots \\ -\infty \quad\quad g_s \\ \hline \\ \underline{D} \end{array}\right] \tag{17-6}$$

In (17-6), the first s rows of $\underline{C}$ constitute an ($s\times s$) "diagonal matrix" having finite elements $g_1,\ldots,g_s$ on the main diagonal and $-\infty$ elsewhere; and $\underline{D}$ is evidently row-G-astic.

Define $\underline{\xi} \in E_s$ by $\{\underline{\xi}\}_j = g_j^{-1}$ ($j=1,\ldots,s$). So $\underline{\xi}$ is finite, and so is $\underline{D} \otimes \underline{\xi}$ since $\underline{D}$ is row-G-astic (Section 12-2). The equation:

$$\underline{C} \otimes \underline{x} = \left[\begin{array}{c} \emptyset \\ \vdots \\ \emptyset \\ \hline \underline{D}\otimes\underline{\xi} \end{array}\right] \tag{17-7}$$

clearly has the solution $\underline{x} = \underline{\xi}$; we assert that this is the only solution.

For any other solution $\underline{y}$ of (17-7) must differ from $\underline{x}$ in at least one coordinate position: $\{\underline{y}\}_j \neq g_j^{-1}$ for some j ($1\leq j\leq s$). Since E_1 is linear, either $\{\underline{y}\}_j > g_j^{-1}$ or $\{\underline{y}\}_j < g_j^{-1}$. But then $\{\underline{C} \otimes \underline{y}\}_j > \emptyset$ or $\{\underline{C} \otimes \underline{y}\}_j < \emptyset$ respectively, and $\underline{y}$ is not a solution to 17-7.

Hence (17-7) has a unique solution, implying by Corollary 16-9 that the columns of $\underline{C}$ are SLI. So $\underline{A}$ has column-rank at least equal to s, and so has $\emptyset$-astic rank at least equal to s.

Hence each doubly G-astic matrix $\underline{A} \in \mathcal{M}_{mn}$ has $\emptyset$-astic rank $r\geq 1$; and since r must, by Theorem 17-7, represent a possible number of rows and a possible number of columns of $\underline{A}$ we infer that $r\leq \min(m,n)$. ●

We are now in a position to show that the dimensional anomalies which, according to Theorems 16-4 and 16-5, arise in relation to linear dependence, are avoided in relation to strong linear dependence.

<u>Theorem 17-10. Let E_1 be a linear blog, and $n\geq 1$ a given integer. Then for each integer m ($1\leq m\leq n$) we can find m n-tuples $\underline{a}(j) \in E_n$ ($j=1,\ldots,m$) which are SLI; but this is impossible for $m>n$.</u>

<u>Proof</u>. For $m\leq n$, define a matrix $\underline{A} \in \mathcal{M}_{mn}$ whose first m rows constitute the "unit matrix" I_m having diagonal elements equal to $\emptyset$ and off-diagonal elements equal to $-\infty$; if $m<n$, let all other elements of $\underline{A}$ be equal to $\emptyset$.

Evidently $\underline{A}$ is doubly G-astic and no columns can be removed from $\underline{A}$ without $\underline{A}$ ceasing to be row-G-astic. Then by the proof of Theorem 17-9, $\underline{A}$ has $\emptyset$-astic rank r with $m \leq r \leq \min(m,n)$ - i.e. $\underline{A}$ has rank m. Hence the columns of $\underline{A}$ furnish m SLI n-tuples.

On the other hand if given n-tuples $\underline{a}(j) \in E_n$ $(j=1,\ldots, m)$ are SLI then the matrix $\underline{A} = [\underline{a}(1),\ldots, \underline{a}(m)]$ has column-rank r=m, and so has $\emptyset$-astic rank r=m. But by Theorem 17-9, $r \leq n$. Hence $m \leq n$. ●

18. SEMINORMS ON E_n

18-1 . Column-Spaces. In previous chapters, we have regarded the relation:

$$\underline{A} \otimes \underline{x} = \underline{b};\ \underline{A} \in \mathcal{M}_{mn};\ \underline{x} \in E_n;\ \underline{b} \in E_m \tag{18-1}$$

as embodying firstly a set of simultaneous equations and secondly a linear dependence among m-tuples. In the present chapter we shall examine (18-1) in yet another light and regard it as expressing the fact that m-tuple $\underline{b}$ belong to a certain space generated by the columns of matrix $\underline{A}$.

Accordingly, for a given belt E_1, let $\underline{A} \in \mathcal{M}_{mn}$ be a given matrix. By the right column-space of $\underline{A}$ we denote the set of all $\underline{b} \in E_m$ for which (18-1) is soluble for $\underline{x}$. And if $\underline{a}(1),\ldots, \underline{a}(n) \in E_m$ are given m-tuples, the notation $\langle \underline{a}(1),\ldots, \underline{a}(n)\rangle$ will denote the right column-space of the matrix $[\underline{a}(1),\ldots, \underline{a}(n)] \in \mathcal{M}_{mn}$ whose columns are $\underline{a}(1),\ldots, \underline{a}(n)$; and this column-space will be called the right space generated by $\underline{a}(1),\ldots, \underline{a}(n)$.

In a manner by now familiar, we can define the left row-space of $\underline{A}$ by considering the relation $\underline{y}\otimes\underline{A}=\underline{c}$ in place of (18-1), and then define the left column-space and right row-space of $\underline{A}$ by reference to the transposed $\underline{A}'$ of $\underline{A}$. Dual definitions of all these concepts are forthcoming by use of dual multiplication in the obvious way and lead to definitions of left or right dual space generated by given m-tuples $\underline{a}(1),\ldots,$ $\underline{a}(n)$. As usual, we shall drop the epithet "right" in most contexts when it will cause no confusion, and speak of the column-space of $\underline{A}$ and the space generated by $\underline{a}(1),\ldots, \underline{a}(n)$.

By making use of the following relations for all $\underline{A} \in \mathcal{M}_{mn}$; $\underline{x},\underline{y} \in E_n$; $\lambda \in E_1$:

$$\underline{A}\otimes(\underline{x}\oplus\underline{y}) = (\underline{A}\otimes\underline{x}) \oplus (\underline{A}\otimes\underline{y})$$

$$A\otimes(x\otimes\lambda) = (A\otimes x)\otimes\lambda$$

we infer:

Proposition 18-1. If E_1 is a given belt then the right column-space of any matrix is a right band-space over E_1. ●

Evidently, we may interpret the diagrams of Figs 8-3 and 8-4 as presenting systems of istone and antitone bijections among the various row and column-spaces of a given matrix $\underline{A}$ and its conjugate $\underline{A}^*$.

Now suppose E_1 is a belt with identity $\emptyset$ (axiom X_8), and let $\delta \in E_1$ satisfy $\delta \leq \emptyset$. We define the matrix $\underline{I}_n(\delta) \in \mathcal{M}_{nn}$:

$$\left.\begin{aligned} \{\underline{I}_n(\delta)\}_{ij} &= \emptyset \text{ if } i=j \\ &= \delta \text{ otherwise} \end{aligned}\right\} \tag{18-2}$$

The column-space of $I_n(\delta)$ we shall call $E_n(\delta)$. Referring back to Theorem 5-8 and the first paragraph of its proof, we infer the following result.

Proposition 18-2. Let E_1 be a belt. If, for $n>1$, $\mathcal{M}_{nn}$ satisfies axiom X_8, then E_1 satisfies axiom X_{10}. The identity element of $\mathcal{M}_{mm}$ is then $I_m(\theta)$, where θ is the (unique) null element of E_1, and $E_m(\theta) = E_m$, for all integers $m \geq 1$. ●

18-2. Seminorms. Let V, E_1 be given commutative bands. A mapping $\tau \varepsilon (E_1)^V$ is said to be sublinear if:

$$\tau(x \oplus y) \leqslant \tau(x) \oplus \tau(y) \qquad (\forall x, y \varepsilon V) \qquad (18\text{-}3)$$

The following results are readily established by the methods of Chapter 2.

Proposition 18-3. Let V, E_1 be given commutative bands. Then the sublinear mappings $\tau \varepsilon (E_1)^V$ form a commutative sub-band of the commutative band $(E_1)^V$. A mapping $\tau \varepsilon (E_1)^V$ is a homomorphism if and only if it is isotone and sublinear. ●

Proposition 18-4. Let V, W, E_1 be given commutative bands. For given integer $k \geqslant 1$, let $\tau_1, \ldots, \tau_k \varepsilon (W)^V$ be sublinear and let $\mu \varepsilon (E_1)^{(W)^k}$ be a homomorphism. Then $\mu(\tau_1, \ldots, \tau_k) \varepsilon (E_1)^{(V)^k}$ is sublinear. ●

If V is a right band-space over E_1 then a sublinear mapping $\tau \varepsilon (E_1)^V$ will be called an E_1-seminorm (on V). We consider now some particular seminorms. First, by taking each τ_i in Proposition 18-4 to be the identity mapping i_{E_1}, and μ to be the summation $\sum_{i=1}^{n}{}_{\oplus}$, we derive the maximum-seminorm on E_n:

$$\sum_{i=1}^{n}{}_{\oplus}\{\underline{x}\}_i \qquad (18\text{-}4)$$

Theorem 18-5. If E_1 is a self-conjugate belt then the mapping τ defined by:

$$\tau(x) = x \oplus x^* \qquad (\forall x \varepsilon E_1) \qquad (18\text{-}5)$$

is an E_1-seminorm on E_1.

Proof. For arbitrary $x, y \varepsilon E_1$ we have:

$$x^* \oplus' y^* \leqslant x^* \oplus y^* \qquad (18\text{-}6)$$

Hence
$$\begin{aligned} \tau(x \oplus y) &= (x \oplus y) \oplus (x \oplus y)^* \\ &= (x \oplus y) \oplus (x^* \oplus' y^*) \quad \text{(by axiom } N_2) \\ &\leqslant (x \oplus y) \oplus (x^* \oplus y^*) \quad \text{(by (18-6) and (2-6))} \\ &= (x \oplus x^*) \oplus (y \oplus y^*) \\ &= \tau(x) \oplus \tau(y) \quad \text{(by definition)} \end{aligned}$$ ●

The seminorm defined by (18-4) may be called the absolute value seminorm; for the principal interpretation, indeed, it yields exactly the absolute value $|x|$.
By taking each τ_i in Proposition 18-4 to be the absolute value seminorm on E_1 and μ to be the summation $\sum_{i=1}^{n}{}_{\oplus}$, we derive the absolute maximum seminorm on E_n:

$$\sum_{i=1}^{n}{}_{\oplus}\{\underline{x} \oplus \underline{x}^*\}_i \qquad (18\text{-}7)$$

For the principal interpretation, (18-7) becomes:

$$\max_{i=1,\ldots,n} |\{\underline{x}\}_i| \tag{18-8}$$

(18-9)

which is the familiar "R_n^∞ -norm".

The following theorem enables us to derive a range of different seminorms applicable to the principal interpretation.

<u>Theorem 18-6. Let E_1 be a linear self-conjugate belt. For given integers $s,t,n\geq 1$ let the given integers $i_1,\ldots, i_s, j_1,\ldots, j_t$ satisfy $1\leq i\leq n$, $1\leq j\leq n$. Let $\mu\varepsilon(E_1)^{E_{t+1}}$ be isotone. Then τ is an E_1-seminorm on E_n, where for each $\underline{x}\ \varepsilon\ E_n$:</u>

$$\tau(\underline{x}) = \mu(\{\underline{x}\}_{i_1} \oplus \ldots \oplus \{\underline{x}\}_{i_s}, (\{\underline{x}\}_{j_1})^*,\ldots, (\{\underline{x}\}_{j_t})^*) \tag{18-10}$$

<u>Proof</u>: Let $\underline{x}$, $\underline{y}\ \varepsilon\ E_n$ be arbitrary. Since E_1 is linear, we may assume without loss of generality that

$$\begin{aligned} \{\underline{x}\oplus\underline{y}\}_{i_1} \oplus \ldots \oplus \{\underline{x}\oplus\underline{y}\}_{i_s} &= \{\underline{x}\oplus\underline{y}\}_{i_1} \qquad \text{(the attained maximum)} \\ &= \text{either } \{\underline{x}\}_{i_1} \text{ or } \{\underline{y}\}_{i_1} \end{aligned}$$

Assuming $\{\underline{x}\oplus\underline{y}\}_{i_1} \oplus \ldots \oplus \{\underline{x}\oplus\underline{y}\}_{i_s} = \{\underline{x}\}_{i_1}$, we have:

$$\begin{aligned} \tau(\underline{x}\oplus\underline{y}) &= \mu(\{\underline{x}\oplus\underline{y}\}_{i_1} \oplus \ldots \oplus \{\underline{x}\oplus\underline{y}\}_{i_s}, (\{\underline{x}\oplus\underline{y}\}_{j_1})^*,\ldots, (\{\underline{x}\oplus\underline{y}\}_{j_t})^*) \\ &= \mu(\{\underline{x}\}_{i_1}, (\{\underline{x}\}_{j_1})^* \oplus' (\{\underline{y}\}_{j_1})^*,\ldots, (\{\underline{x}\}_{j_t})^* \oplus' (\{\underline{y}\}_{j_t})^*) \\ &\leq \mu(\{\underline{x}\}_{i_1} \oplus \ldots \oplus \{\underline{x}\}_{i_s}, (\{\underline{x}\}_{j_1})^*,\ldots, (\{x\}_{j_t})^*) \qquad \text{(because } \mu \text{ is isotone)} \end{aligned}$$

Similarly if $\{\underline{x}\oplus\underline{y}\}_{i_1} \oplus \ldots \oplus \{\underline{x}\oplus\underline{y}\}_{i_s} = \{\underline{y}\}_{i_1}$, we obtain

$$\tau(\underline{x}\oplus\underline{y}) \leq \tau(\underline{y})$$

Hence in any case:

$$\tau(\underline{x}\oplus\underline{y}) \leq \tau(\underline{x}) \oplus \tau(\underline{y}).$$

●

Although it would be notationally inconvenient to include them in the statement of Theorem 18-6, we may associate with (18-10) two other norms obtained, as it were, by taking s=o or t=o. The first of these is:

$$\mu((\{\underline{x}\}_{j_1})^*,\ldots, (\{\underline{x}\}_{j_t})^*) \tag{18-11}$$

And the second is

$$\mu(\{\underline{x}\}_{i_1} \oplus \ldots \oplus \{\underline{x}\}_{i_s}) \tag{18-12}$$

A particular interpretation of Theorem 18-6 is afforded by the E_1-seminorms $\delta_{ij}\ \varepsilon\ (E_1)^{E_n}$ defined for $\underline{x}\ \varepsilon\ E_n$ by:

$$\delta_{ij}(\underline{x}) = \{\underline{x}\}_i \otimes (\{\underline{x}\}_j)^* \tag{18-13}$$

with the meaning $\{\underline{x}\}_i - \{\underline{x}\}_j$ for the principal interpretation.

A seminorm which we shall find particularly useful is the range seminorm $\rho \varepsilon (E_1)^{E_n}$ defined by:

$$\rho(\underline{x}) = (\sum_{j=1}^{n}{}_{\oplus} \{\underline{x}\}_j) \otimes (\sum_{j=1}^{n}{}_{\oplus} (\{\underline{x}\}_i)^*) \tag{18-14}$$

which is clearly a particular case of (18-10). Equivalent forms for $\rho(\underline{x})$ are:

$$\rho(\underline{x}) = \sum_{i,j=1}^{n}{}_{\oplus} (\{\underline{x}\}_j \otimes (\{\underline{x}\}_i)^*)$$

$$\text{or} \quad \rho(\underline{x}) = (\sum_{j=1}^{n}{}_{\oplus} \{\underline{x}\}_j) \otimes (\sum_{i=1}^{n}{}_{\oplus'} \{\underline{x}\}_i)^* \tag{18-15}$$

For the principal interpretation, this last expression is just

$$\max_{j=1,\ldots,n} \{\underline{x}\}_j - \min_{i=1,\ldots,n} \{\underline{x}\}_i$$

which motivates the name "range seminorm". The value $\rho(\underline{x})$ will be called the range of $\underline{x}$.

Finally, for the principal interpretation we may define the deviation seminorm:

$$\max_{i=1,\ldots,n} \{\underline{x}\}_i - \frac{1}{n} \sum_{j=1}^{n} \{\underline{x}\}_j \tag{18-16}$$

which measures the excess of the greatest coordinate of $\underline{x}$ over the mean value of the coordinates. Since the "times $^1/n$" function is monotone, it is easy to see that (18-16) is a special case of (18-10).

18-3. Spaces of Bounded Seminorm

If V,W are right band-spaces over a belt E_1 then a function $\tau \varepsilon (W)^V$ will be called right scale-free if for each $\lambda \varepsilon E_1$ and $x \varepsilon V$ there holds: $\tau(x \otimes \lambda) \leqslant \tau(x)$ (18-17)

As usual, we shall drop the epithet 'right' when not needed for emphasis. The following result motivates the term scale-free.

Lemma 18-7 . Let E_1 be a blog and W a right band-space over E_1. If $\tau \varepsilon (W)^{E_n}$ (for given integer $n \geqslant 1$) is scale-free then there holds: $\tau(x \otimes \lambda) = \tau(x)$ for all $x \varepsilon E_n$ and finite $\lambda \varepsilon E_1$.

Proof. For all $\underline{x} \varepsilon E_n$ and finite $\lambda \varepsilon E_1$ we have:

$$\tau(\underline{x}) = \tau(\underline{x} \otimes \lambda \otimes \lambda^{-1}) \leqslant \tau(\underline{x} \otimes \lambda)$$

which combined with (18-17) yields $\tau(\underline{x}) = \tau(\underline{x} \otimes \lambda)$.

Theorem 18-8. Let E_1 be a pre-residuated belt satisfying axiom X_{12}. Then for each integer n, the function ρ defined on E_n by (18-14) is scale-free.

Proof. Let $\underline{x} \varepsilon E_n$, $\lambda \varepsilon E_1$ be arbitrary. We have:

$$\rho(\underline{x}\otimes\lambda) = (\sum_{j=1}^{n}{}_{\oplus}\{\underline{x}\otimes\lambda\}_j) \otimes (\sum_{i=1}^{n}{}_{\otimes}(\{\underline{x}\otimes\lambda\}_i)^*) \qquad \text{(by (18-14))}$$

$$= (\sum_{j=1}^{n}{}_{\oplus}\{\underline{x}\}_j) \otimes \lambda\otimes(\sum_{i=1}^{n}{}_{\oplus}(\lambda^*\otimes'(\{\underline{x}\}_i)^*)) \qquad \text{(by axioms } X_6', N_3\text{, etc)}$$

$$\leqslant (\sum_{j=1}^{n}{}_{\oplus}\{\underline{x}\}_j) \otimes\lambda\otimes(\lambda^*\otimes' \sum_{i=1}^{n}{}_{\oplus}(\{\underline{x}\}_i)^*) \qquad \text{(by (2-9), (2-17), etc)}$$

$$\leqslant (\sum_{j=1}^{n}{}_{\oplus}\{\underline{x}\}_j) \otimes (\sum_{i=1}^{n}{}_{\oplus}(\{\underline{x}\}_i)^*) \qquad \text{(by Theorem 8-8 for } \mathcal{M}_{11}\text{, and (2-9))}$$

$$= \rho(\underline{x}).$$ ●

It is not too difficult to show similarly that the δ_{ij} norms (18-13) and the deviation norm (18-16) are also scale-free. However, the maximum seminorm (18-4) and the absolute maximum seminorm (18-7) are not scale-free.

Theorem 18-9. Let τ be a right scale-free E_1-seminorm defined on a right band-space V over a belt E_1, and let $\delta\varepsilon E_1$ be given. Then the set:

$$S(\tau,\delta) = \{x \mid x\varepsilon V \text{ and } \tau(x)\leqslant\delta\} \qquad (18\text{-}18)$$

is a (right) subspace of V (or the empty set)

Proof: Let $x,y\varepsilon V$ and $\lambda\varepsilon E_1$ be arbitrary. We have:

$$\tau(x\oplus y) \leqslant \tau(x) \oplus \tau(y) \leqslant \delta\oplus\delta = \delta$$

$$\tau(x\otimes\lambda) \leqslant \tau(x) \leqslant \delta$$

Hence $x \oplus y$ and $x \otimes \lambda$ belong to $S(\tau,\delta)$, which is therefore a right (sub) space over E_1. ●

An immediate inference from Theorem 18-9 is the following.

Proposition 18-10. Let E_1 be a given belt and $\delta \varepsilon E_1$ a given element. Let $n\geqslant 1$ be a given integer and τ a scale-free E_1-seminorm on E_n. If $\underline{a}(1),\ldots, \underline{a}(n) \varepsilon E_n$ all lie in $S(\tau,\delta)$ then so do all elements of the space generated by $\underline{a}(1),\ldots, \underline{a}(n)$. ●

When τ is the range seminorm and E_1 is a linear blog, we can identify the subspace $S(\tau,\delta)$ exactly. First, by direct computations using (18-14) we may confirm the following results.

Proposition 18-11. Let E_1 be a linear blog, $n\geqslant 1$ a given integer and ρ the range seminorm defined on E_n. For any $\underline{x}\varepsilon E_n$, either $\rho(x)\geqslant\emptyset$ or $\rho(\underline{x})=-\infty$. Specifically:

$\rho(\underline{x}) = -\infty$ if and only if either $\underline{x} = \begin{bmatrix}-\infty\\ \vdots\\ -\infty\end{bmatrix}$ or $\underline{x} = \begin{bmatrix}+\infty\\ \vdots\\ +\infty\end{bmatrix}$; otherwise $\rho(\underline{x})\geqslant\emptyset$.

$\rho(\underline{x})$ is finite if and only if $\underline{x}$ is finite

$\rho(\underline{x}) = \emptyset$ if and only if $\underline{x} = \begin{bmatrix}\lambda\\ \vdots\\ \lambda\end{bmatrix}$ for some finite $\lambda\varepsilon E_1$. ●

Theorem 18-12. Let E_1 be a linear blog and let $\delta \geqslant \emptyset$ be a given element of E_1. Let $n \geqslant 1$ be a given integer and let ρ be the range seminorm defined on E_n. Then (with the notation of Section 18-1 and (18-18)):

$$S(\rho,\delta) = E_n(\delta^*)$$

Proof. From the data, E_1 is linear, pre-residuated and satisfies axiom X_{12}, so Theorems 18-6, 18-8 and 18-9 apply. Now by direct use of (18-14) we find that each column of the matrix $\underline{I}_n(\delta^*)$ lies in the subspace $S(\rho,\delta)$ and so therefore does each element of $E_n(\delta^*)$, the column-space of $\underline{I}_n(\delta^*)$.

Conversely, let $\underline{x} \in S(\rho,\delta)$. Now, if $\delta = +\infty$ we have, using Proposition 18-2:

$\underline{x} \in S(\rho,\delta) \subset E_n = E_n(-\infty) = E_n(\delta^*)$.

We must now consider the case where δ is finite. If $\rho(\underline{x}) = -\infty$ then by Proposition 18-11, $\underline{x}$ has all components equal to $-\infty$, or else all equal to $+\infty$. In either case $\underline{I}_n(\delta^*) \otimes \underline{x} = \underline{x}$ so $\underline{x}$ lies in the column-space of $\underline{I}_n(\delta^*)$. The only remaining possibility (with $\delta \geqslant \emptyset$ and finite) is that $\rho(\underline{x})$ is finite, so $\underline{x}$ is finite by Proposition 18-11. Hence we may write:

$$\rho(\underline{x}) = \left(\sum_{j=1}^{n}{}_{\oplus}\{\underline{x}\}_j\right) \otimes \left(\sum_{j=1}^{n}{}_{\oplus}(\{\underline{x}\}_i)^{-1}\right) \leqslant \delta \tag{18-19}$$

Opening w.r.t. indices i and j:

$$\forall\ i,j=1,\ldots,\ n,\ \{\underline{x}\}_j \otimes (\{\underline{x}\}_i)^{-1} \leqslant \delta \tag{18-20}$$

$$\forall\ i,j=1,\ldots,\ n,\ \delta^{-1} \otimes \{\underline{x}\}_j \leqslant \{\underline{x}\}_i \tag{18-21}$$

Now consider $\underline{I}_n(\delta^*) \otimes \underline{x}$. We have for $i=1,\ldots,\ n$:

$$\{\underline{I}_n(\delta^*) \otimes \underline{x}\}_i = \sum_{j\neq i}{}_{\oplus}(\delta^* \otimes \{\underline{x}\}_j) \oplus (\emptyset \otimes \{\underline{x}\}_i) \tag{18-22}$$

$$= \{\underline{x}\}_i \qquad \text{(using (18-21))}$$

Hence $\underline{I}_n(\delta^*) \otimes \underline{x} = \underline{x}$ and $\underline{x}$ lies in the column-space of $\underline{I}_n(\delta^*)$ ●

We may paraphrase Theorem 18-12 by the statement "the space of n-tuples of range bounded by $\delta \geqslant \emptyset$ is exactly the column-space of $\underline{I}_n(\delta^*)$." In the particular case $\delta = \emptyset$, this space consists of all (finite or not) n-tuples whose components are all equal. The space $S(\rho,-\infty)$ is a column-space also. It is generated by the n-tuple having all components equal to $+\infty$ and has as its only other member the n-tuple all of whose elements are equal to $-\infty$. But this is not, of course, a special case of Theorem 18-11, since $\delta < \emptyset$ in this case.

Theorem 18-13. If E_1 is a linear blog and $\delta > \emptyset$ a given element of E_1, then (with the previous notation), the columns of $\underline{I}_n(\delta^*)$ are SLI for each integer $n \geqslant 1$; in fact each $\underline{x} \in E_n$ which satisfies $-\infty < \rho(\underline{x}) < \delta$ has unique expression as a linear combination of the columns of $\underline{I}_n(\delta^*)$.

Proof. The set $T=\{\underline{x}|\underline{x}\varepsilon E_n$ and $-\infty<\rho(\underline{x})<\delta\}$ is not empty; in fact the n-tuple with all components equal to $\emptyset$ lies in T. We remark that $\underline{I}_n(\delta^*)$ is doubly G-astic and strictly doubly $\emptyset$-astic and that each $\underline{x}\varepsilon T$ is finite by Proposition 18-11, since $-\infty<\rho(\underline{x})<+\infty$.

Hence for $\underline{x}\varepsilon T$ we may write (18-19) with strict inequality, and deduce (18-20). But (18-20) cannot hold with equality for any combination of i,j, else we could close w.r.t indices i and j and deduce (18-19) with equality. Hence:

$$\forall i,j=1,\ldots, n, \; \{\underline{x}\}_j \otimes (\{\underline{x}\}_i)^{-1} < \delta \tag{18-23}$$

It is notationally convenient to rearrange (18-23). Recalling that multiplication by finite elements is self-dual we have:

$$\forall i,j=1,\ldots, n, \; \{\underline{x}\}_j \otimes'(\{\underline{x}\}_i)^{-1} < \delta$$

Whence: $(\{x_j\})^{-1} \otimes'\delta\otimes'\{\underline{x}\}_i > \emptyset$

Dualising:

$$(\{\underline{x}\}_i)^{-1} \otimes \delta^* \otimes \{\underline{x}\}_j < \emptyset \qquad (\forall i,j=1,\ldots, n) \tag{18-24}$$

We consider now the matrix $\underline{B}\ \varepsilon \mathcal{M}_{nn}$ given by:

$$\{\underline{B}\}_{ij} = (\{\underline{x}\}_i)^{-1} \otimes \{\underline{I}_n(\delta^*)\}_{ij} \otimes \{\underline{x}\}_j \tag{18-25}$$

If i=j in (18-25) we have:

$$\{\underline{B}\}_{ii} = (\{\underline{x}\}_i)^{-1} \otimes \emptyset \otimes \{x\}_i = \emptyset \tag{18-26}$$

If i≠j in (18-25) we have:

$$\{\underline{B}\}_{ij} = (\{\underline{x}\}_i)^{-1} \otimes \delta^* \otimes \{\underline{x}\}_j < \emptyset \quad \text{(by (18-24))} \tag{18-27}$$

From (18-26) and (18-27) we see that $\underline{B}$ is a strictly doubly $\emptyset$-astic (n×n) matrix, showing by Theorem 15-9 and Corollary 16-9 that $\underline{x}$ has unique expression as a linear combination of the columns of $\underline{I}_n(\delta^*)$ and that these columns are accordingly SLI. ●

19. SOME MATRIX SPACES

19-1. Matrix Seminorms.

By identifying the commutative band $(\mathcal{M}_{mn},\oplus)$ with the space $((E_1)^{mn},\oplus)$ of mn-tuples, we may extend the seminorms of Section 18-2 to matrices. In particular, we have the range seminorm for $\mathcal{M}_{mn}$ defined for $\underline{A}\in\mathcal{M}_{mn}$ by:

$$\rho(\underline{A}) = \Big(\sum_{\substack{i=1,\dots,m\\ j=1,\dots,n}}{}_{\oplus}\{\underline{A}\}_{ij}\Big)\otimes\Big(\sum_{\substack{r=1,\dots,m\\ s=1,\dots,n}}{}_{\oplus}(\{\underline{A}\}_{rs})^*\Big) \qquad (19\text{-}1)$$

We may also define the <u>column-range seminorm</u> ρ_C for $\mathcal{M}_{mn}$ by:

$$\rho_C([\underline{a}(1),\dots,\underline{a}(n)]) = \sum_{j=1}^{n}{}_{\oplus}\rho(\underline{a}(j)) \qquad (19\text{-}2)$$

Intuitively, ρ_C represents "the greatest range of any column of $\underline{A}$."
Using results of the previous chapter, we easily confirm:

<u>Proposition 19-1. The range seminorm and column-range seminorm provide scale-free E_1-seminorms on the right band-space $(\mathcal{M}_{mn},\oplus)$ over a linear blog E_1, for given integers $m,n\geq 1$.</u> ●

In the principal interpretation, let

$$\underline{A} = \begin{bmatrix} 2 & -5 & 1 & 4\\ 3 & 4 & -2 & 4\\ 8 & 7 & -6 & -4\end{bmatrix}$$

The ranges of the columns of $\underline{A}$ are respectively 6,12,7,8, so $\rho_C(\underline{A}) = 12$. However, the greatest element of $\underline{A}$ is 8 and the least is -6, so $\rho(\underline{A}) = 14$. This illustrates the fairly evident fact contained in the next result.

<u>Theorem 19-2. Let E_1 be a linear blog and let ρ, ρ_C be respectively the range and column-range seminorms defined on $\mathcal{M}_{mn}$ for given integers $m,n\geq 1$. Then $\rho_C(\underline{A}) \leq \rho(\underline{A})$ for all $\underline{A}\in\mathcal{M}_{mn}$.</u>

<u>Proof</u>. Opening (19-1) w.r.t. indices i,j,r,s:
$\forall\, i,r=1,\dots,m;\ j,s=1,\dots,n,$

$$\rho(\underline{A}) \geq \{\underline{A}\}_{ij}\otimes(\{\underline{A}\}_{rs})^*$$

Taking s=j, we have:

$$\rho(\underline{A}) \geq \{\underline{A}\}_{ij}\otimes(\{\underline{A}\}_{rj})^*$$

Closing w.r.t indices i and r, we have for $j=1,\dots,n$:

$$\rho(\underline{A}) \geq \sum_{i=1}^{m}{}_{\oplus}\sum_{r=1}^{m}{}_{\oplus}(\{\underline{A}\}_{ij}\otimes(\{\underline{A}\}_{rj})^*)$$

$$= \Big(\sum_{i=1}^{m}{}_{\oplus}\{\underline{A}\}_{ij}\Big)\otimes\Big(\sum_{r=1}^{m}{}_{\oplus}(\{\underline{A}\}_{rj})^*\Big)$$

$$= \rho(\underline{a}(j))$$

where $\underline{a}(j)$ is the jth column of $\underline{A}$. Closing w.r.t index j:

$$\rho(\underline{A}) \geqslant \sum_{j=1}^{n}{}_{\oplus}\rho(\underline{a}(j)) = \rho_{\mathcal{C}}(\underline{A})$$

Of course, the results of Sections 18-2 and 18-3 carry over to $(\mathcal{M}_{mn},\oplus)$ regarded as a right band-space over E_1; but further results hold for matrices, arising from our ability to form matrix products.

<u>Theorem 19-3. Let E_1 be a linear blog and let $\rho_{\mathcal{C}}$ denote indifferently the column-range seminorms defined on $\mathcal{M}_{mn}$ and on $\mathcal{M}_{mp}$ for given integers $m,n,p\geqslant 1$. Let $\underline{A} \in \mathcal{M}_{mn}$ and $\underline{X} \in \mathcal{M}_{np}$. Then :</u>

$$\rho_{\mathcal{C}}(\underline{A}\otimes\underline{X}) \leqslant \rho_{\mathcal{C}}(\underline{A})$$

<u>Proof</u>. Each column of $\underline{A}$ belongs to $S(\rho,\rho_{\mathcal{C}}(\underline{A}))$, the space of n-tuples having range not exceeding $\rho_{\mathcal{C}}(\underline{A})$). By Proposition 18-10 therefore, all columns of $\underline{A}\otimes\underline{X}$ do so Hence $\rho(\underline{y}(j)) \leqslant \rho_{\mathcal{C}}(\underline{A})$ for all columns $\underline{y}(j)$ $(j=1,\dots, p)$ of $\underline{X}$. Closing w.r.t. index j:

$$\rho_{\mathcal{C}}(\underline{A}\otimes\underline{X}) = \sum_{j=1}^{p}{}_{\oplus} (\underline{y}(j)) \leqslant \rho_{\mathcal{C}}(\underline{A})$$

The property of ρ_C asserted by Theorem 19-3 evidently generalises (18-17), and we shall say that Theorem 19-3 asserts that ρ_C is <u>(right) factor-free</u>.

We note that Theorem 19-3 gives us yet another insolubility criterion for the equation $\underline{A}\otimes\underline{x}=\underline{b}$. For example, in the principal interpretation:

$$\underline{A} = \begin{bmatrix} 3 & 20 & -5 & 8 \\ -7 & 11 & -6 & 1 \\ 2 & 13 & -11 & 5 \end{bmatrix} \quad ; \quad \underline{b} = \begin{bmatrix} 9 \\ 7 \\ -2 \end{bmatrix}$$

The ranges of the columns of A are 10, 9, 6, 7, so $\rho_C(\underline{A}) = 10$. However, $\rho(\underline{b}) = 11$, so $\underline{b}$ cannot lie in the column-space of $\underline{A}$ and the equation $\underline{A}\otimes\underline{x}=\underline{b}$ is insoluble.

<u>Theorem 19-4. Let E_1 be a linear blog, and let ρ denote indifferently the range seminorms defined on $\mathcal{M}_{mn}$, $\mathcal{M}_{np}$ and $\mathcal{M}_{mp}$ for given integers $m,n,p\geqslant 1$. Let $\underline{A} \in \mathcal{M}_{mn}$, $B \in \mathcal{M}_{np}$ satisfy $\rho(\underline{A})$, $\rho(\underline{B}) < +\infty$.</u>

<u>Then $\rho(\underline{A}\otimes\underline{B})<+\delta$. Moreover, if E_1 is commutative (i.e. the group G of E_1 is commutative) we have</u>

$$\underline{\rho(\underline{A}\otimes\underline{B}) \leqslant \rho(\underline{A}) \otimes \rho(\underline{B}).}$$

<u>Proof</u>. If either (or both) of $\underline{A}$, $\underline{B}$ has all elements equal to $-\infty$ then so does $\underline{A}\otimes\underline{B}$, so $\rho(\underline{A}\otimes\underline{B}) = -\infty$. so certainly $\rho(\underline{A}\otimes\underline{B})<+\infty$. If $\underline{A},\underline{B}$ are both finite then so is $\underline{A}\otimes\underline{B}$ and so $\rho(\underline{A}\otimes\underline{B})<+\infty$. If one of $\underline{A},\underline{B}$, has all elements equal to $+\infty$ and the other does not have all elements equal to $-\infty$, then $\underline{A}\otimes\underline{B}$ has all elements equal to $+\infty$ and so $\rho(\underline{A}\otimes\underline{B})=-\infty$. These exhaust (by Proposition 18-11) all the cases in which $\rho(\underline{A})$, $\rho(\underline{B})<+\infty$

In all cases $\rho(\underline{A}\otimes\underline{B})<+\infty$, as asserted.

Now let E_1 be commutative. Since $\rho(\underline{A}\otimes\underline{B})<+\infty$ (as just proved), one possibility is that $\rho(\underline{A}\otimes\underline{B})=-\infty$, and then certainly

$$\rho(\underline{A}\otimes\underline{B}) \leq \rho(\underline{A})\otimes\rho(\underline{B})$$

So let $\rho(\underline{A}\otimes\underline{B})$ be finite; then $\underline{A}\otimes\underline{B}$ is finite by Proposition 18-11.
Since E_1 is linear, there exist indices I.J,K,R,S,T such that

$$\rho(\underline{A}\otimes\underline{B}) = \Big(\sum_{\substack{i=1,\ldots,m\\ j=1,\ldots,p}}{}_{\oplus} \sum_{k=1,\ldots,n} (\{\underline{A}\}_{ik} \otimes \{\underline{B}\}_{kj})\Big) \otimes \Big(\sum_{\substack{r=1,\ldots,m\\ s=1,\ldots,p}}{}_{\oplus'} \Big(\sum_{t=1,\ldots,n}{}_{\oplus} (\{\underline{A}\}_{rt}\otimes\{\underline{B}\}_{ts})\Big)\Big)^{*}$$

$$= (\{\underline{A}\}_{IK} \otimes \{\underline{B}\}_{KJ}) \otimes (\{\underline{A}\}_{RT} \otimes \{\underline{B}\}_{TS})^{*}$$

Hence

$$(\underline{A}\otimes\underline{B}) = (\{\underline{A}\}_{IK} \otimes \{\underline{B}\}_{KJ}) \otimes ((\{\underline{B}\}_{TS})^{*} \otimes' (\{\underline{A}\}_{RT})^{*}) \qquad (19\text{-}3)$$

Now the four factors on the right side of (19-3) are all finite since pairwise they form the finite products $\{\underline{A}\otimes\underline{B}\}_{IJ}$, $(\{\underline{A}\otimes\underline{B}\}_{RT})^{*}$. But multiplication is self-dual for finite elements and commutative by hypothesis; hence (19-3) gives:

$$(\underline{A}\otimes\underline{B}) = (\{\underline{A}\}_{IK} \otimes (\{\underline{A}\}_{RT})^{*}) \otimes (\{\underline{B}\}_{KJ} \otimes (\{\underline{B}\}_{TS})^{*})$$

$$\leq \Big(\sum_{\substack{i=1,\ldots,m\\ k=1,\ldots,n}}{}_{\oplus}\{\underline{A}\}_{ik}\Big) \otimes \Big(\sum_{\substack{r=1,\ldots,m\\ t=1,\ldots,n}}{}_{\oplus}(\{\underline{A}\}_{rt})^{*}\Big)$$

$$\otimes \Big(\sum_{\substack{k=1,\ldots,n\\ j=1,\ldots,p}}{}_{\oplus}\{\underline{B}\}_{kj}\Big) \otimes \Big(\sum_{\substack{t=1,\ldots,n\\ s=1,\ldots,p}}{}_{\oplus}(\{\underline{B}\}_{ts})^{*}\Big)$$

$$= \rho(\underline{A}) \otimes \rho(\underline{B})$$ ●

We shall say that the inequality $\rho(\underline{A}\otimes\underline{B})\leq\rho(\underline{A}) \otimes \rho(\underline{B})$ shows ρ to be <u>submultiplicative</u>.

The necessity of the condition $\rho(\underline{A})$, $\rho(\underline{B})<+\infty$ in the proof of foregoing theorem is illustrated by the following example for the principal interpretation:

$$\underline{A} = \begin{bmatrix} +\infty & +\infty \\ -\infty & -\infty \end{bmatrix}; \qquad \underline{B} = \begin{bmatrix} +\infty \\ +\infty \end{bmatrix}, \qquad \underline{A}\otimes\underline{B} = \begin{bmatrix} +\infty \\ -\infty \end{bmatrix}$$

$$\rho(\underline{A})\otimes\rho(\underline{B}) = +\infty\otimes-\infty=-\infty < +\infty = \rho(\underline{A}\otimes\underline{B})$$

19-2. Matrix Spaces

In place of (19-1) let us define the <u>co-range</u> $\bar{\rho}(\underline{A})$ of a matrix $\underline{A}\in\mathcal{M}_{mn}$ by:

$$\bar{\rho}(\underline{A}) = \Big(\sum_{\substack{r=1,\ldots,m\\ s=1,\ldots,n}}{}_{\oplus}(\{\underline{A}\}_{rs})^{*}\Big) \otimes \Big(\sum_{\substack{i=1,\ldots,m\\ j=1,\ldots,n}}{}_{\oplus}\{\underline{A}\}_{ij}\Big) \qquad (19\text{-}4)$$

Evidently this definition enables us to develop "left" versions of the results of Section 18-3. In this connection, we note the following result.

Lemma 19-5. Let E_1 be a linear self-conjugate belt. Then for any given integers $m, n \geq 1$, the co-range function $\bar{\rho}$ defines an E_1- seminorm on $\mathcal{M}_{mn}$ and in particular on $E_n (= \mathcal{M}_{n1})$. If E_1 is pre-residuated and satisfies axiom X_{12} then $\bar{\rho}$ is left scale-free in the sense that $\bar{\rho}(\lambda \otimes \underline{x}) \leq \bar{\rho}(\underline{x})$ for each $\lambda \in E_1$ and $\underline{x} \in E_n$. And if E_1 is a blog and ρ denotes the range seminorm defined on E_n, then for each $\underline{x} \in E_n$ there holds:

(i) Either $\bar{\rho}(\underline{x}) \geq \emptyset$ or $\bar{\rho}(\underline{x}) = -\infty$

(ii) $\bar{\rho}(\underline{x})$ and $\rho(\underline{x})$ both equal $-\infty$, or are both finite, or both equal $+\infty$.

(iii) $\bar{\rho}(x) = \emptyset$ if and only if $\rho(\underline{x}) = \emptyset$

Proof. All assertions up to and including (i) follow by analogy with the properties of the function ρ. Assertion (ii) follows from the fact that we may regard $\bar{\rho}(\underline{x})$ as being of the form $b \otimes a$, where $\rho(\underline{x})$ is $a \otimes b$ $(a, b \in E_1)$; it easily follows from the basic properties of blogs (in particular from Proposition 4-3) that $a \otimes b$ and $b \otimes a$ both equal $-\infty$, or are both finite, or both equal $+\infty$, for any $a, b \in E_1$. Assertion (iii) follows from the fact that $a \otimes b = \emptyset$ if and only if $b = a^{-1}$ if and only if $b \otimes a = \emptyset$. ●

The following is the first of a number of results showing that families of matrices whose ranges satisfy certain inequalities form, as do finite and $\sigma_{\emptyset}$-astic matrices, closed algebraic systems.

Theorem 19-6. Let E_1 be a linear blog and for given integers $m, n \geq 1$ let $\mathfrak{J}_{mn}$ denote the set $\{\underline{A} | \underline{A} \in \mathcal{M}_{mn}$ and $\rho(\underline{A}) < +\infty\}$ where ρ denotes the range seminorm on $\mathcal{M}_{mn}$. Then $\mathfrak{J}_{mn}$ may equivalently be characterised as the set $\{\underline{A} | \underline{A} \in \mathcal{M}_{mn}$ and $\bar{\rho}(\underline{A}) < +\infty\}$ where $\bar{\rho}$ denotes the co-range seminorm on $\mathcal{M}_{mn}$. And the $(\mathfrak{J}, \mathfrak{J}, E_1)$-and $(E_1, \mathfrak{J}, \mathfrak{J})$-hom-rep statements are both true.

Proof. Suppose $\underline{A}, \underline{B} \in \mathfrak{J}_{mn}$. Since E_1 is linear, $\rho(\underline{A}) \oplus \rho(\underline{B})$ equals either $\rho(\underline{A})$ or $\rho(\underline{B})$, whence, using the sublinearity of ρ :

$$\rho(\underline{A} \oplus \underline{B}) \leq \rho(\underline{A}) \oplus \rho(\underline{B}) < +\infty$$

If $\lambda \in E_1$ then since ρ is right scale-free we have:

$$\rho(\underline{A} \otimes \lambda) \leq \rho(\underline{A}) < +\infty$$

Hence $\underline{A} \oplus \underline{B}$, $\underline{A} \otimes \lambda \in \mathfrak{J}_{mn}$, and $(\mathfrak{J}_{mn}, \oplus)$ is a right band-space over E_1. That $(\mathfrak{J}_{mm}, \oplus, \otimes)$ is a belt over which $(\mathfrak{J}_{mn}, \oplus)$ is a left band-space now follows from

Theorem 19-4. The rest of the proof of the $(\mathcal{J}, \mathcal{J}, E_1)$-hom-rep statement is standard.

The alternative characterisation of $\mathcal{J}_{mn}$ follows from Lemma 19-5 (ii), and then the $(E_1, \mathcal{J}, \mathcal{J})$-hom-rep statement follows by the appropriate "left-right variant" of the foregoing argument using $\bar{\rho}$ in the place of ρ. ●

We now introduce another seminorm - the <u>row-range</u> ρ_R, defined for a given matrix $\underline{A} \in \mathcal{M}_{mn}$ as follows:

$$\rho_R(\underline{A}) = \sum_{i=1}^{m}{}_{\oplus} \bar{\rho}(\underline{b}(i)) \quad \text{where } [\underline{b}(1), \ldots, \underline{b}(m)] \text{ is the transposed of matrix } \underline{A},$$

and $\bar{\rho}$ is the co-range seminorm defined on $E_n (= \mathcal{M}_{1n})$. For example, in the principal interpretation, suppose:

$$\underline{A} = \begin{bmatrix} 2 & 5 & -6 & 5 \\ 1 & -4 & 8 & 5 \\ -3 & -4 & 0 & 5 \end{bmatrix} \tag{19-5}$$

The ranges of the columns of $\underline{A}$ are 5, 9, 14, 0 respectively, so $\rho_C(\underline{A}) = 14$.
The (co-) ranges of the rows of $\underline{A}$ are 11, 12, 9 respectively, so $\rho_R(\underline{A}) = 12$.

By analogy with Proposition 19-1 and Theorems 19-2 and 19-3, we record the following

<u>Proposition 19-7. Let E_1 be a linear blog and let ρ_R indifferently denote the row-range functions defined on $\mathcal{M}_{mn}$ and $\mathcal{M}_{qn}$ for given integers $m, n, q \geq 1$, let $\lambda \in E_1$, $\underline{A} \in \mathcal{M}_{mn}$ and $\underline{X} \in \mathcal{M}_{qm}$. Then:</u>

(i) <u>ρ_R is an E_1-seminorm on $\mathcal{M}_{mn}$ (and on $\mathcal{M}_{qn}$)</u>

(ii) <u>$\rho_R(\lambda \otimes \underline{A}) \leq \rho_R(\underline{A})$</u>

(iii) <u>$\rho_R(\underline{X} \otimes \underline{A}) \leq \rho_R(\underline{A})$</u>

<u>Moreover, if $\bar{\rho}$ denotes the range seminorm on $\mathcal{M}_{mn}$, then:</u>

(iv) <u>$\rho_R(\underline{A}) \leq \bar{\rho}(\underline{A})$</u> ●

We shall paraphrase (ii) and (iii) of Proposition 19-7 by saying that ρ_R is <u>left scalar-free</u> and <u>left factor-free</u>.

<u>Theorem 19-8. Let E_1 be a linear blog and δ a given element of E_1. For any given integers $m, n \geq 1$, define the sets:</u>

$$\mathcal{J}_{mn} = \{\underline{A} \mid \underline{A} \in \mathcal{M}_{mn} \text{ and } \rho_C(\underline{A}) \leq \delta\}$$

$\mathcal{G}_{mn} = \{\underline{A} | \underline{A} \in \mathcal{M}_{mn} \text{ and } \rho_R(\underline{A}) \leq \delta\}$

where ρ_C, ρ_R are respectively the column-range and row-range seminorms defined on $\mathcal{M}_{mn}$. Then the $(\mathcal{J}, \mathcal{J}, E_1)$-and $(E_1, \mathcal{G}, \mathcal{G})$-hom-rep statements both hold.

Proof. $\mathcal{J}_{mn}$ always contains the matrix with all elements equal to $-\infty$, and so is non-empty. That $(\mathcal{J}_{mn}, \oplus)$ is a right band-space over E_1 follows from Theorem 18-9; that $(\mathcal{J}_{mn}, \oplus, \otimes)$ is a belt over which $(\mathcal{J}_{mn}, \oplus)$ is a left band-space follows from Theorem 19-3. The rest of the proof of the $(\mathcal{J}, \mathcal{J}, E_1)$-hom-rep statement is standard and the argument for $\mathcal{G}$ is similar. ●

19-3. The Role of Conjugacy

If $\underline{A}$ is as in (19-5) we have:

$$\underline{A}^* = \begin{bmatrix} -2 & -1 & 3 \\ -5 & 4 & 4 \\ 6 & -8 & 0 \\ -5 & -5 & -5 \end{bmatrix}$$

We find $\rho(\underline{A}^*) = 14 = \rho(\underline{A})$. Furthermore, we find the ranges of the columns of $\underline{A}^*$ to be 11, 12, 9 respectively and the (co-) ranges of the rows of $\underline{A}^*$ to be 5, 9, 14, 0 respectively. This illustrates the following result.

Lemma 19-9. Let E_1 be a linear blog and let $\rho_C, \rho_R, \rho, \bar{\rho}$ have their usual significance. Then for any matrix $\underline{A} \in \mathcal{M}_{mn}$ we have $\rho_C(\underline{A}) = \rho_R(\underline{A}^*)$, $\rho_C(\underline{A}^*) = \rho_R(\underline{A})$, and $\rho(\underline{A}^*) = \bar{\rho}(\underline{A})$. If E_1 is commutative we have $\rho(\underline{A}) = \rho(\underline{A}^*)$.

Proof. We have:

$$\rho_C(\underline{A}^*) = \sum_{i=1}^{m}{}_{\oplus} \left(\left(\sum_{j=1}^{n}{}_{\oplus} \{\underline{A}^*\}_{ji}\right) \otimes \left(\sum_{j=1}^{n}{}_{\oplus} (\{\underline{A}^*\}_{ji})^*\right)\right)$$

$$= \sum_{i=1}^{m}{}_{\oplus} \left(\left(\sum_{j=1}^{n}{}_{\oplus} (\{\underline{A}\}_{ij})^*\right) \otimes \left(\sum_{j=1}^{n}{}_{\oplus} \{\underline{A}\}_{ij}\right)\right)$$

$$= \sum_{i=1}^{m}{}_{\oplus} \bar{\rho}(\underline{b}(i))$$

where $\bar{\rho}$ is the co-range seminorm and $[\underline{b}(1), \ldots, \underline{b}(m)]$ the transposed of $\underline{A}$. Hence $\rho_C(\underline{A}^*) = \rho_R(\underline{A})$ and similarly $\rho_C(\underline{A}) = \rho_R(\underline{A}^*)$. Now:

$$\rho(\underline{A}^*) = \left(\sum_{\substack{s=1,\ldots,n \\ r=1,\ldots,m}}{}_{\oplus} \{\underline{A}^*\}_{sr}\right) \otimes \left(\sum_{\substack{j=1,\ldots,n \\ i=1,\ldots,m}}{}_{\oplus} (\{\underline{A}^*\}_{ji})^*\right)$$

$$= \Big(\sum_{\substack{r=1,\ldots,m\\ s=1,\ldots,n}}{}^{\oplus} (\{\underline{A}\}_{rs})^{*}\Big) \otimes \Big(\sum_{\substack{i=1,\ldots,m\\ j=1,\ldots,n}}{}^{\oplus} \{\underline{A}\}_{ij}\Big) = \bar{\rho}(\underline{A})$$

$$= \Big(\sum_{\substack{i=1,\ldots,m\\ j=1,\ldots,n}}{}^{\oplus} \{\underline{A}\}_{ij}\Big) \otimes \Big(\sum_{\substack{r=1,\ldots,m\\ s=1,\ldots,n}}{}^{\oplus} (\{\underline{A}\}_{rs})^{*}\Big) \qquad \text{(if } E_1 \text{ is commutative)}$$

$$= \rho(\underline{A})$$

●

Lemma 19-9 allows us to prove the perhaps surprising result that our seminorms ρ_C, ρ_R and ρ enjoy certain properties dual to those which we have already established, as set out in our next results. First let us note that by use of the dual operations $\oplus'$ and $\otimes'$ in place of $\oplus$ and $\otimes$ respectively, we may introduce the terms dually sublinear, dually submultiplicative and dual E_1-seminorm; and further, the terms left and right dually scale-free and left and right dually factor-free.

Theorem 19-10. Let E_1 be a linear blog and let ρ_c, ρ_R, ρ have their usual significance. Then:

(i) ρ is (right) dually scale-free and dually factor-free.

(ii) ρ_R is left dually scale-free and left dually factor-free.

(iii) If E_1 is commutative then ρ is dually submultiplicative.

Proof. For case (i) let $\underline{A} \in \mathcal{M}_{mn}$ and $\underline{X} \in \mathcal{M}_{nq}$. We have:

$$\rho_C(\underline{A} \otimes' \underline{X}) = \rho_C((\underline{X}^{*} \otimes \underline{A}^{*})^{*})$$

$$= \rho_R(\underline{X}^{*} \otimes \underline{A}^{*}) \qquad \text{(by Lemma 19-9)}$$

$$\leqslant \rho_R(\underline{A}^{*}) \qquad \text{(by Proposition 19-5)}$$

$$= \rho_C(\underline{A}) \qquad \text{(by Lemma 19-9)}$$

Hence ρ_C is dually factor-free. The other assertions are proved similarly. ●

We remarked earlier that our matrices of bounded range form closed algebraic systems, and illustrated this fact in Theorems 19-6 and 19-8, which are analogous to Theorem 12-3 (for α-σ_ϕ-astic matrices) and Theorem 12-4 (for finite matrices). We now add further results, in the spirit of Theorems 13-17 and 13-19.

Theorem 19-11. Let E_1 be a linear blog and for any given integers $m, n \geqslant 1$ let $\mathcal{F}_{mn}$ denote $\{\underline{A} | \underline{A} \in \mathcal{M}_{mn}$ and $\rho(\underline{A}) < +\infty\}$ where ρ denotes the range seminorm on $\mathcal{M}_{mn}$. Then $(\mathcal{F}_{mn}, \oplus, \oplus')$ and $(\mathcal{F}_{nm}, \oplus, \oplus')$ are commutative bands with duality, which are conjugate; and for any given integer $p \geqslant 1$, we have:

$$\mathcal{L}\mathcal{J}_{mn} : \mathcal{J}_{np} \Longleftrightarrow \mathcal{J}_{mp} : \mathcal{L}^*\mathcal{J}^*_{mn}, \qquad (19\text{-}6)$$

and $\mathcal{J}_{mn}$ has a faithful representation as the commutative belt $\mathcal{L}\mathcal{J}_{mn}$ of residuomorphisms of $\mathcal{J}_{np}$ to $\mathcal{J}_{mp}$. In particular, $(\mathcal{J}_{mm}, \oplus, \otimes, \oplus', \otimes')$ is a belt with duality which is self-conjugate and has a faithful representation as a belt of endo-residuomorphisms of $\mathcal{J}_{mp}$.

Proof. That $(\mathcal{J}_{mn}, \oplus)$ is a commutative band is already asserted under Theorem 19-7. Now, let $\underline{A}, \underline{B} \in \mathcal{J}_{mn}$. Then $\underline{A}, \underline{B}$ have $\rho < +\infty$ so $\underline{A}^*, \underline{B}^*$ have $\bar{\rho} < +\infty$ by Lemma 19-9, whence $\underline{A}^*, \underline{B}^* \in \mathcal{J}_{nm}$ by Theorem 19-6, so $\underline{A}^* \oplus \underline{B}^* \in \mathcal{J}_{nm}$ because $\mathcal{J}_{nm}$ is a commutative band. Hence $\bar{\rho}(\underline{A}^* \oplus \underline{B}^*) < +\infty$ by Theorem 19-6, so $\rho((\underline{A}^* \oplus \underline{B}^*)^*) < +\infty$. But this says that $\underline{A} \oplus' \underline{B} \in \mathcal{J}_{mn}$, whence $(\mathcal{J}_{mn}, \oplus, \oplus')$ is a commutative band with duality. Similarly, so is $(\mathcal{J}_{nm}, \oplus, \oplus')$, and by Lemma 19-9 and Theorem 19-6 we know that $\underline{A} \in \mathcal{J}_{mn}$ if and only if $\underline{A}^* \in \mathcal{J}_{nm}$. Clearly, the conjugacy $\underline{A} \leftrightarrow \underline{A}^*$ between $\mathcal{M}_{mn}$ and $\mathcal{M}_{nm}$ operates between $\mathcal{J}_{mn}$ and $\mathcal{J}_{nm}$. In particular, if $n = m$ in the foregoing argument then $\underline{A}^*, \underline{B}^* \in \mathcal{J}_{mm}$, so $\rho(\underline{B}^* \otimes \underline{A}^*) < +\infty$ by Theorem 19-4. Then by Lemma 19-9, $\bar{\rho}(\underline{A} \otimes' \underline{B}) < +\infty$, whence $\underline{A} \otimes' \underline{B} \in \mathcal{J}_{mm}$ by Theorem 19-6. Hence $(\mathcal{J}_{mm}, \oplus', \otimes')$ is evidently a belt. That $(\mathcal{J}_{mm}, \oplus, \otimes)$ is a belt is already asserted under Theorem 19-7, and the self-conjugacy $\underline{A} \leftrightarrow \underline{A}^*$ of $\mathcal{M}_{mm}$ clearly operates in $\mathcal{J}_{mm}$.

Using the $(\mathcal{J}, \mathcal{J}, E_1)$-hom-rep statement, (19-6) now follows routinely. The mapping $\tau : \underline{A} \mapsto g_{\underline{A}}$ (in the usual notation) gives, according to Theorem 5-10 a bijection between $\mathcal{J}_{mn}$ and $\mathcal{L}\mathcal{J}_{mn}$, quâ subset of $\mathrm{Hom}_{E_1}(\mathcal{M}_{np}, \mathcal{M}_{mp})$. However, our theorem asserts something a little stronger, namely that τ is bijective if we replace each $g_{\underline{A}}$ by its restriction $g_{\underline{A}}/\mathcal{J}_{np}$. This follows, in fact, by the same reasoning as leads to Proposition 10-13. The rest follows on taking $n = m$. ●

By closely similar reasoning we arrive at the following results.

Proposition 19-12. Let E_1 be a linear blog and δ a given element of E_1. For given integers $m, n \geqslant 1$, and with the usual notation, let:

$$\mathcal{J}_{mn} \text{ denote } \{\underline{A} | \underline{A} \in \mathcal{M}_{mn} \text{ and } \rho_C(\underline{A}) \leqslant \delta\}$$

$$\mathcal{G}_{nm} \text{ denote } \{\underline{A} | \underline{A} \in \mathcal{M}_{nm} \text{ and } \rho_R(\underline{A}) \leqslant \delta\}$$

Then $(\mathcal{J}_{mn}, \oplus, \oplus')$ and $(\mathcal{G}_{nm}, \oplus, \oplus')$ are both commutative bands with duality, which are conjugate. Moreover $(\mathcal{J}_{mm}, \oplus, \otimes, \oplus', \otimes')$ and $(\mathcal{G}_{mm}, \oplus, \otimes, \oplus', \otimes')$ are both belts with duality, which are conjugate. If $\mathcal{H}_{mn}$ denotes $\mathcal{J}_{mn} \cap \mathcal{G}_{mn}$ then there holds:

$$\mathcal{L}\mathcal{H}_{mn} : \mathcal{H}_{np} \Longleftrightarrow \mathcal{H}_{mp} : \mathcal{L}^*\mathcal{H}^*_{mn} \qquad ●$$

20-1. The Principal Solution

Suppose we have a one-cycle activity of the kind discussed in Chapter 9, defined by an (n×n) matrix $\underline{A}$ and a given n-tuple $\underline{b}$ of required finishing times. Overrunning the given finishing time on any activity is prohibited so we speak of a zero lateness problem, but there is also (we assume) a penalty associated with earliness - i.e. finishing an activity too early - say the cost of capital tied up in goods which are finished but cannot yet be moved.

It is evident that we should like to choose our vector of start-times $\underline{x}$ to satisfy $\underline{A}\otimes\underline{x} = \underline{b}$, and we have discussed in previous chapters some properties of this equation and in particular the principal solution $\underline{x} = \underline{A}^*\otimes'\underline{b}$. However, if the given $\underline{b}$ does not happen to lie in the column-space of $\underline{A}$, we must choose $\underline{x}$ so as to satisfy $\underline{A}\otimes\underline{x}\leq\underline{b}$ and simultaneously to minimise the penalty for earliness.

Depending how this penalty is computed, different start-time vectors will be preferred. We shall show, in fact, that the expression $\underline{A}^*\otimes'\underline{b}$ gives the preferred start-time vector relative to a number of criteria. Some of these criteria are expressible by the use of some of the seminorms discussed in the previous chapter, so our discussion acquires something of the character of a norm-minimisation problem in conventional algebra. The virtues of the expression $\underline{A}^*\otimes'\underline{b}$ derive from the following result.

Lemma 20-1. Let E_1 be a pre-residuated belt satisfying axiom X_{12} and let $\underline{A}\in\mathcal{M}_{mn}$, $\underline{b}\in E_m$. If $\underline{u}\in E_n$ satisfies $\underline{A}\otimes\underline{u}\leq\underline{b}$, then $\underline{u}\leq\underline{A}^*\otimes'\underline{b}$ and:

$$\underline{A}\otimes\underline{u}\leq\underline{A}\otimes(\underline{A}^*\otimes'\underline{b})\leq\underline{b} \qquad (20\text{-}1)$$

Proof. By Lemma 9-1, if $\underline{A}\otimes\underline{u}\leq\underline{b}$ then $\underline{u}\leq\underline{A}^*\otimes'\underline{b}$ so (matrix multiplications being isotone):

$$\underline{A}\otimes\underline{u}\leq\underline{A}\otimes(\underline{A}^*\otimes'\underline{b})\leq\underline{b} \qquad \text{(by Theorem 8-8)}$$ ●

We may paraphrase (20-1) by saying that $\underline{A}\otimes(\underline{A}^*\otimes'\underline{b})$ is componentwise closer to $\underline{b}$ than any other n-tuple lying in the column-space of $\underline{A}$.

Suppose now that E_1 is a blog and $\underline{b}\in E_m$. We are given $\underline{c}\in E_m$ such that $\underline{c}\leq\underline{b}$. The m-tuple $\underline{c}$ is supposed to be an approximation to $\underline{b}$: how good an approximation is it? In the principal interpretation, for example, let:

$$\underline{b} = \begin{bmatrix} 1 \\ -3 \\ 2 \end{bmatrix} \quad ; \qquad \underline{c} = \begin{bmatrix} -4 \\ -5 \\ 1 \end{bmatrix}$$

If we subtract $\underline{c}$ componentwise from $\underline{b}$ we get

$$\begin{bmatrix} 5 \\ 2 \\ 1 \end{bmatrix}$$

The greatest component in this difference-vector is 5, i.e. 5 is the greatest componentwise deviation of $\underline{c}$ from $\underline{b}$ and so measures the badness of the approximation.

In more general terms, let E_1 be a self-conjugate belt, and let $\underline{b},\underline{c}\varepsilon E_m$ with $\underline{c}\leq\underline{b}$. Then we may form the m-tuple:

$$\underline{\alpha} = \underline{\alpha}(\underline{c},\underline{b}) = \begin{bmatrix} (\{\underline{c}\}_1)^* \otimes \{\underline{b}\}_1 \\ \\ (\{\underline{c}\}_m)^* \otimes \{\underline{b}\}_m \end{bmatrix} \tag{20-2}$$

and apply to $\underline{\alpha}$ some function $\tau\varepsilon(E_1)^{E_m}$ and regard $\tau(\underline{\alpha})$ as measuring how badly $\underline{c}$ approximates $\underline{b}$. If we are constrained to choose $\underline{c}$ from some set S, we shall attempt to choose $\underline{c}\varepsilon S$ so as to minimise $\tau(\underline{\alpha})$.

Relating this to our zero-lateness problem, we arrive at the following problem: Given $\underline{A}\ \varepsilon \mathcal{M}_{mn}$ and $\underline{b}\ \varepsilon\ E_m$, find $\underline{x}\ \varepsilon\ E_n$ such that $\tau(\underline{\alpha}(\underline{A}\otimes\underline{x},\underline{b}))$ is minimised, subject to $\underline{A}\otimes\underline{x}\leq\underline{b}$ (20-3)

Theorem 20-2. Let E_1 be a pre-residuated belt satisfying axiom X_{12} and let τ be the maximum-seminorm (18-4) defined on E_m. Then problem (20-3) is solved by taking $\underline{x}=\underline{A}^*\otimes'\underline{b}$.

Proof. If τ is the maximum-seminorm, then for any $\underline{y},\underline{z}\varepsilon E_m$:

$$\tau(\underline{\alpha}(\underline{y},\underline{z})) = \underline{y}^*\otimes\underline{z}, \tag{20-4}$$

because $$\tau(\underline{\alpha}(\underline{y},\underline{z})) = \sum_{i=1}^{m}{}_{\oplus}((\{\underline{y}\}_i)^* \otimes \{\underline{z}\}_i).$$

Now if $\underline{u}$ satisfies $\underline{A}\otimes\underline{u}\leq\underline{b}$ then $\underline{u}\leq\underline{A}^*\otimes'\underline{b}$, so $\underline{u}^*\geq\underline{b}^*\otimes\underline{A}$ and then by isotonicity of multiplication:

$$(\underline{u}^*\otimes'\underline{A}^*) \otimes \underline{b} \geq ((\underline{b}^*\otimes\underline{A}) \otimes'\underline{A}^*) \otimes \underline{b}$$

$$\text{i.e. } (\underline{A}\otimes\underline{u})^* \otimes \underline{b} \geq (\underline{A} \otimes (\underline{A}^*\otimes'\underline{b}))^* \otimes \underline{b} \tag{20-5}$$

But from (20-4), relation (20-5) is just:

$$\tau(\underline{\alpha}(\underline{A}\otimes\underline{u},\underline{b})) \geq \tau(\underline{\alpha}(\underline{A}\otimes(\underline{A}^*\otimes'\underline{b}), \underline{b}).$$ ●

Theorem 20-2 thus gives, for the principal interpretation, the solution to the Chebychev problem posed under (1-12). In fact, because it makes statements only about the "worst component", Theorem 20-2 is a good deal weaker than (20-1) which makes statements about all components. We can take account of this by introducing, for the principal interpretation, the R_m^1-norm defined by

$$\sum_{i=1}^{m} |\{\underline{x}\}_i| \tag{20-6}$$

This is not an E_1-seminorm in the sense of Chapter 18 but, as is well-known, defines a norm on the m-dimensional real vector space R_m in the usual sense of analysis.

Theorem 20-3. Under the principal interpretation, if τ is the R_m^1-norm then problem (20-3) is solved by taking $\underline{x}=\underline{A}^*\otimes'\underline{b}$.

Proof. Direct application of (20-1).

Now the R^1_m-norm (20-6) in effect gives equal weight to earliness in each coordinate position. But in practice we may wish to penalise earliness differently (and non-linearly) for each component. We assume merely that the penalty is a monotone increasing function of the earliness in each coordinate position.

Theorem 20-4. Let E_1 be a pre-residuated belt satisfying axiom X_{12} and let $\tau\varepsilon(E_1)^{E_m}$ be isotone. Then problem (20-3) is solved by taking $\underline{x}=\underline{A}^*\otimes'\underline{b}$.

Proof. From (20-1), for any $\underline{u}\varepsilon E_n$ satisfying $\underline{A}\otimes\underline{u}=\underline{b}$:

$$\{\underline{A}\otimes\underline{u}\}_i \leqslant \{\underline{A}\otimes(\underline{A}^*\otimes'\underline{b}\}_i \qquad (i=1,\ldots, m)$$

Hence $(\{\underline{A}\otimes\underline{u}\}_i)^* \otimes \{\underline{b}\}_i \geqslant (\{\underline{A}\otimes(\underline{A}^*\otimes'\underline{b})\}_i)^* \otimes \{\underline{b}\}_i \quad (i=1,\ldots, m)$

So $\tau(\underline{\alpha}(\underline{A}\otimes\underline{u},\underline{b})) \geqslant \tau(\underline{\alpha}(\underline{A}\otimes(\underline{A}^*\otimes'\underline{b}),\underline{b})$ since τ is isotone.

20-2. Cases of Equality

Since $\underline{A}^*\otimes'\underline{b}$ gives us the Chebychev-best underapproximation to $\underline{b}$ lying in the column-space of $\underline{A}$, it is intuitive that $\underline{A}^*\otimes'\underline{b}$ must equal $\underline{b}$ in at least one coordinate position, otherwise we could take $\underline{u}$ "a little greater" than $\underline{A}^*\otimes'\underline{b}$ and obtain a "better solution" to problem (20-3). The following result gives conditions wich are sufficient to guarantee this intuitive result.

Theorem 20-5. Let E_1 be a linear blog and let $\underline{A}\ \varepsilon\mathcal{M}_{mn}$, $\underline{b}\ \varepsilon\ E_m$. If at least one component of $\underline{A}^*\otimes'\underline{b}$ is finite then $\underline{A}\otimes(\underline{A}^*\otimes'\underline{b})$ has the same value as $\underline{b}$ in at least one coordinate position, the common value being finite.

Proof. Suppose in fact that $\{\underline{A}^*\otimes'\underline{b}\}_J$ is finite ($1\leqslant J\leqslant n$), i.e.:

$$\sum_{k=1}^{m}{}_{\oplus'}(\{\underline{A}^*\}_{Jk}\otimes'\{\underline{b}\}_k) \text{ is finite.}$$

So, since E_1 is linear, there is an index K ($1\leqslant k\leqslant m$) such that:

$$\{\underline{A}^*\}_{JK}\otimes'\{\underline{b}\}_K = \{\underline{A}^*\otimes'\underline{b}\}_J \text{ is finite.} \tag{20-7}$$

It follows from this that both $\{\underline{A}^*\}_{JK}$ and $\{\underline{b}\}_K$ are finite. Since multiplication is self-dual for finite elements, we may rewrite (20-7) as:

$$(\{\underline{A}\}_{KJ})^{-1}\otimes\{\underline{b}\}_K = \{\underline{A}^*\otimes'\underline{b}\}_J \tag{20-8}$$

Now, $\{\underline{A}\otimes(\underline{A}^*\otimes'\underline{b})\}_K$

$$= \sum_{j=1}^{n}{}_{\oplus}(\{\underline{A}\}_{Kj}\otimes\{\underline{A}^*\otimes'\underline{b}\}_j)$$

$$\geqslant\{\underline{A}\}_{KJ}\otimes\{\underline{A}\otimes'\underline{b}\}_J \qquad \text{(by (3-15))}$$

$$= \{\underline{A}\}_{KJ} \otimes (\{\underline{A}\}_{KJ})^{-1} \otimes \{\underline{b}\}_K$$

$$= \{\underline{b}\}_K \tag{20-9}$$

Hence $\underline{A}\otimes(\underline{A}^*\otimes'\underline{b})$ at least equals $\underline{b}$ at its Kth coordinate position, which coupled with (20-1) shows that equality occurs at the Kth coordinate position.

We remark that in the "standard" situation where $\underline{A}$ is doubly G-astic and $\underline{b}$ is finite, the product $\underline{A}^*\otimes'\underline{b}$ is finite and so the conclusion of Theorem 20-5 holds.

Our main motive for developing Theorem 20-5 is to discuss the concept of the <u>critical path</u> in Section 20-3. However, it is useful to notice that Theorem 20-5 allows us, for the principal interpretation, to solve the Chebychev problem which we get on dropping the zero-lateness constraint $\underline{A}\otimes\underline{x}\leq\underline{b}$ namely:

$$\begin{array}{l}\text{Given } \underline{A} \in \mathcal{M}_{mn} \text{ and } \underline{b} \in E_m, \text{ find } \underline{x} \in E_n \text{ such that} \\ \max\limits_{i=1,\dots,m} (\{\underline{b}\}_i - \{\underline{A} \otimes \underline{x}\}_i) \quad \text{is minimised.}\end{array} \tag{20-10}$$

We simply find the principal solution $\underline{A}\otimes(\underline{A}^*\otimes'\underline{b})$ and then increase it in all components by one-half of the maximum deviation. An example will make this clear. We take:

$$\underline{A} = \begin{bmatrix} 3 & -2 \\ 4 & 1 \\ -1 & 2 \end{bmatrix} \quad ; \quad \underline{b} = \begin{bmatrix} 5 \\ -2 \\ -1 \end{bmatrix}$$

Since $\rho_c(\underline{A}) = 5$ whilst $\rho_c(\underline{b}) = 7$, Theorem 19-3 tells us that $\underline{b}$ cannot lie in the column-space of $\underline{A}$. Let us solve problem (20-10) for these data. Now we find

$$\underline{A} \otimes (\underline{A}^*\otimes'\underline{b}) = \begin{bmatrix} -3 \\ -2 \\ -1 \end{bmatrix}$$

This agrees with $\underline{b}$ in components two and three but underestimates by 8 in component one. If we add $4 = \frac{1}{2}(8)$ to all components, we have as our solution:

$$\begin{bmatrix} 1 \\ 2 \\ 3 \end{bmatrix}$$

This 3-tuple lies in the column-space of $\underline{A}$ and has a maximum absolute deviation from $\underline{b}$ equal to 4 (in component one). No other element $\underline{v}$ of the column-space of $\underline{A}$ can do better, else by subtracting a constant (i.e. making a scalar multiplication in max algebra) we should be able to derive an element $\underline{u}$ of the column-space of $\underline{A}$ violating Lemma 20-1.

Of course, the procedure generalises to more abstract blogs, but extra axiomatisation is necessary and this would take us too far afield in the context of the present memorandum.

20-3. Critical Paths. Let us reconsider the multiple-cycle example of Section 9-3 in the light of Theorem 20-5.

Taking

$$\underline{A} = \begin{bmatrix} 3 & -\infty & 2 \\ -\infty & 1 & 1 \\ 2 & 3 & 2 \end{bmatrix}, \qquad \underline{z} = \begin{bmatrix} 17 \\ 11 \\ 15 \end{bmatrix},$$

we compute successively $\underline{A}^{*}\otimes'\underline{z}$, $\underline{A}^{2*}\otimes'\underline{z}$, $\underline{A}^{3*}\otimes'\underline{z}$, $A^{4*}\otimes'\underline{z}$, this last being $\begin{bmatrix} 2 \\ 3 \\ 3 \end{bmatrix}$,

We now choose $\underline{x}\leqslant\underline{A}^{4*}\otimes'\underline{z}$, say $\underline{x} = \begin{bmatrix} 2 \\ 1 \\ 2 \end{bmatrix}$.

As discussed in Chapter 9, the 3-tuple $\underline{x}$ gives start-times consistent with the target completion times $\underline{z}$. Suppose in fact we adopt the start-times $\underline{x}$ and we compute successively $\underline{A}\otimes\underline{x}$, $\underline{A}^{2}\otimes\underline{x}$, $\underline{A}^{3}\otimes\underline{x}$ and $\underline{A}^{4}\otimes\underline{x}$. Fig 20-1 sets out all the relevant information.

r	0	1	2	3	4
	(2)	(5)	(8)	11	14
$\underline{A}^{r}\otimes\underline{x}$	1	3	5	8	(11)
	2	4	7	(10)	13
	2	5	8	13	17
$\underline{A}^{(4-r)*}\otimes'\underline{z}$	3	5	7	10	11
	3	6	8	10	15

Fig 20-1: Double Orbit Table

Let us now examine Fig 20-1 to discover whether there are activities which have zero float; as discussed in Chapter 9 these can be identified by components which have equal value in the two 3-tuples in a particular column. Thus in the column r=4 the middle component of the two 3-tuples has the common value 11. For convenience we have ringed all such components in the top row of 3-tuples in Fig 20-1; in all other cases, of course, elements in the top row are strictly less than corresponding elements in the bottom row.

We shall call a table constructed in the manner of Fig 20-1 a double orbit table. The following result is relevant to the properties of such tables.

Theorem 20-6. Let E_1 be a linear blog, let $\underline{A}\in\mathcal{M}_{mn}$ be doubly G-astic, $\underline{b}\in E_m$ be finite and $\underline{x}\in E_n$ satisfy $\underline{A}\otimes\underline{x}\leqslant\underline{b}$. If for some i $(1\leqslant i\leqslant m)$ there holds $\{\underline{A}\otimes\underline{x}\}_i = \{\underline{b}\}_i$ then for some j $(1\leqslant j\leqslant n)$ there hold:

$\{\underline{x}\}_j = \{\underline{A}^*\otimes'\underline{b}\}_j$ and $\{\underline{b}\}_i = \{\underline{A}\}_{ij} \otimes \{\underline{x}\}_j$.

Proof. By hypothesis, $\{\underline{b}\}_i = \sum_{k=1}^{n}{}_{\oplus}\{\underline{A}\}_{ik} \otimes \{\underline{x}\}_k$

and by the linearity of E_1 there is an index j $(1\leq j\leq n)$ such that:

$$\{\underline{b}\}_i = \{\underline{A}\}_{ij} \otimes \{\underline{x}\}_j$$

Moreover, $\{\underline{A}\}_{ij}$ and $\{\underline{x}\}_j$ are finite, having finite product. Hence using the fact that multiplication is self-dual for finite elements:

$$\{\underline{x}\}_j = (\{\underline{A}\}_{ij})^{-1} \otimes' \{\underline{b}\}_i = \{\underline{A}^*\}_{ji} \otimes' \{\underline{b}\}_i$$

$$\geq \sum_{k=1}^{m}{}_{\oplus'}(\{\underline{A}^*\}_{ji} \otimes' \{\underline{b}\}_i)$$

$$= \{\underline{A}^* \otimes' \underline{b}\}_j$$

But $\underline{x} \leq \underline{A}^* \otimes' \underline{b}$, since $\underline{A} \otimes \underline{x} \leq \underline{b}$ by hypothesis (Lemma 9-1). Hence $\{\underline{x}\}_j = \{\underline{A}^*\otimes'\underline{b}\}_j$. ●

Corollary 20-7. Let E_1 be a linear blog and $\underline{A} \in \mathcal{M}_{nn}$ be doubly G-astic. For given integer $N\geq 1$ let $\underline{x},\underline{z}\in E_n$ be compatible for $(\underline{A},N)$. In the corresponding double orbit table, either there is no element which can be ringed to denote zero float, or there is at least one such element in each column.

Proof. Theorem 20-6 implies that if some element may be ringed in column $r\geq 1$ of the table then some element may be ringed in column $(r-1)$. And the dual of Theorem 20-6 implies that if some element may be ringed in column $(r-1) \leq (N-1)$, then some element may be ringed in column r. Hence ringing is possible in all columns, or in none. ●

If element $\{\underline{A}^r\otimes\underline{x}\}_i$ $(r\geq 1)$ of a double orbit table is ringed, then according to Theorem 20-6 we can find some ringed element $\{\underline{A}^{r-1}\otimes\underline{x}\}_j$ in column $(r-1)$ such that:

$$\{\underline{A}\}_{ij} \otimes \{\underline{A}^{r-1}\otimes\underline{x}\}_j = \{\underline{A}^r\otimes\underline{x}\}_i$$

Let us join all such pairs of elements with a line, as shown in Fig 20-1. Then starting from any ringed element in column N, we may trace a continuous path along such lines, moving left until we arrive at a ringed element in Column 0. Such a path is known as a critical path. In terms of project planning, we may say that no activity on the critical path may be delayed without causing the activity at the right-hand end of the path to be late, thus violating the zero-lateness condition. Hence in a practical situation the activities on the critical path would attract great management attention.

There may be more than one critical path. For example, if we choose $\underline{x}=\underline{A}^{4*}\otimes'\underline{z}$ in the preceding example we get the double orbit table shown in Fig. 20-2.

r	0	1	2	3	4
$\underline{A}^r \otimes \underline{x}$	(2)	(5)	(8)	11	14
	(3)	4	(7)	9	(11)
	(3)	(6)	(8)	(10)	13
$\underline{A}^{(4-r)*} \otimes ' \underline{z}$	2	5	8	13	17
	3	5	7	10	11
	3	6	8	10	15

Fig 20-2. Critical Paths

Fig 20-2 contains five critical paths. It is easy to see that Theorem 20-6 together with its dual implies that all and only ringed elements in column 0 begin critical paths; that all and only ringed elements in column N end critical paths; that all and only ringed elements in other columns are intermediate elements of critical paths.

21. PROJECTIONS

21-1. Congruence Classes

Let E_1 be a pre-residuated belt satisfying axiom X_{12} and let $\underline{A} \in \mathcal{M}_{mn}$ for given integers $m,n\geq 1$. As usual, denote by $g_{\underline{A}}$, $g^*_{\underline{A}}$ the functions:

$$\left.\begin{aligned} g_{\underline{A}}: E_n \to E_m \quad \text{where } g_{\underline{A}}(\underline{x}) = \underline{A} \otimes \underline{x} \quad (\underline{x} \in E_n) \\ g^*_{\underline{A}}: E_m \to E_n \quad \text{where } g^*_{\underline{A}}(\underline{y}) = \underline{A}^* \otimes' \underline{y} \quad (\underline{y} \in E_m) \end{aligned}\right\} \qquad (21\text{-}1)$$

In the terminology of sections 8-4 and 18-1, (Ran $g_{\underline{A}}$,$\oplus$) is the column-space of $\underline{A}$ and (Ran $g^*_{\underline{A}}$,$\oplus'$) the dual column-space of $\underline{A}^*$.

If $\underline{b} \in E_m$ is given, then the results of Chapter 20 indicate that $(g_{\underline{A}} o g^*_{\underline{A}})(\underline{b}) = \underline{A}\otimes(\underline{A}^*\otimes'\underline{b})$ is, relative to several different criteria, an element of Ran $g_{\underline{A}}$ which is "nearest" to $\underline{b}$. It is natural to call $(g_{\underline{A}} o g^*_{\underline{A}})(\underline{b})$ the projection of $\underline{b}$ on Ran $g_{\underline{A}}$; similarly for $\underline{c} \in E_n$ we should call $(g^*_{\underline{A}} o g_{\underline{A}})(\underline{c})$ the dual projection of $\underline{c}$ on Ran $g^*_{\underline{A}}$.

We introduce the notations:

$$\pi_{\underline{A}} = g_{\underline{A}} o g^*_{\underline{A}} \; ; \; \pi^*_{\underline{A}} = g^*_{\underline{A}} o g_{\underline{A}} \qquad (21\text{-}2)$$

From the results of Section 10-15:

$$\left.\begin{aligned} g^*_{\underline{A}} \circ \pi_{\underline{A}} = \pi^*_{\underline{A}} \circ g^*_{\underline{A}} = g^*_{\underline{A}} \\ g_{\underline{A}} \circ \pi^*_{\underline{A}} = \pi_{\underline{A}} \circ g_{\underline{A}} = g_{\underline{A}} \\ \pi_{\underline{A}} \circ \pi_{\underline{A}} = \pi_{\underline{A}} \; ; \; \pi^*_{\underline{A}} \circ \pi^*_{\underline{A}} = \pi^*_{\underline{A}} \end{aligned}\right\} \qquad (21\text{-}3)$$

Now we know from Lemma 8-10 that $g_{\underline{A}}\backslash$Ran $g^*_{\underline{A}}$ and $g^*_{\underline{A}}\backslash$Ran $g_{\underline{A}}$ are mutually inverse bijections; but we cannot say that they constitute isomorphisms since it is not at once clear how to endow Ran $g^*_{\underline{A}}$ and Ran $g_{\underline{A}}$ with the structure of commutative belts with duality. It is clear that (Ran $g_{\underline{A}}$,$\oplus$) is a commutative band and (Ran $g^*_{\underline{A}}$,$\oplus'$) a (dual) commutative band, but in general the dualities in $(E_n,\oplus,\oplus')$, $(E_m,\oplus,\oplus')$ do not extend to Ran $g^*_{\underline{A}}$ and Ran $g_{\underline{A}}$. Thus, although it is true to say:

$$g_{\underline{A}}(\underline{x}\oplus\underline{y}) = g_{\underline{A}}(\underline{x})\oplus g_{\underline{A}}(\underline{y}) \qquad (21\text{-}4)$$

for arbitrary $\underline{x}$, $\underline{y} \in E_n$, we cannot apply (21-4) to Ran $g^*_{\underline{A}}$ because in general $\underline{x}\oplus\underline{y} \notin$ Ran $g^*_{\underline{A}}$ for given $\underline{x},\underline{y} \in$ Ran $g^*_{\underline{A}}$.

It is to this problem that we now address ourselves. First, in the light of (21-4) and its dual, we record the following.

Proposition 21-1. Let E_1 be a pre-residuated belt satisfying axiom X_{12} and let $\underline{A} \in \mathcal{M}_{mn}$ for given integers $m,n\geq 1$. Using the notation of (21-1), define binary relations $\sim$, $\overset{*}{\sim}$ on E_n, E_m respectively by:

$$\left.\begin{array}{l} \text{If } \underline{x},\underline{y} \varepsilon E_n, \text{ then } \underline{x} \sim \underline{y} \text{ if and only if } g_{\underline{A}}(\underline{x}) = g_{\underline{A}}(\underline{y}) \\ \text{If } \underline{u},\underline{v} \varepsilon E_m \text{ then } \underline{u} \overset{*}{\sim} \underline{v} \text{ if and only if } g^*_{\underline{A}}(\underline{u}) = g^*_{\underline{A}}(\underline{v}) \end{array}\right\} \quad (21\text{-}5)$$

Then $\sim$, $\overset{*}{\sim}$ constitute congruence relations on $(E_n,\oplus)$, $(E_m,\oplus')$ respectively

Let us call the congruence classes established by (21-5) the $g_{\underline{A}}$-congruence classes and and $g^*_{\underline{A}}$-congruence classes respectively.

Theorem 21-2. Let E_1 be a pre-residuated belt satisfying axiom X_{12} and let $\underline{A} \varepsilon \mathcal{M}_{mn}$ for given integers $m,n\geq 1$. Then each $g_{\underline{A}}$-congruence class of E_n contains exactly one element of $\text{Rang}^*_{\underline{A}}$ and each $g^*_{\underline{A}}$-congruence class of E_m contains exactly one element of $\text{Rang}_{\underline{A}}$.

Proof. From (21-3), $(g_{\underline{A}} o \pi^*_{\underline{A}})(\underline{x}) = g_{\underline{A}}(\underline{x})$ for all $\underline{x} \varepsilon E_n$

$$\text{i.e.} \quad \underline{x} \sim \pi^*_{\underline{A}}(\underline{x}) = (g^*_{\underline{A}} o g_{\underline{A}})(\underline{x}) \varepsilon \text{Rang}^*_{\underline{A}} \qquad (21\text{-}6)$$

Hence the congruence class containing any $\underline{x} \varepsilon E_n$ also contains an element of $\text{Rang}^*_{\underline{A}}$. Suppose in fact $g^*_{\underline{A}}(u)$, $g^*_{\underline{A}}(\underline{v})$ both belong to the same $g_{\underline{A}}$-congruence class, i.e. $g_{\underline{A}}(g^*_{\underline{A}}(\underline{u})) = g_{\underline{A}}(g^*_{\underline{A}}(\underline{v}))$. Then $g^*_{\underline{A}}(\underline{u}) = (g^*_{\underline{A}} o g_{\underline{A}} o g^*_{\underline{A}})(u) = (g^*_{\underline{A}} o g_{\underline{A}} o g^*_{\underline{A}})(v) = g^*_{\underline{A}}(\underline{v})$. The argument for the $g^*_{\underline{A}}$-congruence classes is dual.

Corollary 21-3. Let E_1 be a pre-residuated belt satisfying axiom X_{12}, let $\underline{A} \varepsilon \mathcal{M}_{mn}$ for given integers $m,n\geq 1$, and let $\underline{x},\underline{y} \varepsilon E_n$, $\underline{u},\underline{v} \varepsilon E_m$ Then $\underline{x} \sim \underline{y}$ if and only if $\pi^*_{\underline{A}}(\underline{x}) = \pi^*_{\underline{A}}(\underline{y})$; and $\underline{u} \overset{*}{\sim} \underline{v}$ if and only if $\pi_{\underline{A}}(\underline{u}) = \pi_{\underline{A}}(\underline{v})$.

Proof. Since $\underline{x} \sim \pi^*_{\underline{A}}(x)$ ((21-6)), it follows that $\underline{x} \sim \underline{y}$ if and only if $\pi^*_{\underline{A}}(\underline{x}) \sim \pi^*_{\underline{A}}(\underline{y})$. But a $g_{\underline{A}}$-congruence class contains only one element of $\text{Rang}^*_{\underline{A}}$, whence $\pi^*_{\underline{A}}(\underline{x}) \sim \pi^*_{\underline{A}}(\underline{y})$ if and only if $\pi^*_{\underline{A}}(\underline{x}) = \pi^*_{\underline{A}}(\underline{y})$. The argument for $\underline{u}$ and $\underline{v}$ is dual.

We may paraphrase the preceding results by saying that each $g^*_{\underline{A}}$-congruence class consists exactly of the elements of E_m having a common projection on $\text{Rang}_{\underline{A}}$, and includes that projection itself (by (21-3)); and the dual holds.

The situation is represented diagramatically in Fig 21-1.

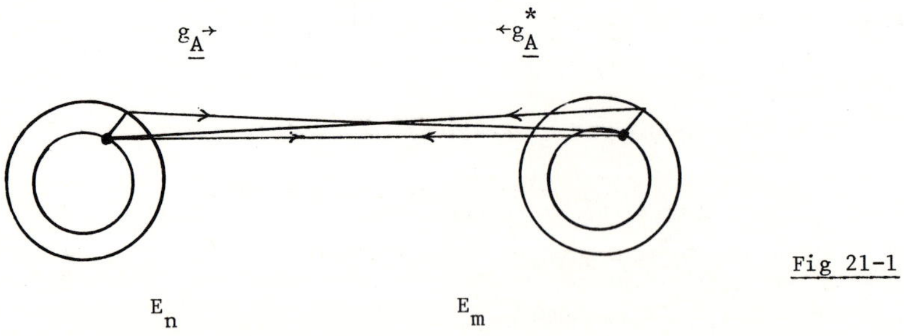

Fig 21-1

The left-hand annulus represents E_n, each radius representing a $g_{\underline{A}}$-congruence class. The inner circumference represents $\text{Rang}^*_{\underline{A}}$, intersecting each radius in a unique point, which we may call its base point. The projection $\pi^*_{\underline{A}}$ condenses each radius onto its intersection with the inner circumference. The right-hand annulus represents E_m similarly. The functions $g_{\underline{A}}, g^*_{\underline{A}}$ take whole radii in one annulus into base points of radii in the other annulus, and act as mutually inverse bijections between inner circumferences.

21-2. Operations in $\text{Rang}_{\underline{A}}$. We are now in a position to introduce a dual addition in $\text{Rang}_{\underline{A}}$. We define:

$$\underline{u}\ \dot{\oplus}'\ \underline{v} = \pi_{\underline{A}}(\underline{u}\oplus'\underline{v}) \quad (\underline{u},\ \underline{v} \in \text{Rang}_{\underline{A}}) \tag{21-7}$$

Theorem 21-4. Let E_1 be a pre-residuated belt satisfying axiom X_{12} and let $\underline{A} \in \mathcal{M}_{mn}$ for given integers $m,n \geq 1$. Then $(\text{Rang}_{\underline{A}}, \oplus, \dot{\oplus}')$ is a commutative band with duality, when $\dot{\oplus}'$ is as in (21-7).

Proof. Certainly $(\text{Rang}_{\underline{A}}, \oplus)$ is a commutative band. Now let $\underline{u}, \underline{v}, \underline{w} \in \text{Rang}_{\underline{A}}$ throughout. Then $\underline{u} = \pi_{\underline{A}}(\underline{u})$, $\underline{v} = \pi_{\underline{A}}(\underline{v})$ and $\underline{w} = \pi_{\underline{A}}(\underline{w})$.

$$\text{Now, } \underline{u}\ \dot{\oplus}'\ \underline{u} = \pi_{\underline{A}}(\underline{u}\oplus'\underline{u}) = \pi_{\underline{A}}(\underline{u}) = \underline{u} \tag{21-8}$$

$$\text{And } \underline{u}\ \dot{\oplus}'\ \underline{v} = \pi_{\underline{A}}(\underline{u}\oplus'\underline{v}) = \pi_{\underline{A}}(\underline{v}\oplus'\underline{u}) = \underline{v}\dot{\oplus}'\underline{u}. \tag{21-9}$$

We now prove the associativity of the operation $\dot{\oplus}'$. From (21-3) and (21-7),

$$\underline{u}\ \dot{\oplus}'\ \underline{v} \overset{*}{\sim} \underline{u}\ \oplus'\ \underline{v}$$

Since $\overset{*}{\sim}$ is a congruence relation on $(E_m, \oplus')$ we infer:

$$(\underline{u}\dot{\oplus}'\underline{v})\ \oplus'\ \underline{w} \overset{*}{\sim} \underline{u}\ \oplus'\ \underline{v}\ \oplus'\ \underline{w} \tag{21-10}$$

Whence $(\underline{u}\dot{\oplus}'\underline{v})\ \dot{\oplus}'\ \underline{w} = \pi_{\underline{A}}((\underline{u}\dot{\oplus}'\underline{v})\ \oplus'\ \underline{w})$ (by (21-7))

$= \pi_{\underline{A}}(\underline{u}\oplus'\underline{v}\oplus'\underline{w})$ (by Corollary 21-3 and (21-10))

Similarly $\underline{u}\dot{\oplus}'(\underline{v}\dot{\oplus}'\underline{w}) = \pi_{\underline{A}}(\underline{u}\oplus'\underline{v}\oplus'\underline{w})$

and associativity is established. Hence $(\text{Rang}_{\underline{A}}, \dot{\oplus}')$ is a (dual) commutative band. We must now establish the lattice absorbtion laws L_3.

Now $\underline{u}\oplus(\underline{v}\dot{\oplus}'\underline{u}) = \pi_{\underline{A}}(\underline{u}) \oplus \pi_{\underline{A}}(\underline{v}\oplus'\underline{u})$ (using $\underline{u} = \pi_{\underline{A}}(\underline{u})$ and (21-7))

$= (g_{\underline{A}}\circ g^*_{\underline{A}})(\underline{u})\oplus(g_{\underline{A}}\circ g^*_{\underline{A}})(\underline{v}\oplus'\underline{u})$

$= g_{\underline{A}}(g^*_{\underline{A}}(\underline{u})\oplus g^*_{\underline{A}}(\underline{v}\oplus'\underline{u}))$ (by linearity of $g_{\underline{A}}$)

$= g_{\underline{A}}(g^*_{\underline{A}}(\underline{u})\oplus(g^*_{\underline{A}}(\underline{v})\oplus' g^*_{\underline{A}}(\underline{u})))$ (by dual linearity of $g^*_{\underline{A}}$)

$= g_{\underline{A}}(g^*_{\underline{A}}(\underline{u}))$ (by L_3)

$= \pi_{\underline{A}}(\underline{u})$

$= \underline{u}$

Again $(\underline{u}\oplus\underline{v})\dot{\oplus}'\underline{u}$ $= \pi_{\underline{A}}((\underline{u}\oplus\underline{v})\oplus'\underline{u})$ (by (21-7))

$= \pi_{\underline{A}}(\underline{u})$ (by L_3)

$= \underline{u}.$

The proof is now complete.

Obviously $(\text{Rang}_{\underline{A}},\oplus)$ admits the same right scalar multiplication as $(E_m,\oplus)$ and is therefore a subspace of $(E_m,\oplus)$. Now let us define a right multiplication:

$$\underline{u}\ \dot{\otimes}'\ \lambda = \pi_{\underline{A}}(\underline{u}\otimes'\lambda)\quad (\underline{u}\in\text{Rang}_{\underline{A}},\ \lambda\in E_1) \tag{21-11}$$

Theorem 21-5. Let E_1 be a pre-residuated belt satisfying axiom X_{12} and let $\underline{A}\in\mathcal{M}_{mn}$ for given integers $m,n\geqslant 1$. Then $(\text{Rang}_{\underline{A}},\dot{\oplus}')$ is a (dual) right band-space over E_1 when $\dot{\oplus}'$ is as in (21-7) and dual scalar multiplication $\dot{\otimes}'$ is as in (21-11).

Proof. Let $\underline{u},\underline{v}\in\text{Rang}_{\underline{A}}$; $\lambda,\mu\in E_1$. We confirm (appropriate duals of) S_2, S_3, S_4 of Section 5-1, as follows.

Consider $(\underline{u}\ \dot{\otimes}'\ \mu)\ \dot{\otimes}'\lambda = \pi_{\underline{A}}\ ((\underline{u}\ \dot{\otimes}'\ \mu)\ \otimes'\lambda)$ (by 21-11))

$= (g_{\underline{A}}\circ g^*_{\underline{A}})\ ((\pi_{\underline{A}}(\underline{u}\otimes'\mu))\otimes'\lambda)$ (by 21-2), (21-11))

$= g_{\underline{A}}(((g^*_{\underline{A}}\circ\pi_{\underline{A}})(\underline{u}\otimes'\mu))\otimes'\lambda)$ (by dual of (5-1))

$= g_{\underline{A}}((g^*_{\underline{A}}(\underline{u}\otimes'\mu))\otimes'\lambda)$ (by 21-3)

$= g_{\underline{A}}(g^*_{\underline{A}}((\underline{u}\otimes'\mu)\otimes'\lambda))$ (by dual of (5-1))

$= \pi_{\underline{A}}(\underline{u}\otimes'(\mu\otimes'\lambda))$ (by (21-2) and dual of S_2 for E_1)

$= \underline{u}\ \dot{\otimes}'\ (\mu\otimes'\lambda)$ (by (21-11)

This confirms the dual of S_2. Now consider:

$(\underline{u}\dot{\oplus}'\underline{v})\dot{\otimes}'\lambda = \pi_{\underline{A}}((\underline{u}\dot{\oplus}'\underline{v})\otimes'\lambda)$ (by 21-11))

$= g_{\underline{A}}((g^*_{\underline{A}}\circ\pi_{\underline{A}}(\underline{u}\oplus'\underline{v}))\otimes'\lambda)$ (by (21-7) and dual of (5-1))

$= g_{\underline{A}}((g^*_{\underline{A}}(\underline{u}\oplus'\underline{v}))\otimes'\lambda)$ (by 21-3))

$= g_{\underline{A}}(g^*_{\underline{A}}((\underline{u}\oplus'\underline{v})\otimes'\lambda))$ (by dual of (5-1))

$= g_{\underline{A}}(g^*_{\underline{A}}((\underline{u}\otimes'\lambda)\oplus'(\underline{v}\otimes'\lambda)))$ (by dual of S_3 for E_1)

$= g_{\underline{A}}(g^*_{\underline{A}}(\underline{u}\otimes'\lambda)\oplus' g^*_{\underline{A}}(\underline{v}\otimes'\lambda))$ (by dual linearity of $g^*_{\underline{A}}$)

$= g_{\underline{A}}((g^*_{\underline{A}} o \pi_{\underline{A}})(\underline{u}\otimes'\lambda)\oplus'(g^*_{\underline{A}} o \pi_{\underline{A}})(\underline{v}\otimes'\lambda))$ (by (21-3))

$= g_{\underline{A}}(g^*_{\underline{A}}(\underline{u}\dot{\otimes}'\lambda)\oplus' g^*_{\underline{A}}(\underline{v}\dot{\otimes}'\lambda))$ (by (21-11))

$=(g_{\underline{A}} o g^*_{\underline{A}})((\underline{u}\dot{\otimes}'\lambda)\oplus'(\underline{v}\dot{\otimes}'\lambda))$ (by dual linearity of $g^*_{\underline{A}}$)

$= (\underline{u}\dot{\otimes}'\lambda)\dot{\oplus}'(\underline{v}\dot{\otimes}'\lambda)$ (by (21-2) and (21-7))

This confirms the dual of S_3. Finally, consider:

$\underline{u}\dot{\otimes}'(\lambda\oplus'\mu) = \pi_{\underline{A}}(\underline{u}\otimes'(\lambda\oplus'\mu))$ (by (21-7))

$= (g_{\underline{A}} o g^*_{\underline{A}})((\underline{u}\otimes'\lambda)\oplus'(\underline{u}\otimes'\mu))$ (by (21-2) and dual of S_4 for E_1)

$= g_{\underline{A}}(g^*_{\underline{A}}(\underline{u}\otimes'\lambda)\oplus' g^*_{\underline{A}}(\underline{u}\otimes'\mu))$ (by dual linearity of $g^*_{\underline{A}}$)

$= g_{\underline{A}}((g^*_{\underline{A}} o \pi_{\underline{A}})(\underline{u}\otimes'\lambda)\oplus'(g^*_{\underline{A}} o \pi_{\underline{A}})(\underline{u}\otimes'\mu))$ (by (21-3))

$= g_{\underline{A}}(g^*_{\underline{A}}(\underline{u}\dot{\otimes}'\lambda)\oplus' g^*_{\underline{A}}(\underline{u}\dot{\otimes}'\mu))$ (by (21-11))

$= (g_{\underline{A}} o g^*_{\underline{A}})((\underline{u}\dot{\otimes}'\lambda)\oplus'(\underline{u}\dot{\otimes}'\mu))$ (by dual linearity of $g^*_{\underline{A}}$)

$= (\underline{u}\dot{\otimes}'\lambda)\dot{\oplus}'(\underline{u}\dot{\otimes}'\mu)$ (by (21-11))

This confirms the dual of S_4 and the proof is complete.

The following theorem now establishes the isomorphism discussed in Section 21-1.

Theorem 21-6. Let E_1 be a pre-residuated belt satisfying axiom X_{12} and let $\underline{A} \in \mathcal{M}_{mn}$ for given integers $m,n \geq 1$. Then $(Rang_{\underline{A}}, \oplus, \dot{\oplus}')$ and $(Rang^*_{\underline{A}}, \dot{\oplus}, \oplus')$ are isomorphic as right spaces with duality over E_1 when $\dot{\oplus}'$, $\dot{\oplus}$ are as in (21-7) and its dual respectively, and dual scalar multiplication for $Rang_{\underline{A}}$ and scalar multiplication for $Rang^*_{\underline{A}}$ are as in (21-11) and its dual respectively.

Proof. That $(Rang_{\underline{A}}, \oplus, \dot{\oplus}')$ and $(Rang^*_{\underline{A}}, \dot{\oplus}, \oplus')$ are spaces with duality is established in the preceding results. Moreover, $g_{\underline{A}}$ and $g^*_{\underline{A}}$ provide mutually inverse bijections. Now for all $\underline{x},\underline{y} \in Rang^*_{\underline{A}}$, we have:

$g_{\underline{A}}(\underline{x}\dot{\oplus}\underline{y}) = g_{\underline{A}} o \pi^*_{\underline{A}}(\underline{x}\oplus\underline{y})$ (by dual of (21-7))

$$= \underline{g}_{\underline{A}}(\underline{x}\oplus\underline{y}) \qquad \text{(by (21-3))}$$

$$= g_{\underline{A}}(\underline{x})\oplus g_{\underline{A}}(\underline{y}) \qquad \text{(by linearity of } g_{\underline{A}})$$

Also: $g_{\underline{A}}(\underline{x}\oplus'\underline{y}) = g_{\underline{A}}(g^*_{\underline{A}}(g_{\underline{A}}(\underline{x}))\oplus' g^*_{\underline{A}}(g_{\underline{A}}(\underline{y})))$

$$= (g_{\underline{A}} o g^*_{\underline{A}})(g_{\underline{A}}(\underline{x})\oplus' g_{\underline{A}}(\underline{y}) \qquad \text{(by dual linearity of } g^*_{\underline{A}})$$

$$= g_{\underline{A}}(\underline{x})\dot{\oplus}' g_{\underline{A}}(\underline{y}) \qquad \text{(by(21-7))}.$$

Finally, for all $\underline{x} \in \text{Rang}^*_{\underline{A}}$ and $\lambda\varepsilon E_1$, we have:

$$g_{\underline{A}}(\underline{x}\otimes'\lambda) = g_{\underline{A}}(\pi^*_{\underline{A}}(\underline{x})\otimes'\lambda) \qquad \text{(because } \underline{x} = \pi^*_{\underline{A}}(\underline{x}) \text{ if } \underline{x}\varepsilon\text{Rang}^*_{\underline{A}})$$

$$= g_{\underline{A}}(g^*_{\underline{A}}(g_{\underline{A}}(\underline{x}))\otimes'\lambda) \qquad \text{(by (21-2))}$$

$$= g_{\underline{A}}(g^*_{\underline{A}}(g_{\underline{A}}(\underline{x})\otimes'\lambda) \qquad \text{(by dual of (5-1))}$$

$$= (g_{\underline{A}} o g^*_{\underline{A}})(g_{\underline{A}}(\underline{x})\otimes'\lambda)$$

$$= g_{\underline{A}}(\underline{x})\dot{\otimes}'\lambda \qquad \text{(by (21-11))}$$

The arguments for $g^*_{\underline{A}}$ proceed dually and the theorem is proved. ●

By virtue of Theorem 21-6 we can carry algebraic relations between $\text{Rang}_{\underline{A}}$ and $\text{Rang}^*_{\underline{A}}$ by means of $g_{\underline{A}}$, $g_{\underline{A}*}$. Thus, if a linear dependence:

$$\underline{b} = \sum_{j=1}^{s}{}_{\oplus}(\underline{u}(j)\otimes\lambda_j)$$

holds in $\text{Rang}_{\underline{A}}$, we can infer:

$$\underline{A}^*\otimes'\underline{b} = \sum_{j=1}^{s}{}_{\dot{\oplus}}((\underline{A}^*\otimes'\underline{u}(j))\dot{\otimes}\lambda_j)$$

21-3. Projection matrices. In Section 21-1 we introduced the projection operators $\pi_{\underline{A}}$, $\pi^*_{\underline{A}}$ for a given matrix $\underline{A} \in \mathcal{M}_{mn}$. If n=1, consider the projection operators $\pi_{\underline{\xi}}$, $\pi^*_{\underline{\xi}*}$ for given $\underline{\xi} \in E_m$. For arbitrary $\underline{x} \in E_m$ there holds:

$$\pi_{\underline{\xi}}(\underline{x}) = \underline{\xi}\otimes(\underline{\xi}^*\otimes'\underline{x}); \; \pi^*_{\underline{\xi}*}(\underline{x}) = \underline{\xi}\otimes'(\underline{\xi}^*\otimes\underline{x}): \qquad (21\text{-}12)$$

Now, if E_1 is a given self-conjugate belt, let $\underline{\xi}\varepsilon E_m$, and consider the matrices:

$$\left.\begin{aligned} \underline{P}(\underline{\xi}) &= \underline{\xi}\otimes'\underline{\xi}^* \in \mathcal{M}_{mm} \\ \underline{P}^*(\underline{\xi}) &= \underline{\xi}\otimes\xi^* \in \mathcal{M}_{mm} \end{aligned}\right\} \qquad (21\text{-}13)$$

We shall call $\underline{P}(\underline{\xi})$ the projection matrix associated with $\underline{\xi}$ and $\underline{P}^*(\underline{\xi})$ the dual projection matrix associated with $\underline{\xi}$. (These definitions may appear to be the wrong way round, but the following result motivates the choice of names).

Theorem 21-7. If E_1 is a blog, and $\underline{\xi} \in E_m$ is finite, (for given integer $m \geq 1$) then for any $\underline{x} \in E_m$ there hold:

$$\left.\begin{aligned} \underline{P}(\underline{\xi}) \otimes' \underline{x} &= \underline{\xi} \otimes (\underline{\xi}^* \otimes' \underline{x}) = \pi_{\underline{\xi}}(\underline{x}) \\ \underline{P}^*(\underline{\xi}) \otimes \underline{x} &= \underline{\xi} \otimes' (\underline{\xi}^* \otimes \underline{x}) \end{aligned}\right\} \qquad (21\text{-}14)$$

Proof. For i=1,..., m we have:

$$\{\underline{P}(\underline{\xi}) \otimes' \underline{x}\}_i = \sum_{j=1}^{m}{}_{\oplus'} \left((\{\underline{\xi}\}_i \otimes' (\{\underline{\xi}\}_j)^*) \otimes' \{\underline{x}\}_j\right)$$

$$= \{\underline{\xi}\}_i \otimes' \sum_{j=1}^{m}{}_{\oplus'} ((\{\underline{\xi}\}_j)^* \otimes' \{\underline{x}\}_j)$$

$$= \{\underline{\xi}\}_i \otimes' (\underline{\xi}^* \otimes' \underline{x})$$

However, if $\underline{\xi}$ is finite then multiplication by $\{\underline{\xi}\}_i$ is self-dual, and the first line of (21-14) follows. The second line is proved similarly. ●

In other words, (21-14) asserts that $\underline{P}(\underline{\xi})$ can be used (when $\underline{\xi}$ is finite) to map any arbitrary vector into a scalar multiple of $\underline{\xi}$, namely $\pi_{\underline{\xi}}(\underline{x})$.

The following result is the exact analogue of a classical result for projection matrices - namely that they are just the idempotent self-conjugate square matrices.

Theorem 21-8. Let E_1 be a blog, and let $\underline{P} \in \mathcal{M}_{mm}$ (for given integer $m \geq 1$) be finite. Then necessary and sufficient conditions that $\underline{P}$ be the projection matrix associated with some $\underline{\xi} \in E_m$ are:

(i) $\underline{P} = \underline{P}^*$ and
(ii) $\underline{P}^2 = \underline{P}$ (so $\underline{P}^{2*} = \underline{P}^*$)

Proof. Suppose firstly that $\underline{P} = \underline{P}(\underline{\xi})$. Then $\underline{\xi}$ is finite, since each product $\{\underline{\xi}\}_i \otimes' (\{\underline{\xi}\}_j)^*$ is finite. Hence:

(i) $\{\underline{P}^*\}_{ij} = (\{\underline{P}\}_{ji})^* = (\{\underline{\xi}\}_j \otimes' (\{\underline{\xi}\}_i)^*)^*$

$$= \{\underline{\xi}\}_i \otimes (\{\underline{\xi}\}_j)^*$$

$$= \{\underline{\xi}\}_i \otimes' (\{\underline{\xi}\}_j)^* \qquad \text{(because } \underline{\xi} \text{ finite)}$$

$$= \{\underline{P}\}_{ij}$$

(ii) $\{\underline{P}^2\}_{ij} = \sum_{k=1}^{m} {}_{\oplus}\left((\{\underline{\xi}\}_i \otimes' (\{\underline{\xi}\}_k)^*) \otimes (\{\underline{\xi}\}_k \otimes' (\{\underline{\xi}\}_j)^*)\right)$

But since $\underline{\xi}$ is finite, so that $(\{\underline{\xi}\}_k)^* = (\{\underline{\xi}\}_k)^{-1}$ and multiplication is self-dual, we may infer:

$$\{\underline{P}^2\}_{ij} = \sum_{k=1}^{m} {}_{\oplus} (\{\underline{\xi}\}_i \otimes' (\{\underline{\xi}\}_j)^*)$$

$$= \{\underline{\xi}\}_i \otimes' (\{\underline{\xi}\}_j)^*$$

$$= \{\underline{P}\}_{ij}$$

Hence $\underline{P}^2 = \underline{P}$ and so $\underline{P}^{2*} = \underline{P}^*$

Conversely suppose that (i) and (ii) hold. Then:

$$\underline{P} \otimes \underline{P}^* = \underline{P}^2 = \underline{P} = \underline{P}^* = \underline{P}^{2*} = \underline{P} \otimes' \underline{P}^*$$

Now for given i,j ($1 \leqslant i \leqslant m$; $1 \leqslant j \leqslant m$), we have:

$$\{\underline{P} \otimes \underline{P}^*\}_{ij} = \sum_{k=1}^{m} {}_{\oplus} (\{\underline{P}\}_{ik} \otimes (\{\underline{P}\}_{jk})^*)$$

$$\geqslant \{\underline{P}\}_{i1} \otimes (\{\underline{P}\}_{j1})^* \qquad \text{(by (3-15))}$$

$$= \{\underline{P}\}_{i1} \otimes' (\{\underline{P}\}_{j1})^* \qquad \text{(since } \underline{P} \text{ is finite)}$$

$$\geqslant \sum_{k=1}^{m} {}_{\oplus'} (\{\underline{P}\}_{ik} \otimes' (\{\underline{P}\}_{jk})^*) \qquad \text{(by (3-17))}$$

$$= \{\underline{P} \otimes' \underline{P}^*\}_{ij}$$

So, since in fact we know $\underline{P} \otimes \underline{P}^* = \underline{P} \otimes' \underline{P}^*$, we infer from the foregoing inequalities that $\{\underline{P} \otimes' \underline{P}^*\}_{ij} = \{\underline{P}\}_{i1} \otimes' (\{\underline{P}\}_{j1})^*$, i.e.:

$$\underline{P} \otimes' \underline{P}^* = \underline{\xi} \otimes' \underline{\xi}^* \text{ where } \underline{\xi} = \begin{bmatrix} \{\underline{P}\}_{11} \\ \vdots \\ \{\underline{P}\}_{m1} \end{bmatrix} \varepsilon\ E_m.$$

But $\underline{P} = \underline{P} \otimes' \underline{P}^*$. Hence $\underline{P} = \underline{\xi} \otimes' \underline{\xi}^* = \underline{P}(\underline{\xi})$

We consider now some properties of the matrix $\underline{P}(\underline{\xi})$ when $\underline{\xi}$ is chosen from the column-space of a given matrix.

Lemma 21-9. Let E_1 be a pre-residuated belt satisfying axiom X_{12}, let $\underline{A}\ \varepsilon \mathcal{M}_{mn}$ (for given integers $m,n \geqslant 1$), have columns $\underline{a}(1),\ldots,\underline{a}(n)$ and let $\underline{u}\ \varepsilon\ E_n$. Then $\underline{P}(\underline{\xi}) \geqslant \sum_{k=1}^{n} {}_{\oplus'} \underline{P}(\underline{a}(k)) = \underline{A} \otimes' \underline{A}^*$, if $\underline{\xi}$ is either $\underline{A} \otimes \underline{u}$ or $\underline{A} \otimes' \underline{u}$.

Proof. If $\underline{\xi}$ is $\underline{A}\otimes\underline{u}$ then $\underline{P}(\underline{\xi}) = (\underline{A}\otimes\underline{u}) \otimes' (\underline{A}\otimes\underline{u})^*$

$$= (\underline{A}\otimes\underline{u})\otimes'\underline{u}^*\otimes'\underline{A}^*$$

However from Theorem 8-8 with $\underline{u}$ and $\underline{A}$ in the roles of $\underline{A}$ and $\underline{X}$ respectively:

$$(\underline{A}\otimes\underline{u}) \otimes' \underline{u}^* \geq \underline{A}$$

After an (isotone) right dual multiplication by $\underline{A}^*$, we derive: $\underline{P}(\underline{\xi}) \geq \underline{A}\otimes'\underline{A}^*$

Similarly, if ξ is $\underline{A}\otimes'\underline{u}$ then $\underline{P}(\underline{\xi}) = (\underline{A}\otimes'\underline{u})\otimes'(\underline{A}\otimes'\underline{u})^*$

$$= \underline{A}\otimes'\underline{u}\otimes'(\underline{u}^*\otimes\underline{A}^*)$$

$$\geq \underline{A}\otimes'\underline{A}^*$$

Now, $\{\underline{A}\otimes'\underline{A}^*\}_{ij} = \sum_{k=1}^{n}{}_{\oplus'}\{\underline{A}\}_{ik}\otimes'\{\underline{A}^*\}_{kj}$

$$= \sum_{k=1}^{n}{}_{\oplus'}(\{\underline{a}(k)\}_i\otimes'(\{\underline{a}(k)\}_j)^*)$$

$$= \{ \sum_{k=1}^{n}{}_{\oplus'} (\underline{a}(k) \otimes' (\underline{a}(k))^*)\}_{ij}$$

$$= \{ \sum_{k=1}^{n}{}_{\oplus'}\underline{P}(\underline{a}(k))\}_{ij}$$

Thus $\underline{A}\otimes'\underline{A}^*$ may alternatively be written $\sum_{k=1}^{n}{}_{\oplus'}\underline{P}(\underline{a}(k))$. ●

The foregoing result sets the lower bound $\underline{A}\otimes'\underline{A}^*$ on $\underline{P}(\underline{\xi})$; we now seek an upper bound.

Corollary 21-10. Let E_1 be a blog, let $\underline{A} \in \mathcal{M}_{mn}$ (for given integers $m,n\geq 1$) be finite and let $\underline{\xi} \in E_m$ be finite, where $\underline{\xi}$ is either $\underline{A}\otimes\underline{u}$ or $\underline{A}\otimes'\underline{u}$ for some $\underline{u}\in E_n$. Then

$$\underline{P}(\underline{\xi}) \leq \sum_{k=1}^{n}{}_{\oplus} \underline{P}(\underline{a}(k)) = \underline{A} \otimes \underline{A}^*.$$

Proof. Taking the conjugate of Lemma 21-9:

$$\underline{P}^*(\underline{\xi}) \leq \sum_{k=1}^{n}{}_{\oplus} \underline{P}^*(\underline{a}(k)) = \underline{A} \otimes \underline{A}^*$$

But if $\underline{\xi}$, $\underline{a}(k)$ are finite then by Theorem 21-8, $\underline{P}^*(\underline{\xi}) = \underline{P}(\underline{\xi})$ and $\underline{P}^*(\underline{a}(k)) = \underline{P}(\underline{a}(k))$ and the result follows. ●

We shall make use of the properties of projection matrices later, in Chapter 26.

22. DEFINITE AND METRIC MATRICES

22-1. Some Graph Theory

The topic which we propose to discuss in the following chapters is the eigenvector-eigenvalue problem for square matrices, introduced in Section 1-2.1. As a preliminary, however, it will be convenient to develop some graph-theoretical tools, and establish the properties of certain classes of matrices.

A directed graph or as we shall simply say, a graph, is a pair (N,F), where N is a finite set consisting of $n \geq 1$ elements, called nodes, denoted by the numerals $1, \ldots, n$, and F is a mapping which associates to each node $i \varepsilon N$ a subset F(i) of N. The graph is said to be of order n.

A pair of nodes (i,j) with $j \varepsilon F(i)$ is said to be an arc. If to each arc (i,j) we associate a weight $a_{ij} \varepsilon E_1$ where $(E_1, \oplus, \otimes)$ is a given belt, we say that (N,F) is a weighted graph. A path (from $i_0 \varepsilon N$ to $i_t \varepsilon N$) is a sequence:

$$(i_0, i_1, \ldots, i_t) \tag{22-1}$$

of nodes $i_k \varepsilon N$ such that $i_k \varepsilon F(i_{k-1})$, $(k=1,2,\ldots, t)$. The nodes $i_0, \ldots, i_t$ will be said to be on the path (22-1), and the path will be said to pass through them.
The length of the path (22-1) is t.
The path (22-1) is said to be an elementary path if the nodes $i_0, \ldots, i_t$ are pairwise distinct. The length of an elementary path is thus at most n-1.
A path (22-1) is called a circuit, if $i_0 = i_t$, and it is an elementary circuit if $i_0 = i_t$ and the path $(i_0, i_1, \ldots, i_{t-1})$ is elementary.
If the circuit consists of a single arc, then it is called a loop. The length of a circuit is its length as a path; in particular, loops have length one. Elementary circuits have length at most n. Notice that the definition of a circuit does not regard e.g. (i_0, i_1, i_2, i_0) as being the same circuit as e.g. (i_1, i_2, i_0, i_1).

To each path $(i_0, i_1, i_2, \ldots, i_t)$ of a weighted graph (N,F), we may associate a (path) product (if $i_0 = i_t$ then (circuit) product) defined as:

$$a_{i_0 i_1} \otimes \ldots \otimes a_{i_{t-1} i_t} \tag{22-2}$$

If E_1 is a belt with duality then path dual product and circuit dual product are defined in the obvious way.

For example, in Fig 22-1, where an abstract graph is given its usual visual presentation and the weights are from the principal interpretation:

(1,2,3) is an elementary path from 1 to 3. The path product is $a_{12} \otimes a_{23} = (-2) + (-4) = -6$.

(1,2,3,1) is an elementary circuit. The circuit product p is given by $a_{12} \otimes a_{23} \otimes a_{31} = (-2) + (-4) + 0 = -6$. The length of this circuit is t=3.

A graph (N,F) is called strongly complete if for every pair of nodes $i, j \varepsilon F$, there holds $j \varepsilon F(i)$. For a strongly complete weighted graph we can define the associated matrix $\underline{A} \varepsilon \mathcal{M}_{nn}$ by:

$$\{\underline{A}\}_{ij} = [a_{ij}] \qquad (i,j=1,\ldots, n) \tag{22-3}$$

Conversely, given such a square matrix over E_1, we can obviously define the associated (strongly complete, weighted) graph by first constructing a strongly complete graph of order n by setting $N=\{1,2,\ldots, n\}$ and $F(i)=N$ for all $i\varepsilon N$, and then introducing the weights a_{ij} as in (22-3).

For example, in the principal interpretation, consider the matrix

$$\underline{A} = \begin{bmatrix} 5 & -2 & 1 \\ -3 & 6 & -4 \\ 0 & 2 & 7 \end{bmatrix}$$

The associated graph is shown in Fig 22-1.

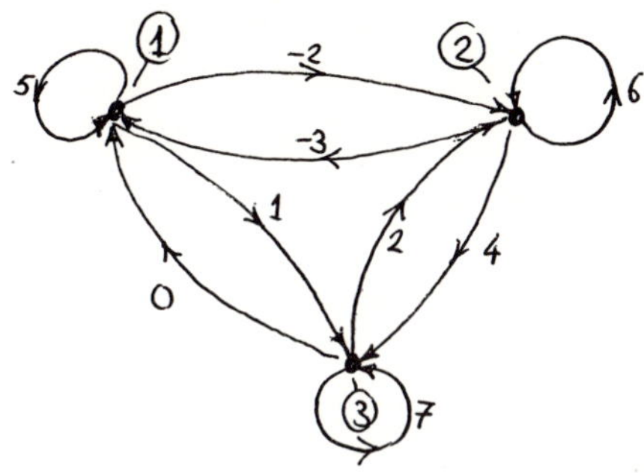

Fig 22-1. A strongly complete weighted graph

As a standard notation, we shall use $\Delta(\underline{A})$ to denote the graph associated with a given matrix $\underline{A}$.

In the sequel, we shall make frequent use of the following simple considerations. If $\underline{A}$ is a given matrix then:

$$\{\underline{A}^2\}_{ij} = \sum_{k=1}^{n}{}_{\oplus}(\{\underline{A}\}_{ik} \otimes \{\underline{A}\}_{kj}) \qquad (i,j=1,\ldots, n)$$

so each element of $\underline{A}^2$ is a summation of path products of paths of length two in $\Delta(\underline{A})$. Similarly:

$$\{\underline{A}^3\}_{ij} = \sum_{k=1}^{n}{}_{\oplus} \sum_{h=1}^{n}{}_{\oplus} (\{\underline{A}\}_{ik} \otimes \{\underline{A}\}_{kh} \otimes \{\underline{A}\}_{hj})$$

which is a summation of all possible path products of paths from i to j of length 3 in $\Delta(\underline{A})$. Summarising:

Proposition 22-1. Let E_1 be a belt and $\underline{A}\ \varepsilon\ \mathcal{M}_{nn}$ for given integer $n\geqslant 1$. Then for all integers i,j $(1\leqslant j\leqslant n;\ 1\leqslant j\leqslant n)$ and $r\geqslant 1$, we have:

$$\{\underline{A}^r\}_{ij} = \sum_{k}{}_{\oplus} p_k$$

where each p_k is a path product for a path of length r from i to j in $\Delta(\underline{A})$ and the summation is over all possible such paths. Hence also $\{\underline{A}^r\}_{ij}\geqslant p_k$ for every such path. ●

<u>22-2 . Definite Matrices</u>. Suppose now that E_1 is a blog, and $\underline{B} \in \mathcal{M}_{nn}$. Adapting a term from Carré [17] we shall say that $\underline{B}$ is <u>definite</u> if it satisfies the following condition:

Each circuit σ in $\Delta(\underline{B})$ has circuit product $p(\sigma)\leq\emptyset$, and at least one such circuit has $p(\sigma)=\emptyset$ (22-4)

<u>Lemma 22-2. Let E_1 be a blog and let $\underline{B} \in \mathcal{M}_{nn}$ for given integer $n\geq 1$. Then $\underline{B}$ is definite if and only if there holds:</u>

<u>Each elementary circuit σ in $\Delta(\underline{B})$ has circuit product $p(\sigma)\leq\emptyset$, and at least one such circuit has $p(\sigma)=\emptyset$</u> (22-5)

<u>Proof</u>. Let $\underline{B}$ merely satisfy the condition:

Each elementary circuit σ in $\Delta(\underline{B})$ has circuit-product $p(\sigma)\leq\emptyset$ (22-6)

Now make the following induction hypothesis, with induction variable k:

For each circuit τ of $\Delta(\underline{B})$ of length $t(\tau)\leq k$, there exists an elementary circuit δ of $\Delta(\underline{B})$ such that $p(\tau)\leq p(\delta)$ (22-7)

Certainly, (22-7) is true for k=1, since if $t(\tau)=1$ then τ is a loop and so elementary, whence $p(\tau)\leq p(\delta)$ holds with $\delta=\tau$.

Suppose hypothesis (22-7) then true for a particular value of $k\geq 1$ and let σ be any non-elementary circuit having $t(\sigma)=k+1$. (Such circuits certainly exist - e.g. $(i_0,i_0,\ldots, i_0)$).

Suppose σ is $(i_0,i_1,\ldots, i_k,i_0)$. Now, as we trace σ starting from i_0 and visiting in turn $i_1,\ldots, i_k$, let j be the first node to be encountered for the second time. Then σ contains an elementary sub-circuit η which runs from the first occurrence of j to the second occurrence of j. Let τ be the circuit obtained by tracing σ from i_0 to the first occurrence of j, and then from the nodes following the second occurrence of j to the end.

Then $p(\eta)\leq\emptyset$ by (22-6). Now $p(\sigma)$ is calculated as a product of terms consisting of those making up $p(\tau)$, interrupted by (or preceded by or followed by) those making up η. So by isotonicity, $p(\sigma)\leq p(\tau)$. But evidently $t(\tau)<k$, so by (22-7) there exists an elementary circuit δ of $\Delta(\underline{B})$ such that $p(\tau)\leq p(\delta)$. Hence also $p(\sigma)\leq p(\delta)$.

By induction, therefore, it follows that if $\underline{B}$ satisfies (22-6) then it also satisfies the following:

For each circuit τ of $\Delta(\underline{B})$, there exists an elementary circuit δ of $\Delta(\underline{B})$ such that $p(\tau)\leq p(\delta)$ (22-8)

In the light of (22-8) it is clear that (22-4) and (22-5) are equivalent.

<u>Theorem 22-3. Let E_1 be a blog and let $\underline{B} \in \mathcal{M}_{nn}$ (for given integer $n\geq 1$) be either row-$\emptyset$-astic or column-$\emptyset$-astic. Then $\underline{B}$ is definite.</u>

<u>Proof</u>. Suppose $\underline{B}$ is row-$\emptyset$-astic. Obviously $p(\sigma)\leq\emptyset$ for each cycle σ in $\Delta(\underline{B})$ since $p(\sigma)$ is a product of elements $\{\underline{B}\}_{ij}\leq\emptyset$. Now for each index i=1,..., n, let c(i) be

the lowest index $(1\leqslant c(i)\leqslant n)$ such that $\{\underline{B}\}_{ic(i)}=\emptyset$, and consider the path:

$$(1,c(1),c^2(1),\ldots,c^n(1)) \tag{22-9}$$

where $c^2(1) = c(c(1))$ etc. Since the path (22-9) contains (n+1) terms drawn from a set of n indices, it must contain a repeated term, and thus a circuit σ (say). Evidently $p(\sigma)=\emptyset$. The argument when $\underline{B}$ is column-$\emptyset$-astic is similar.

Theorem 22-4. Let E_1 be a blog and let $\underline{B}\in\mathcal{M}_{nn}$ for given integer $n\geqslant 1$. If $\underline{B}$ is definite then so is $\underline{B}^r$ for any integer $r\geqslant 0$.

Proof. By convention, $\underline{B}^o$ is $\underline{I}_n$, the $(n\times n)$ identity matrix whose diagonal elements are all $\emptyset$, and whose off diagonal elements are all $-\infty$. Evidently, $\underline{I}_n$ is definite. So assume $r\geqslant 1$.

The circuit product p of a given circuit σ of length t in $\Delta(\underline{B}^r)$ has the form:

$$p(\sigma) = b_{i_0 i_1} \otimes \ldots \otimes b_{i_{t-1} i_t} \qquad \text{with } i_t = i_0 \tag{22-10}$$

In (22-10), each $b_{i_{k-1} i_k}$ is $\{\underline{B}^r\}_{i_{k-1} i_k}$ and therefore by Proposition 22-1 is of the form $\sum_{h_k}{}_{\oplus}\, p_{kh_k}$, where each p_{kh_k} is a path product of a path from i_{k-1} to i_k of length r in $\Delta(\underline{B})$, the summation with respect to the dummy variable h_k being over all such paths. Hence, for all $h_1,\ldots, h_t$:

$$p_{1h_1} \otimes \ldots \otimes p_{th_t}$$

is a circuit product of length rt for some circuit in $\Delta(\underline{B})$, so by hypothesis:

$$p_{1h_1} \otimes \ldots \otimes p_{th_t} \leqslant \emptyset$$

Closing w.r.t. the indices $h_1,\ldots, h_t$, we obtain

$$p(\sigma) \leqslant \emptyset \tag{22-11}$$

On the other hand the graph $\Delta(\underline{B})$ contains, by hypothesis, a circuit τ of length s (say) and of circuit product $p(\tau) = \emptyset$. Suppose τ is $(j_0,j_1,\ldots, j_s)$ where $j_s=j_o$ and consider a product of r identical terms:

$$q = (\{\underline{B}\}_{j_0 j_1} \otimes \ldots \otimes \{\underline{B}\}_{j_{s-1} j_s})\otimes \ldots \otimes (\{\underline{B}\}_{j_0 j_1} \otimes \ldots \otimes \{\underline{B}\}_{j_{s-1} j_s}) \tag{22-12}$$

Evidently $q=(\emptyset)^r=\emptyset$. On the other hand q may be re-associated, without changing the order of the factors, into a product of s brackets, each containing a path product for a path of length r from the graph $\Delta(\underline{B})$. Thus

$$q = C_{d_0 d_1} \otimes \ldots \otimes C_{d_{s-1} d_s}$$

where $d_o=j_o$, $d_s=j_s=j_o$, and for $k=1,\ldots, s$:

$$C_{d_{k-1} d_k} = \text{path product for a path of length r from } d_{k-1} \text{ to } d_k \text{ in } \Delta(\underline{B})$$

$\leqslant \sum_{\oplus}$ (such path products) $= \{\underline{B}^r\}_{d_{k-1}\, d_k}$ (by Proposition 22-1)

Hence $\quad \phi = q \leqslant \{\underline{B}^r\}_{d_0 d_1} \otimes \ldots \otimes \{\underline{B}^r\}_{d_{s-1} d_s}$ (22-13)

Obviously (22-13) exhibits a circuit in $\Delta(\underline{B}^r)$, having circuit product at least equal to ϕ. Together with (22-11), this shows that $\underline{B}^r$ is definite. ●

22-3. Metric Matrices

For any given matrix $\underline{B} \in \mathcal{M}_{nn}$, the metric matrix generated by $\underline{B}$ is by definition:

$$\underline{\Gamma}(\underline{B}) = \underline{B} \oplus \underline{B}^2 \oplus \ldots \oplus \underline{B}^n \tag{22-14}$$

And if E_1 has a duality, we define the dual metric matrix generated by $\underline{B}$ as:

$$\underline{\Gamma}^*(\underline{B}) = \underline{B}^* \oplus' \underline{B}^{2*} \oplus' \ldots \oplus' \underline{B}^{n*} \tag{22-15}$$

The notations $\underline{\Gamma}$, $\underline{\Gamma}^*$ will be standard throughout the sequel.

Lemma 22-5. Let E_1 be a blog, and $\underline{B} \in {}_{nn}$ for given integer $n>1$. Then $\underline{\Gamma}(\underline{B}) = \underline{B} \otimes (\underline{I} \oplus \underline{B})^{n-1}$ where $\underline{I}_n$ is the $(n\times n)$ identity matrix.

Proof. If $n=1$ then $(\underline{I}\oplus\underline{B})^{n-1}$ is by convention $\underline{I}$ and the result follows. Otherwise, since $\underline{I},\underline{B}$ commute, we may carry out the iterated multiplication $(\underline{I}\oplus\underline{B})\otimes \ldots \otimes(\underline{I}\oplus\underline{B})$ to obtain $I\oplus\sum_{\oplus}$(Powers of $\underline{B}$). Each power of $\underline{B}$ occurs at least once up to and including $\underline{B}^{n-1}$ and since $\underline{B}^r \oplus \underline{B}^r = \underline{B}^r$, we have

$$(\underline{I}\oplus\underline{B})^{n-1} = \underline{I}\oplus\underline{B}\oplus \ldots \oplus \underline{B}^{n-1}$$

and the result follows. ●

Theorem 22-6. Let E_1 be a blog and let $\underline{B} \in \mathcal{M}_{nn}$ for given integer $n \geqslant 1$. If $\underline{B}$ is definite then $\underline{B}^r \leqslant \underline{\Gamma}(\underline{B})$ for every integer $r \geqslant 1$.

Proof. Fix i,j ($1\leqslant i\leqslant n$; $1\leqslant j\leqslant n$). Each path η of length $n+1$ from i to j in $\Delta(\underline{B})$, the graph associated with $\underline{B}$, must contain a circuit σ, say, since some node must recur. The path product $p(\eta)$ is therefore a product of the circuit product $p(\sigma)$ for this circuit, together with factors making up a path product $p(\tau)$ for a path τ of length $t(\tau)\leqslant n$. But $p(\sigma)\leqslant\phi$ by hypothesis, so $p(\eta)\leqslant p(\tau)$ by isotonicity.

Now $\quad p(\tau) \leqslant \{\underline{B}^{t(\tau)}\}_{ij}$ (by Proposition 22-1)

$\leqslant \{\underline{B} \oplus \ldots \oplus B^n\}_{ij}$ (since $1 < t(\tau) \leqslant n$)

$= \{\underline{\Gamma}(\underline{B})\}_{ij}$

Hence $\quad p(\eta) \leqslant \{\underline{\Gamma}(\underline{B})\}_{ij}$

Closing w.r.t. η (regarded as indexing all paths of length $(n+1)$ from i to j):

$$\{\underline{B}^{n+1}\}_{ij} = \Sigma_{\oplus} p(\eta) \qquad \text{(by Proposition 22-1)}$$

$$\leq \{\underline{\Gamma}(\underline{B})\}_{ij} \tag{22-16}$$

Hence $\underline{B}^{n+1} \leq \underline{\Gamma}(\underline{B})$. Assume now that $\underline{B}^{n+s} \leq \underline{\Gamma}(\underline{B})$ for some integer $s \geq 1$. Then by isotonicity,

$$\underline{B}^{n+s+1} \leq \underline{B} \otimes \underline{\Gamma}(\underline{B}) = \underline{B}^2 \oplus \ldots \oplus \underline{B}^n \oplus \underline{B}^{n+1}$$

$$\leq \underline{\Gamma}(\underline{B}) \quad \text{(by (22-14) and (22-16))} \tag{22-17}$$

Hence the result holds by induction for $r>n$, and is clearly trivial for $1 \leq r \leq n$. ●

<u>Corollary 22-7. Let E_1 be a blog and let $\underline{B} \in \mathcal{M}_{nn}$ for given integer $n \geq 1$. If $\underline{B}$ is definite then:</u>

$$\underline{(\underline{\Gamma}(\underline{B}))^r \leq \underline{\Gamma}(\underline{B})} \qquad (r \geq 1)$$

<u>and $\underline{\Gamma}(\underline{B}) = B \otimes (\underline{I}_n \oplus \underline{B})^r$ for any $r \geq (n-1)$</u>

<u>where I_n is the (n×n) identity matrix.</u>

<u>Proof</u>. $(\underline{\Gamma}(\underline{B}))^2 = (\underline{B} \oplus \ldots \oplus \underline{B}^n)^2$

$$= \underline{B}^2 \oplus \ldots \oplus \underline{B}^{2n}$$

$$\leq \underline{\Gamma}(\underline{B}) \quad \text{(by Theorem 22-6)}$$

By isotonicity $(\underline{\Gamma}(\underline{B}))^3 \leq (\underline{\Gamma}(\underline{B}))^2 \leq \underline{\Gamma}(\underline{B})$ and the result follows by iteration

Moreover:

$$\underline{B} \otimes (\underline{I}_n \oplus \underline{B})^r = \underline{B} \oplus \ldots \oplus \underline{B}^{r+1} \leq \underline{\Gamma}(\underline{B}) \qquad \text{(by Theorem 22-6)}$$

On the other hand, if $r \geq (n-1)$ then:

$$\underline{B} \oplus \ldots \oplus \underline{B}^{r+1} = (\underline{B} \oplus \ldots \oplus B^n) \oplus (\underline{B}^n \oplus \ldots \oplus \underline{B}^{r+1})$$

$$\geq \underline{\Gamma}(\underline{B}) \qquad \text{(from (22-14))}$$

Hence $\underline{B} \otimes (\underline{I}_n \oplus \underline{B})^r = \underline{\Gamma}(\underline{B})$ when $r \geq (n-1)$ ●

<u>22-4 . The Shortest Distance Matrix</u>

The dual of the first result in Corollary 22-7 motivates the use of the term <u>metric matrix</u>. In fact, in the principal interpretation if element $\{\underline{B}^*\}_{ij}$ represents the direct distance from node i to node j of a transportation network, then as discussed in Section 1-2.2, $\underline{\Gamma}'(\underline{B}^*)$ represents the <u>shortest distance matrix</u>. Hence the elements of $\underline{\Gamma}'(\underline{B}^*)$ satisfy the <u>triangle law for distances</u>:

$$\{\underline{\Gamma}'(\underline{B}^*)\}_{ij} + \{\underline{\Gamma}'(\underline{B}^*)\}_{jk} \geq \{\underline{\Gamma}'(\underline{B}^*)\}_{ik}$$

In min algebra notation, this is just:

$$\{\underline{\Gamma}'(\underline{B}^*)\}_{ij} \otimes' \{\underline{\Gamma}'(B^*)\}_{jk} \geq \{\underline{\Gamma}'(B^*)\}_{ik} \tag{22-18}$$

Now we can readily confirm that $\underline{\Gamma}'(\underline{B}^*)$ is just $(\underline{\Gamma}(\underline{B}))^*$, so (22-18) may be written equivalently: $(\underline{\Gamma}(\underline{B}))^{2*} \geq (\underline{\Gamma}(\underline{B}))^*$, which is the dual of the first result in Corollary 22-7, for r=2.

We conclude with an example (in the principal interpretation). In the (5×5) matrix $\underline{B}^*$ given below, $\{\underline{B}^*\}_{ij}$ represents the time taken to get from city i to city j of a transportation network using the direct connection (an ∞ indicates 'no direct connection'). We wish to compute the matrix of shortest possible intercity journey times.

$$\underline{B}^* = \begin{bmatrix} 0 & 7 & 3 & 5 & \infty \\ 7 & 0 & 3 & \infty & 1 \\ 3 & 3 & 0 & \infty & 6 \\ 5 & \infty & \infty & 0 & 3 \\ \infty & 1 & 6 & 3 & 0 \end{bmatrix} = (\underline{I}_5 \oplus \underline{B})^*$$

$$(\underline{I}_5 \oplus \underline{B})^{2*} = \begin{bmatrix} 0 & 6 & 3 & 5 & 8 \\ 6 & 0 & 3 & 4 & 1 \\ 3 & 3 & 0 & 8 & 4 \\ 5 & 4 & 8 & 0 & 3 \\ 8 & 1 & 4 & 3 & 0 \end{bmatrix}$$

$$(\underline{I}_5 \oplus \underline{B})^{4*} = \begin{bmatrix} 0 & 6 & 3 & 5 & 7 \\ 6 & 0 & 3 & 4 & 1 \\ 3 & 3 & 0 & 7 & 4 \\ 5 & 4 & 7 & 0 & 3 \\ 7 & 1 & 4 & 3 & 0 \end{bmatrix}$$

Hence the shortest time matrix is

$$(\underline{\Gamma}(\underline{B}))^* = \underline{B}^* \otimes' (\underline{I}_5 \oplus \underline{B})^{4*} = \begin{bmatrix} 0 & 6 & 3 & 5 & 7 \\ 6 & 0 & 3 & 4 & 1 \\ 3 & 3 & 0 & 7 & 4 \\ 5 & 4 & 7 & 0 & 3 \\ 7 & 1 & 4 & 3 & 0 \end{bmatrix}$$

As an alternative notation to $(\underline{\Gamma}(\underline{B}))^*$ we may more conveniently write $\underline{\Gamma}^*(\underline{B})$.

Thus $\underline{\Gamma}^*(\underline{B}) = (\underline{\Gamma}(\underline{B}))^* = \underline{\Gamma}'(\underline{B}^*)$ (22-19)

The preceding discussion leaves out any consideration of computational efficiency, since our present aim is the logically prior one of establishing algebraic structure. But see further [17] and [39].

23. FUNDAMENTAL EIGENVECTORS

23-1 · The Eigenproblem

In Section 1-2.1, we described a problem in the control of industrial processes which gives rise to the eigenvector-eigenvalue problem or eigenproblem for short. We shall now begin our study of this problem, for the case when E_1 is a (linear) blog. If $\underline{A} \in \mathcal{M}_{nn}$ is a square matrix, we shall say that $\underline{x} \in E_n$ is an eigenvector of $\underline{A}$ and $\lambda \in E_1$ a corresponding eigenvalue of $\underline{A}$ if there holds:

$$\underline{A} \otimes \underline{x} = \lambda \otimes \underline{x} \tag{23-1}$$

As an illustration, we have the following result, which follows directly from Theorem 8-9.

Proposition 23-1. Let E_1 be a blog, and let $\underline{A} \in \mathcal{M}_{mn}$ for given integers $m,n \geq 1$. Then there exists a matrix $\underline{B} \in \mathcal{M}_{mm}$ for which every m-tuple in the column-space of $\underline{A}$ is an eigenvector with corresponding eigenvalue $\emptyset$. Namely, $\underline{B}=\underline{A} \otimes' \underline{A}^*$ ●

In conventional linear algebra, it is necessary to stipulate that an eigenvector be non-zero by definition, in order to avoid trivialities. Analogously, we shall say that (23-1) is finitely soluble (when E_1 is a blog), if we can find finite λ and $\underline{x}$ satisfying (23-1).

Lemma 23-2. Let E_1 be a blog and let $\underline{B} \in \mathcal{M}_{nn}$, for given integer $n \geq 1$. A necessary condition that the eigenproblem for $\underline{B}$ be finitely soluble is that $\underline{B}$ be row-G-astic; a sufficient condition is that $\underline{B}$ be row-$\emptyset$-astic.

Proof. If $\underline{B} \otimes \underline{x} = \lambda \otimes \underline{x}$ with λ, $\underline{x}$ finite then, using Proposition 4-3, since:

$$\sum_{j=1}^{n}{}_{\oplus}(\{\underline{B}\}_{ij} \otimes \{\underline{x}\}_j) = \lambda \otimes \{\underline{x}\}_i \text{ is finite, } (i=1,\ldots, n), \text{ it is clear that no}$$

$\{\underline{B}\}_{ij}$ can be $+\infty$ and no row of $\underline{B}$ can consist entirely of $-\infty$'s. Hence $\underline{B}$ is row-G-astic.

If $\underline{B}$ is row-$\emptyset$-astic, let $\underline{x} \in E_n$ be defined by $\{\underline{x}\}_i=\emptyset$ $(i=1,\ldots, n)$ and let $\lambda=\emptyset$. Then:

$$\begin{aligned}\{\underline{B} \otimes \underline{x}\}_i &= \sum_{j=1}^{n}{}_{\oplus}(\{\underline{B}\}_{ij} \otimes \emptyset) \\ &= \emptyset \qquad \text{(by (12-1))} \\ &= \{\lambda \otimes \underline{x}\}_i\end{aligned}$$

●

The following lemma will be useful in subsequent arguments.

Lemma 23-3. Let E_1 be a blog, and let $\underline{B} \in \mathcal{M}_{nn}$ for given integer $n \geq 1$. Let σ be a circuit in $\Delta(\underline{B})$, of circuit product $p(\sigma)$ and let τ, of circuit product $p(\tau)$, be any circuit obtained from σ by cyclic permutation of the nodes on σ. Then $p(\tau) \sim \emptyset$ if and only if $p(\sigma) \sim \emptyset$, where $\sim$ is any given one of the symbols $=$, $\geq$, $\leq$.

Proof. Suppose σ is $(i_0, i_1, \ldots, i_{t-1}, i_0)$. If $p(\sigma)$ is not finite then at least one of $\{\underline{B}\}_{i_0 i_1}, \ldots, \{\underline{B}\}_{i_{t-1} i_0}$ is infinite, and rearrangement of the factors of $p(\sigma)$ will produce the same (infinite) value for $p(\tau)$.

Now suppose all factors of $p(\sigma)$ are finite and that:

$$\{\underline{B}\}_{i_0 i_1} \otimes \ldots \otimes \{\underline{B}\}_{i_{t-2} i_{t-1}} \otimes \{\underline{B}\}_{i_{t-1} i_0} \geqslant \emptyset$$

Pre-multiplying by $\{\underline{B}\}_{i_{t-1} i_0}$ and post-multiplying by $(\{\underline{B}\}_{i_{t-1} i_0})^{-1}$ we have by isotonicity:

$$\{\underline{B}\}_{i_{t-1} i_o} \otimes \{\underline{B}\}_{i_0 i_1} \otimes \ldots \otimes \{\underline{B}\}_{i_{t-2} i_{t-1}} \geqslant \emptyset$$

And evidently we may proceed to generate all cyclic permutations in this way. Moreover, the argument goes through similarly when $\sim$ is $=$ or $\leqslant$.

Our study of (23-1) will extend over several chapters. First we shall establish some results for definite matrices. These will then enable us to prove some fundamental results regarding the eigenproblem for row-$\emptyset$-astic matrices. Finally, we shall extend these results to all matrices for which (23-1) is finitely soluble, by finding a transformation of such matrices into row-$\emptyset$-astic matrices.

So suppose a certain matrix $\underline{B}$ over a blog E_1 is definite. Then $\Delta(\underline{B})$ contains at least one circuit with circuit-product equal to $\emptyset$. An eigen-node of $\Delta(\underline{B})$ is any node on such a circuit. Two eigen-nodes are equivalent if they are both on any one such circuit. Lemma 23-3 implies that this defines an equivalence relation.

Lemma 23-4. Let E_1 be a blog and let $\underline{B} \in \mathcal{M}_{nn}$, for given integer $n \geqslant 1$, be definite. Then $\underline{\Gamma}(\underline{B})$ is definite and if j is an eigen-node of $\Delta(\underline{B})$, then:

$$\{\underline{\Gamma}(\underline{B})\}_{jj} = \emptyset$$

Conversely, if E_1 is linear, and $\{\underline{\Gamma}(\underline{B})\}_{jj} = \emptyset$ for some index j $(1 \leqslant j \leqslant n)$ then j is an eigen-node.

Proof. If $k (1 \leqslant k \leqslant n)$ is any index, then each circuit from k to k of length $r \geqslant 1$ in $\Delta(\underline{B})$ has circuit product $\leqslant \emptyset$ by hypothesis. Closing w.r.t that class of circuits we have by Proposition 22-1 that $\{\underline{B}^r\}_{kk} \leqslant \emptyset$ and hence by definition (22-14):

$$\{\underline{\Gamma}(\underline{B})\}_{kk} \leqslant \emptyset \qquad (k=1, \ldots, n) \tag{23-2}$$

On the other hand, j is on some circuit in $\Delta(\underline{B})$ of length r (say), with circuit product $\emptyset$, by hypothesis. By Lemma 23-3 we can assume this circuit to run from j to j. Hence by Proposition 22-1, $\{\underline{B}^r\}_{jj} \geqslant \emptyset$, so by Theorem 22-6:

$$\{\underline{\Gamma}(\underline{B})\}_{jj} \geqslant \emptyset$$

Hence $\{\underline{\Gamma}(\underline{B})\}_{jj} = \emptyset$. Now if $k (1 \leqslant k \leqslant n)$ is any index, then any circuit σ of length $r \geqslant 1$ from k to k in $\Delta(\underline{\Gamma}(\underline{B}))$ has a circuit product $p(\sigma)$ which by Proposition 22-1 satisfies

$p(\sigma) \leqslant \{(\underline{\Gamma}(\underline{B}))^r\}_{kk}$. But by Corollary 22-7, $\{(\underline{\Gamma}(\underline{B}))^r\}_{kk} \leqslant \{\underline{\Gamma}(\underline{B})\}_{kk}$. Hence, using (23-2): $p(\sigma) \leqslant \emptyset$. So, since $\underline{\Gamma}(\underline{B})$ has at least one circuit-product equal to $\emptyset$ (namely $\{\underline{\Gamma}(\underline{B})\}_{jj} = \emptyset$ for eigen-node j), $\underline{\Gamma}(\underline{B})$ is definite.

Conversely, if E_1 is linear and $\{\underline{\Gamma}(\underline{B})\}_{jj} = \emptyset$ for some index j $(1 \leqslant j \leqslant n)$ then $\{\underline{B}^r\}_{jj} = \emptyset$ for some r $(1 \leqslant r \leqslant n)$, by (22-14) and the linearity of E_1. So by Proposition 22-1 and the linearity of E_1 there is a circuit (of length r) from j to j in $\Delta(\underline{B})$ with circuit product equal to $\emptyset$. But this means that j is an eigen-node.

<u>Theorem 23-5. Let E_1 be a blog and let $\underline{B} \in \mathcal{M}_{nn}$, for given integer $n \geqslant 1$, be definite. If j is an eigen-node of $\Delta(\underline{B})$, then $\underline{B} \otimes \underline{\xi}(j) = \underline{\xi}(j)$ where $\underline{\xi}(j)$ is the jth column of $\underline{\Gamma}(\underline{B})$.</u>

<u>Proof</u>. From (22-17), $\underline{B} \otimes \underline{\Gamma}(\underline{B}) \leqslant \underline{\Gamma}(\underline{B})$ so for the jth column:

$$\underline{B} \otimes \underline{\xi}(j) \leqslant \underline{\xi}(j) \tag{23-3}$$

On the other hand,

$$\{\underline{B} \otimes \underline{\Gamma}(\underline{B})\}_{ij} = \{\underline{B}^2 \oplus \ldots \oplus \underline{B}^n \oplus \underline{B}^{n+1}\}_{ij} \quad (i=1,\ldots,n) \tag{23-4}$$

$$\geqslant \{\underline{B}^2 \oplus \ldots \oplus \underline{B}^n\}_{ij} \quad (i=1,\ldots,n)$$

Moreover, $\{\underline{B} \otimes \underline{\Gamma}(\underline{B})\}_{ij} = \sum_{k=1}^{n}{}_{\oplus} \{\underline{B}\}_{ik} \otimes \{\underline{\Gamma}(\underline{B})\}_{kj} \quad (i=1,\ldots,n)$

$$\geqslant \{\underline{B}\}_{ij} \otimes \{\underline{\Gamma}(\underline{B})\}_{jj} \quad (i=1,\ldots,n) \text{ (by (3-15))}$$

Hence $\{\underline{B} \otimes \underline{\Gamma}(\underline{B})\}_{ij} \geqslant \{\underline{B}\}_{ij} \quad (i=1,\ldots,n)$ (by Lemma 23-4) (23-5)

Evidently (23-4) and (23-5) together imply:

$$\{\underline{B} \otimes \underline{\Gamma}(\underline{B})\}_{ij} \geqslant \{\underline{B} \oplus \underline{B}^2 \oplus \ldots \oplus \underline{B}^n\}_{ij} \quad (i=1,\ldots,n)$$

i.e. $\underline{B} \otimes \underline{\xi}(j) \geqslant \underline{\xi}(j)$

which together with (23-3) implies the required result.

In other words, Theorem 23-5 states that columns of $\underline{\Gamma}(\underline{B})$ which correspond to eigen-nodes furnish eigenvectors for $\underline{B}$, with corresponding eigenvalue $\emptyset$. We call such columns of $\underline{\Gamma}(\underline{B})$ the <u>fundamental eigenvectors of $\underline{B}$.</u> Such eigenvectors need not be finite, even for a row-$\emptyset$-astic matrix. Consider for example:

$$\underline{B} = \begin{bmatrix} \emptyset & -\infty \\ -\infty & \emptyset \end{bmatrix}$$

$\underline{B}$ is row-$\emptyset$-astic, and $\underline{\Gamma}(\underline{B}) = \underline{B}$. Both columns of $\underline{B}$ are fundamental eigenvectors, but neither is finite.

<u>23-2 . Blocked Matrices</u>

<u>Theorem 23-6. Let E_1 be a blog and let $\underline{B} \in \mathcal{M}_{nn}$, for given integer $n \geqslant 1$ be definite. If $\underline{\xi}(h), \underline{\xi}(k) \in E_n$ are fundamental eigenvectors of $\underline{B}$ corresponding to equivalent eigen-nodes h and k respectively then $\underline{\xi}(h), \underline{\xi}(k)$ are finite scalar multiples of one</u>

another.

Proof. For i=1,...,n we have:

$$\{\underline{\Gamma}(\underline{B})\}_{ih} \geqslant \{(\underline{\Gamma}(\underline{B}))^2\}_{ih} \qquad \text{(by Corollary 22-7)}$$

$$= \sum_{j=1}^{n}{}_{\oplus}(\{\underline{\Gamma}(\underline{B})\}_{ij} \otimes \{\underline{\Gamma}(\underline{B})\}_{jh})$$

$$\geqslant \{\underline{\Gamma}(\underline{B})\}_{ik} \otimes \{\underline{\Gamma}(\underline{B})\}_{kh} \qquad \text{(by (3-15))} \tag{23-6}$$

$$\geqslant \{(\underline{\Gamma}(\underline{B}))^2\}_{ik} \otimes \{\underline{\Gamma}(\underline{B})\}_{kh} \qquad \text{(by Corollary 22-7)}$$

$$= (\sum_{j=1}^{n}{}_{\oplus}(\{\underline{\Gamma}(\underline{B})\}_{ij} \otimes \{\underline{\Gamma}(\underline{B})\}_{jk})) \otimes \{\underline{\Gamma}(\underline{B})\}_{kh}$$

$$\geqslant \{\underline{\Gamma}(\underline{B})\}_{ih} \otimes \{\underline{\Gamma}(\underline{B})\}_{hk} \otimes \{\underline{\Gamma}(\underline{B})\}_{kh} \qquad \text{(by (3-15))} \tag{23-7}$$

Now, by hypothesis, nodes h and k are both on some circuit σ in $\Delta(\underline{B})$ of circuit product $p(\sigma)=\emptyset$.

By Lemma 23-3 we may assume that σ consits of a (not necessarily elementary) path α from h to k of length $t(\alpha)$ and path product $p(\alpha)$ (say), followed by a (not necessarily elementary) path β of length $t(\beta)$ and path product $p(\beta)$ (say) from k to h. Evidently:

$$p(\alpha) \otimes p(\beta) = p(\sigma) = \emptyset \tag{23-8}$$

Now by Proposition 22-1:

$$p(\alpha) \leqslant \{\underline{B}^{t(\alpha)}\}_{hk} \text{ and } p(\beta) \leqslant \{\underline{B}^{t(\beta)}\}_{kh}$$

Hence $p(\alpha) \otimes p(\beta) \leqslant \{\underline{B}^{t(\alpha)}\}_{hk} \otimes \{B^{t(\beta)}\}_{kh}$

$$\leqslant \{\underline{\Gamma}(B)\}_{hk} \otimes \{\underline{\Gamma}(\underline{B})\}_{kh} \qquad \text{(by Theorem 22-6)} \tag{23-9}$$

From (23-6), (23-7), (23-8) and (23-9):

$$\{\underline{\Gamma}(\underline{B})\}_{ih} \geqslant \{\underline{\Gamma}(\underline{B})\}_{ik} \otimes \{\underline{\Gamma}(\underline{B})\}_{kh} \geqslant \{\underline{\Gamma}(\underline{B})\}_{ih}$$

In other words: $\{\underline{\xi}(h)\}_i = \{\underline{\xi}(k)\}_i \otimes \mu \qquad (i=1,\ldots, n)$ (23-10)

where $\mu = \{\underline{\Gamma}(\underline{B})\}_{kh} \varepsilon E_1$ is a constant independent of i. Thus:

$$\underline{\xi}(h) = \underline{\xi}(k) \otimes \mu \tag{23-11}$$

Evidently μ is finite because from (23-10) with i=h, we have

$$\{\underline{\xi}(h)\}_h = \{\underline{\xi}(k)\}_h \otimes \mu \qquad (23\text{-}12)$$

But $\{\underline{\xi}(h)\}_h = \{\underline{\Gamma}(\underline{B})\}_{hh} = \emptyset$ by Lemma 23-4, so μ must be finite. Evidently also

$$\underline{\xi}(k) = \underline{\xi}(h) \otimes \mu^{-1}$$

Theorem 23-6 shows that two equivalent eigen-nodes of $\Delta(\underline{B})$ determine essentially the same eigenvectors of $\underline{B}$. We call two fundamental eigenvectors $\underline{\xi}(h)$, $\underline{\xi}(k)$ equivalent if nodes h and k are equivalent, otherwise we say they are non-equivalent.

In future arguments it will sometimes be convenient to assume that the indices i=1,..., n (and therefore the nodes in $\Delta(\underline{B})$ have been (re)allocated in such a way that equivalent nodes are numbered consecutively. Specifically, if there are q eigen-nodes, which fall into r equivalence classes containing $s_1, s_2, \ldots, s_r$ eigen-nodes each, with $s_1 + \ldots + s_r = q \leq n$, then we shall say that $\underline{B}$ is blocked if indices are (re)allocated in such a way that nodes 1,..., s_1 are equivalent eigen-nodes,..., nodes $(s_1 + \ldots + s_{r-1}+1), \ldots, (s_1 + \ldots + s_r)$ are equivalent eigen-nodes, and nodes $(s_1 + \ldots + s_r+1)$ to n (if any) are not eigen-nodes.

For example, in the principal interpretation, let

$$\underline{B} = \begin{bmatrix} -1 & -3 & -1 & -2 \\ 3 & -2 & -5 & 0 \\ -6 & -2 & 0 & -3 \\ -5 & -5 & -4 & -2 \end{bmatrix}$$

$$\text{Then } \underline{\Gamma}(\underline{B}) = \left[\begin{array}{cc|c|c} 0 & -3 & -1 & -2 \\ 3 & 0 & 2 & 1 \\ 1 & -2 & 0 & -1 \\ -2 & -5 & -3 & -2 \end{array}\right]$$

Equivalent eigenvectors (columns 1, 2)   Eigenvector (column 3)   Non-eigenvector (column 4)

There are two circuits in $\Delta(\underline{B})$ having circuit product =0, namely;

(1,2,1) and (3,3)

Hence the eigen-nodes are 1,2 and 3, with 1 equivalent to 2. So $\underline{B}$ is blocked. If we ring the elements in $\underline{B}$ which contribute to circuit products of value $\emptyset$, we see that they fall within blocks around the principal diagonal, corresponding to equivalence classes of eigen-nodes:

$$\left[\begin{array}{cc|c|c} -1 & \textcircled{-3} & -1 & -2 \\ \textcircled{3} & -2 & -5 & 0 \\ \hline -6 & -2 & \textcircled{0} & -3 \\ \hline -5 & -5 & -4 & -2 \end{array}\right]$$

1st block (rows/columns 1, 2)   2nd block (row/column 3)   Non-eigen-node (row/column 4)

Evidently $\Gamma(\underline{B})$ may be partitioned in the same way. We note en passant that the first column of $\Gamma(\underline{B})$ may be obtained from the second column by adding 3 to each component, thus illustrating Theorem 23-6.

23-3. Ø-astic Definite Matrices

We examine now the possibility of linear dependence among non-equivalent fundamental eigenvectors when E_1 is a linear blog. We do this first for a particular class of matrices; it is then a simple matter to derive results for general matrices. Accordingly, if E_1 is a blog and $\underline{B} \varepsilon \mathcal{M}_{nn}$ for given integer $n \geq 1$, we shall say that $\underline{B}$ is Ø-astic definite if:

$$\underline{B} \text{ is definite, and } \{\underline{B}\}_{ij} \leq \emptyset \quad (i,j=1,\ldots, n)$$

Lemma 23-7. Let E_1 be a blog and let $\underline{B} \varepsilon \mathcal{M}_{nn}$ for given integer $n \geq 1$. Then $\underline{B}$ is Ø-astic definite if and only if: B is definite and the elements of $\underline{B}$ form a Ø-astic set. In particular, if $\underline{B}$ is row-Ø-astic or column-Ø-astic, then $\underline{B}$ is Ø-astic definite.

Proof. If $\underline{B}$ is definite then some product of elements of $\underline{B}$ (namely a circuit product) equals Ø. If also $\{\underline{B}\}_{ij} \leq \emptyset$ $(i,j=1,\ldots, n)$ then Corollary 4-12 shows that the elements making up the circuit all equal Ø. Hence the elements of $\underline{B}$ form a Ø-astic set. The converse is trivial, and the rest follows from Theorem 22-3.

Theorem 23-8. Let E_1 be a blog and let $\underline{B} \varepsilon \mathcal{M}_{nn}$, for given integer $n \geq 1$, be Ø-astic definite. Then:

(i) $\Gamma(\underline{B})$ is Ø-astic definite, and $\underline{B}^r$ is Ø-astic definite for each integer $r \geq 1$.

(ii) If $\underline{\xi}(h)$, $\underline{\xi}(k)$ are equivalent fundamental eigenvectors of $\underline{B}$, corresponding to equivalent eigen-nodes h,k respectively, then $\underline{\xi}(h) = \underline{\xi}(k)$.

(iii) Each fundamental eigenvector of $\underline{B}$ is column-Ø-astic.

(iv) The fundamental eigenvector $\underline{\xi}(j)$ corresponding to a given eigen-node j of $\Delta(\underline{B})$ contains components equal to Ø in all positions corresponding to eigen-nodes equivalent to j.

Proof. (i) For $i,j=1,\ldots, n$:

$$\{\underline{B}^2\}_{ij} = \sum_{k=1}^{n}{}_{\oplus}(\{\underline{B}\}_{ik} \otimes \{\underline{B}\}_{kj})$$

$$\leq \emptyset \quad (\text{since } \{\underline{B}\}_{ik}, \{\underline{B}\}_{kj} \leq \emptyset).$$

Similarly $\{\underline{B}^r\}_{ij} \leq \emptyset$ $(i,j=1,\ldots, n)$ and so $\{\Gamma(\underline{B})\}_{ij} \leq \emptyset$ $(i,j=1,\ldots, n)$. But $\underline{B}^r$ is definite by Theorem 22-4, and $\Gamma(\underline{B})$ is definite by Lemma 23-4.

(ii) From (i), $\{\Gamma(\underline{B})\}_{kh} \leq \emptyset$, so from (23-10) in the proof of Theorem 23-6, $\underline{\xi}(h) \leq \underline{\xi}(k)$. Similarly $\underline{\xi}(k) \leq \underline{\xi}(h)$ since equivalence is a symmetric relation.

(iii) If $\underline{\xi}(j)$, the jth column of $\Gamma(\underline{B})$, is a fundamental eigenvector, then $\{\underline{\xi}(j)\}_i \leq \emptyset$ by (i), but $\{\underline{\xi}(j)\}_j = \emptyset$ by Lemma 23-4.

(iv) If eigenvector $\underline{\xi}(k)$ corresponds to eigen-node k, equivalent to eigen-node j,

then $\{\underline{\xi}(j)\}_k = \{\underline{\xi}(k)\}_k$ (by (ii))

$= \emptyset$ (by Lemma 23-4). ●

Lemma 23-9 Let E_1 be a linear blog and let $\underline{B} \in \mathcal{M}_{nn}$, for given integer $n\geq 1$, be $\emptyset$-astic definite. If $\underline{\xi}(j)$, $\underline{\xi}(k)$ are fundamental eigenvectors corresponding to eigen-nodes j, k of $\Delta(\underline{B})$ respectively, and j,k are not equivalent, then: either $\{\underline{\xi}(j)\}_k \lneq \emptyset$ or $\{\underline{\xi}(k)\}_j \lneq \emptyset$

Proof. If the lemma is false then in the light of Theorem 23-8 (iii) the only possibility is $\{\underline{\xi}(j)\}_k = \emptyset = \{\underline{\xi}(k)\}_j$ (23-13)

Now $\{\underline{\xi}(j)\}_k$ is $\{\underline{\Gamma}(\underline{B})\}_{kj} = \{\underline{B} \oplus \ldots \oplus \underline{B}^n\}_{kj}$ and is therefore a sum of path products of all paths of lengths $1,\ldots,n$ from k to j in $\Delta(\underline{B})$. From the linearity of E_1, $\{\underline{\xi}(j)\}_k$ must actually equal one such product. We may argue analogously for $\{\underline{\xi}(k)\}_j$. Thus there are integers r,s ($1\leq r\leq n$; $1\leq s\leq n$) such that

$\{\underline{\xi}(j)\}_k \otimes \{\underline{\xi}(k)\}_j$ is a circuit product for some circuit of length (r+s) in $\Delta(B)$ on which nodes j and k both occur. But in the light of (23-13), this would imply equivalence of j and k. ●

Theorem 23-10. Let E_1 be a linear blog and let $\underline{B} \in \mathcal{M}_{nn}$, for given integer $n\geq 1$ be $\emptyset$-astic definite. If $\underline{\xi}(j_1),\ldots, \underline{\xi}(j_s)$ are fundamental eigenvectors corresponding to pairwise non-equivalent eigen-nodes $j_1,\ldots, j_s$ respectively, then no one of $\underline{\xi}(j_1),\ldots, \underline{\xi}(j_s)$ is linearly dependent on the others.

Proof. Suppose on the contrary that

$$\underline{\xi}(j_s) = \sum_{k=1}^{s-1}{}_{\oplus}\, \underline{\xi}(j_k) \otimes \alpha_k \quad (\alpha_k \in E_1,\ k=1,\ldots, s-1) \tag{23-14}$$

Considering row j_s, in (23-14), we have (using Lemma 23-4):

$$\emptyset = \{\underline{\xi}(j_s)\}_{j_s} = \sum_{k=1}^{s-1}{}_{\oplus} \{\underline{\xi}(j_k)\}_{j_s} \otimes \alpha_k$$

By the linearity of E_1, therefore, there is an index h ($1\leq h\leq s-1$) such that:

$$\emptyset = \{\underline{\xi}(j_h)\}_{j_s} \otimes \alpha_h$$

Hence α_h is finite and $\alpha_h^{-1} = \{\underline{\xi}(j_h)\}_{j_s} \lneq \emptyset$ (since $\underline{\Gamma}(\underline{B})$ is $\emptyset$-astic definite) (23-15)

However, $\emptyset \geq \{\underline{\xi}(j_s)\}_{j_h}$ (since $\underline{\Gamma}(\underline{B})$ is $\emptyset$-astic definite)

$\geq \{\underline{\xi}(j_h)\}_{j_h} \otimes \alpha_h$ (from row j_h of (23-14))

$= \alpha_h$ (by Lemma 23-4) (23-16)

Evidently from (23-15) and (23-16) the only possibility is $\alpha_h=\emptyset$, giving also $\{\underline{\xi}(j_s)\}_{j_h} = \emptyset = \{\underline{\xi}(j_h)\}_{j_s}$ (from the previous working).

But by Lemma 23-9 this contradicts the assumption that j_h and j_s are not equivalent. ●

We conclude with the following lemma, which we shall require later.

<u>Lemma 23-11. Let E_1 be a blog and let $\underline{B} \in \mathcal{M}_{nn}$, for given integer $n \geq 1$, be row-$\emptyset$-astic. Let $\underline{\xi}(j_1),\ldots, \underline{\xi}(j_s)$ be a maximal set of non-equivalent fundamental eigenvectors for $\underline{B}$ and let $\underline{\Phi}(B) \in \mathcal{M}_{ns}$ be the matrix whose columns are $\underline{\xi}(j_1),\ldots, \underline{\xi}(j_s)$ in that order. Then $\underline{\Phi}(B)$ is doubly $\emptyset$-astic.</u>

<u>Proof</u>. It follows from Theorem 22-3 and Theorem 23-8 (iii) that $\underline{\Phi}(\underline{B})$ is column-$\emptyset$-astic. So:

$$\{\underline{\Phi}(\underline{B})\}_{jk} \leq \emptyset \qquad (j=1,\ldots, n;\ k=1,\ldots, s). \tag{23-17}$$

Moreover, if j is an eigen-node of $\Delta(\underline{B})$ then j is equivalent to one of $j_1,\ldots, j_s$: to j_k, say. Then:

$$\{\underline{\Phi}(\underline{B})\}_{jj_h} = \emptyset \qquad \text{(by Theorem 23-8(iv))} \tag{23-18}$$

Now suppose j is a non-eigen-node of $\Delta(\underline{B})$. Since $\underline{B}$ is row-$\emptyset$-astic, there is for each index $i=1,\ldots, n$ a least index $c(i)$ $(1\leq i\leq n)$ such that $\{\underline{B}\}_{ic(i)} = \emptyset$.

Consider the path $(j, c(j), c^2(j),\ldots, c^n(j))$ in $\Delta(\underline{B})$, where $c^2(j) = c(c(j))$ etc. This cannot consist exclusively of non-eigen-nodes since it must contain a repeated index and would therefore identify a circuit consisting of non-eigen-nodes but having circuit product $\emptyset$ - a contradiction. Hence some eigen-node d occurs on this path. Thus there is a path p, of length t (say), from j to d, having path-product $\emptyset$. So by Proposition 22-1 and Theorem 22-6:

$$\emptyset \leq \{\underline{B}^t\}_{jd} \leq \{\underline{\Gamma}(\underline{B})\}_{jd} \tag{23-19}$$

But since d is an eigen-node, it is equivalent to one of $j_1,\ldots, j_s$: to j_h say. Then by Theorem 23-8 (ii) column d of $\underline{\Gamma}(\underline{B})$ equals $\xi(j_h)$. Hence (23-19) gives:

$$\{\underline{\Phi}(\underline{B})\}_{jj_h} = \{\underline{\xi}(j_h)\}_j \geq \emptyset \tag{23-20}$$

Summarising, we see that (23-17) and (23-18) imply that row j of $\underline{\Phi}(\underline{B})$ is $\emptyset$-astic when j is an eigen-node, whilst (23-17) and (23-20) imply the same when j is a non-eigen-node. Hence $\underline{\Phi}(\underline{B})$ is row-$\emptyset$-astic and so doubly $\emptyset$-astic. ●

24. ASPECTS OF THE EIGENPROBLEM

24-1. The Eigenspace

If $\underline{B}$ is a definite square matrix over a blog E_1, let $\underline{\xi}(j_1),\ldots, \underline{\xi}(j_s)$ be a maximal set of non-equivalent fundamental eigenvectors of $\underline{B}$. The space $<\underline{\xi}(j_1),\ldots, \underline{\xi}(j_s)>$ generated by these eigenvectors will be called the <u>eigenspace of $\underline{B}$.</u>

<u>Lemma 24-1. Let E_1 be a blog and let $\underline{B} \in \mathcal{M}_{nn}$, for given integer $n\geq 1$, be definite. Then each element of the eigenspace of $\underline{B}$ is an eigenvector of $\underline{B}$ with corresponding eigenvalue equal to $\emptyset$.</u> Any maximal non-equivalent $\underline{\xi}(j_s),\ldots, \underline{\xi}(j_s)$ defines the same eigenspace.

<u>Proof.</u> Any element $\underline{u}$ of the eigenspace may be written: $\sum_{\oplus\, k=1}^{s}(\underline{\xi}(j_k) \otimes \alpha_k)$ $(\alpha_k \in E_1,\ k=1,\ldots, s)$; Theorem 23-6 shows that each non-equivalent $\underline{\xi}(j_1,\ldots, \underline{\xi}(j_s)$ give the same eigenspace.

$$\text{Now, } \underline{B} \otimes \underline{u} = \underline{B} \otimes \left(\sum_{\oplus\, k=1}^{s}(\underline{\xi}(j_k) \otimes \alpha_k)\right) = \left(\sum_{\oplus\, k=1}^{s}(\underline{B} \otimes \underline{\xi}(j_k)) \otimes \alpha_k\right)$$

$$= \sum_{\oplus\, k=1}^{s}(\underline{\xi}(j_k) \otimes \alpha_k) \qquad \text{(by Theorem 23-5)}$$

$$= \emptyset \otimes \underline{u} = \underline{u}$$

It is easy to see that the total set of eigenvectors of $\underline{B}$ which have corresponding eigenvalue $\emptyset$ is a space, containing the eigenspace of $\underline{B}$. Our next aim is to derive sufficient conditions for all finite elements of this space actually to lie in the eigenspace of $\underline{B}$.

<u>Lemma 24-2 Let E_1 be a blog and let $\underline{B} \in \mathcal{M}_{nn}$, for given integer $n\geq 1$, be $\emptyset$-astic definite. Let $\underline{\eta} \in E_n$ be an eigenvector of $\underline{B}$ having corresponding eigenvalue $\emptyset$. If h, k are (distinct) equivalent eigen-nodes in $\Delta(\underline{B})$, then $\{\underline{\eta}\}_h = \{\underline{\eta}\}_k$.</u>

<u>Proof.</u> By hypothesis and Lemma 23-3, there is some circuit starting at node h and passing through node k, with circuit product $\emptyset$. Let this circuit be $(h,h_1,\ldots, h_{t-1},h)$. Then

$$\{\underline{B}\}_{hh_1} \otimes \ldots \otimes \{\underline{B}\}_{t-1\ h} = \emptyset \tag{24-1}$$

and since $\{\underline{B}\}_{hh_1},\ldots, \{\underline{B}\}_{t-1h} \leq \emptyset$ but (by (24-1)) are all finite Corollary 4-12 shows that $\{\underline{B}\}_{hh_1} = \ldots = \{\underline{B}\}_{t-1h} = \emptyset$. Then, since $\underline{\eta} = \underline{B} \otimes \underline{\eta}$, we have by iteration:

$$\{\underline{\eta}\}_h = \sum_{\oplus\, j=1}^{n} \underline{B}_{hj} \otimes \{\underline{\eta}\}_j \geq \underline{B}_{hh_1} \otimes \{\underline{\eta}\}_{h_1} \qquad \text{(by (3-15))}$$

$$= \{\underline{\eta}\}_{h_1} \geq \ldots\ldots\ldots\ldots\ldots\ldots$$

$$\ldots\ldots\ldots\ldots\ldots\ldots \geq \{\underline{B}\}_{t-1h} \otimes \{\underline{\eta}\}_h$$

$$= \{\underline{\eta}\}_h$$

Since $\{\underline{\eta}\}_k$ occurs somewhere in this sequence, the result follows.

If we assume the matrix $\underline{B}$ is blocked, we can display the situation as in Fig 24-1.

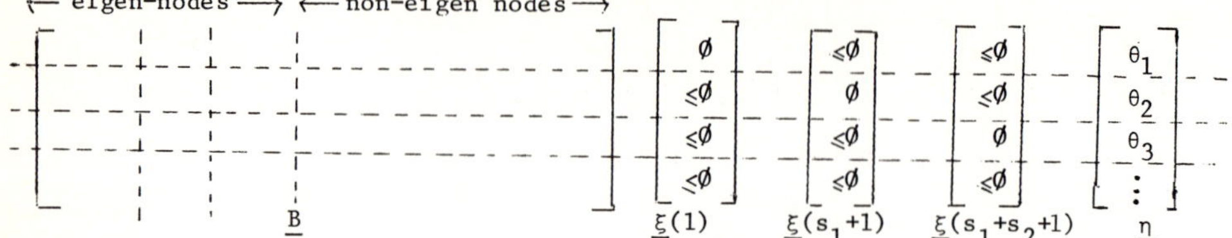

Fig. 24-1. BLOCKED MATRIX

In Fig 24-1 we assume 3 equivalence classes of nodes, containing s_1, s_2, s_3 nodes respectively and with corresponding fundamental eigenvectors $\underline{\xi}(1), \underline{\xi}(s_1+1), \underline{\xi}(s_1+s_2+1)$. These eigenvectors contain blocks of components equal to $\emptyset$ as shown (by Theorem 23-8(iv)) A given eigenvector $\underline{\eta}$ with corresponding eigenvalue $\emptyset$ is also shown, with blocks of equal components. The last $(n-s_1-s_3-s_3)$ components have not yet been discussed. We discuss them now.

Lemma 24-3. Let E_1 be a linear blog and let $\underline{B} \in \mathcal{M}_{nn}$, for given integer $n \geq 1$, be $\emptyset$-astic definite. Let $\underline{\eta} \in E_n$ be a finite eigenvector of $\underline{B}$ having corresponding eigenvalue $\emptyset$. If j is a non-eigen-node in $\Delta(\underline{B})$ then there is an eigen-node d of $\Delta(\underline{B})$ such that $\{\underline{\eta}\}_j = \{\Gamma(\underline{B})\}_{jd} \otimes \{\underline{\eta}\}_d$.

Proof. By hypothesis, $\underline{B} \otimes \underline{\eta} = \underline{\eta}$, i.e.

$$\sum_{s=1}^{n}{}_{\oplus} \{\underline{B}\}_{rs} \otimes \{\underline{\eta}\}_s = \{\underline{\eta}\}_r \quad (r=1,\ldots, n) \tag{24-2}$$

By the linearity of E_1, there is for each r (r=1,..., n) a (least) index c(r). $(1 \leq c(r) \leq n)$ such that:

$$\{\underline{\eta}\}_r = \{\underline{B}\}_{rc(r)} \otimes \{\underline{\eta}\}_{c(r)} \tag{24-3}$$

For the given non-eigen-node j consider the path:

$$(j, c(j), c^2(j), \ldots, c^n(j)) \tag{24-4}$$

where $c^2(j) = c(c(j))$ etc. We assert that this cannot consist exclusively of non-eigen-nodes. For it must contain a repeated index, i.e. $c^{\alpha}(j) = c^{\alpha+\beta}(j)$ for integers $\alpha \geq o$, $\beta \geq 1$, (where $c^{o}(j) = j$). But then by repeated use of (24-2):

$$\begin{aligned}\{\underline{\eta}\}_{c^{\alpha}(j)} &= \{\underline{B}\}_{c^{\alpha}(j)c^{\alpha+1}(j)} \otimes \{\underline{\eta}\}_{c^{\alpha+1}(j)} \\ &= \ldots \\ &= \{\underline{B}\}_{c^{\alpha}(j)c^{\alpha+1}(j)} \otimes \ldots \otimes \{\underline{B}\}_{c^{\alpha+\beta-1}(j)c\ (j)} \otimes \{\underline{\eta}\}_{c^{\alpha}(j)}\end{aligned}$$

But since $\underline{\eta}$ is finite, this would give

$$\{\underline{B}\}_{c^{\alpha}(j)\ c^{\alpha+1}(j)} \otimes \ldots \otimes \{\underline{B}\}_{c^{\alpha+\beta-1}(j)c^{\alpha}(j)} = \emptyset$$

But this would indicate a cycle composed exclusively of non-eigen-nodes and having cycle-product $\emptyset$, a contradiction.

Hence some eigen-node d occurs in (24-4) and there holds by repeated use of (24-2):

$$\{\underline{\eta}\}_j = \{\underline{B}\}_{jc(j)} \otimes \ldots \otimes \{\underline{B}\}_{c^{\gamma-1}(j)d} \otimes \{\underline{\eta}\}_d \qquad (24\text{-}5)$$

for some integer $\gamma \geq 1$.

$$\text{Now } \underline{\Gamma}(\underline{B}) \otimes \underline{\eta} = (\underline{B} \oplus \ldots \oplus \underline{B}^n) \otimes \underline{\eta} = \underline{\eta} \oplus \ldots \oplus \underline{\eta} = \underline{\eta} \qquad (24\text{-}6)$$

$$\text{Hence } \{\underline{\eta}\}_j = \sum_{k=1}^{n}{}_{\oplus} \{\underline{\Gamma}(\underline{B})\}_{jk} \otimes \{\underline{\eta}\}_k \qquad \text{(from (24-6))} \qquad (24\text{-}7)$$

$$\geq \{\underline{\Gamma}(\underline{B})\}_{jd} \otimes \{\underline{\eta}\}_d \qquad \text{(by (3-15))} \qquad (24\text{-}8)$$

$$\geq \{\underline{B}^{\gamma}\}_{jd} \otimes \{\underline{\eta}\}_d \qquad \text{(by Theorem 22-6)}$$

$$\geq \{\underline{B}\}_{jc(j)} \otimes \ldots \otimes \{\underline{B}\}_{c^{\gamma-1}(j)d} \otimes \{\underline{\eta}\}_d \qquad \text{(by Proposition 22-1)}$$

$$= \{\underline{\eta}\}_j \qquad \text{(by (24-5))}.$$

Hence equality applies throughout, and to (24-8) in particular. ●

<u>Theorem 24-4. Let E_1 be a linear blog and let $\underline{B} \in \mathcal{M}_{nn}$, for given integer $n \geq 1$, be ϕ-astic definite. If $\underline{\eta} \in E_n$ is a finite eigenvector having corresponding eigenvalue ϕ, then $\underline{\eta}$ lies in the eigenspace of $\underline{B}$.</u>

<u>Proof</u>. Let $\underline{\xi}(j_1), \ldots, \underline{\xi}(j_s)$ be a maximal set of non-equivalent eigenvectors for $\underline{B}$. We shall prove:

$$\underline{\eta} = \sum_{k=1}^{s}{}_{\oplus} (\underline{\xi}(j_k) \otimes \{\underline{\eta}\}_{j_k})$$

Now, since $\underline{\xi}(j_1), \ldots, \underline{\xi}(j_s)$ form a subset of the columns of $\underline{\Gamma}(\underline{B})$:

$$\sum_{k=1}^{s}{}_{\oplus} (\underline{\xi}(j_k) \otimes \{\underline{\eta}\}_{j_k}) \leq \underline{\Gamma}(\underline{B}) \otimes \underline{\eta} = \underline{\eta} \qquad \text{(by (24-6))} \qquad (24\text{-}9)$$

Now let j be an eigen-node of $\Delta(\underline{B})$. Then one of $j_1, \ldots, j_s$ is equivalent to j: say j_h is. We have:

$$\{\sum_{k=1}^{s}{}_{\oplus} (\underline{\xi}(j_k) \otimes \{\underline{\eta}\}_{j_k})\}_j \geq \{\underline{\xi}(j_h)\}_j \otimes \{\underline{\eta}\}_{j_h} \qquad \text{(by (3-15))}$$

$$= \phi \otimes \{\underline{\eta}\}_j \qquad \text{(by Theorem 23-8 (iv) and Lemma 24-2)}$$

$$= \{\underline{\eta}\}_j \qquad (24\text{-}10)$$

Finally, let j be a non-eigen-node of $\Delta(\underline{B})$. Then by Lemma 24-3, there is an eigen-node d such that:

$$\{\underline{\eta}\}_j = \{\underline{\Gamma}(\underline{B})\}_{jd} \otimes \{\underline{\eta}\}_d \qquad (24\text{-}11)$$

Now, one of $j_1, \ldots, j_s$ is equivalent to d: say j_h is. Then:

$$\{\sum_{k=1}^{s}{}_{\oplus}(\underline{\xi}(j_k) \otimes \{\underline{\eta}\}_{j_k})\}_j \geq \{\underline{\xi}(j_h)\}_j \otimes \{\underline{\eta}\}_{j_h} \qquad \text{(by (3-15))}$$

$$= \{\underline{\Gamma}(\underline{B})\}_{jd} \otimes \{\underline{\eta}\}_d \qquad \text{(by Theorem 23-8(iv) and Lemma 24-2)}$$

$$= \{\underline{\eta}\}_j \qquad \text{(by (24-11))} \qquad (24\text{-}12)$$

Evidently (24-9), (24-10) and (24-12) imply the desired result. ●

Taking Theorem 24-4 with Theorem 23-10 we see that every finite eigenvector of $\underline{B}$ having corresponding eigenvalue $\emptyset$ is a linear combination of certain linearly independent fundamental eigenvectors.

24-2. Directly Similar Matrices

Theorem 24-4 is the essential key to the eigenproblem for more general square matrices. The following discussion establishes the connection. We shall say that two matrices $\underline{A}, \underline{B} \in \mathcal{M}_{nn}$ over a blog are directly similar if there are n finite elements $\eta_1, \ldots, \eta_n \in E_1$ such that $\{\underline{B}\}_{ij} = \eta_i^{-1} \otimes \{\underline{A}\}_{ij} \otimes \eta_j$. We shall write: $\underline{A} \simeq \underline{B}$.

Lemma 24-5. Let E_1 be a blog. Then $\simeq$ is an equivalence relation on $\mathcal{M}_{nn}$ for any given integer $n \geq 1$. If $\underline{A}, \underline{B} \in \mathcal{M}_{nn}$ and $\underline{A} \simeq \underline{B}$ then there are $\underline{P}, \underline{Q} \in \mathcal{M}_{nn}$ such that $\underline{B} = \underline{P} \otimes \underline{A} \otimes \underline{Q}$ and $\underline{P} \otimes \underline{Q} = \underline{Q} \otimes \underline{P} = \underline{I}_n$, the $(n \times n)$ identity matrix. Then $\underline{B}^r = \underline{P} \otimes \underline{A}^r \otimes \underline{Q}$ for all integers $r \geq 0$ and $\underline{\Gamma}(\underline{B}) = \underline{P} \otimes \underline{\Gamma}(\underline{A}) \otimes \underline{Q}$.

Proof. Given finite elements $\eta_1, \ldots, \eta_n \in E_1$ define diag $(\eta_1, \ldots, \eta_n) \in \mathcal{M}_{nn}$ to be a matrix having diagonal elements $\eta_1, \ldots, \eta_n$ in that order and off-diagonal elements $-\infty$.

If $\underline{P} = \text{diag}\,(\eta_1^{-1}, \ldots, \eta_n^{-1})$ and $\underline{Q} = \text{diag}\,(\eta_1, \ldots, \eta_n)$, we readily verify that $\underline{P} \otimes \underline{Q} = \underline{Q} \otimes \underline{P} = \underline{I}_n$ and that $\underline{A} \simeq \underline{B}$ if and only if $\underline{B} = \underline{P} \otimes \underline{A} \otimes \underline{Q}$ for suitable $\eta_1, \ldots, \eta_n$. Now $\underline{A} = \underline{I}_n \otimes \underline{A} \otimes \underline{I}_n$ so $\simeq$ is reflexive. And $\underline{Q} \otimes (\underline{P} \otimes \underline{A} \otimes \underline{Q}) \otimes \underline{P} = \underline{A}$ so $\simeq$ is symmetric. And with an obvious notation, $\underline{R} \otimes (\underline{P} \otimes \underline{A} \otimes \underline{Q}) \otimes \underline{S} = (\underline{R} \otimes \underline{P}) \otimes \underline{A} \otimes (\underline{Q} \otimes \underline{S})$ where $(\underline{R} \otimes \underline{P}) \otimes (\underline{Q} \otimes \underline{S}) = \underline{I}_n$ so $\simeq$ is transitive.

Further, $(\underline{P} \otimes \underline{A} \otimes \underline{Q})^r = \underline{P} \otimes \underline{A} \otimes (\underline{Q} \otimes \underline{P}) \otimes \underline{A} \otimes \ldots \otimes (\underline{Q} \otimes \underline{P}) \otimes \underline{A} \otimes \underline{Q}$

$$= \underline{P} \otimes \underline{A}^r \otimes \underline{Q} \qquad (r \geq 1)$$

And $\quad \Gamma(\underline{P} \otimes \underline{A} \otimes \underline{Q}) = \sum_{r=1}^{n}{}_{\oplus}(\underline{P} \otimes \underline{A} \otimes \underline{Q})^r$

$$= P \otimes \left(\sum_{r=1}^{n}{}_{\oplus}\underline{A}^r\right) \otimes Q = \underline{P} \otimes \Gamma(\underline{A}) \otimes \underline{Q}.$$ ●

Lemma 24-6. Let E_1 be a blog and let $\underline{A}, \underline{B} \in \mathcal{M}_{nn}$, for given integer $n \geq 1$, be directly similar. Then $\underline{A}$ is definite if and only if $\underline{B}$ is definite; and then a set of indices $j_1, \ldots, j_s$ give a set of non-equivalent eigen-nodes in $\Delta(\underline{A})$ if and only if they give a set of non-equivalent eigen-nodes in $\Delta(\underline{B})$.

Proof. Let a set of indices $(i_0, i_1, \ldots, i_{t-1}, i_0)$ constitute a circuit σ in $\Delta(\underline{A})$

with corresponding circuit products $p(\sigma)$ and $p(\tau)$. We have, for suitable finite elements $\eta_1,\ldots, \eta_n \in E_1$:

$$p(\tau) = (\eta_{i_o})^{-1} \otimes \{\underline{A}\}_{i_o i_1} \otimes \eta_{i_1}$$

$$\otimes (\eta_{i_1})^{-1} \otimes \{\underline{A}\}_{i_1 i_2} \otimes \eta_{i_2}$$
$$\ldots$$
$$\otimes (\eta_{i_{t-1}})^{-1} \otimes \{\underline{A}\}_{i_{t-1} i_o} \otimes \eta_{i_o}$$

$$= (\eta_{i_o})^{-1} \otimes p(\sigma) \otimes \eta_{i_o}$$

It follows that $p(\tau) \sim \emptyset$ if and only if $p(\sigma) \sim \emptyset$, where $\sim$ is any of the relations $\leqslant,=,\geqslant$. Hence $\underline{A}$ is definite if and only if $\underline{B}$ is definite. Hence also, if $\underline{A}$ and $\underline{B}$ are definite, then the same indices are eigen-nodes in $\Delta(\underline{A})$ and in $\Delta(\underline{B})$, and two indices give equivalent eigen-nodes in $\Delta(\underline{A})$ if and only if they give equivalent eigen-nodes in $\Delta(\underline{B})$. ●

Lemma 24-7. Let E_1 be a blog and let $\underline{A}, \underline{B} \in \mathcal{M}_{nn}$ for given integer $n\geqslant 1$ be such that $\underline{A} \simeq \underline{B}$. Let $j_1,\ldots, j_{s+1}$ $(s\geqslant 1)$ be indices $(1\leqslant j_k \leqslant n;\ k=1,\ldots, s+1)$. Then column j_{s+1} in $\underline{B}$ is linearly dependent on columns $j_1,\ldots, j_s$ in $\underline{B}$ if and only if the same holds in $\underline{A}$. An analogous result holds for the rows.

Proof. If such a dependence holds in $\underline{B}$ then for certain scalars β_k $(k=1,\ldots, s)$ there holds:

$$\{\underline{B}\}_{ij_{s+1}} = \sum_{k=1}^{s}{}_{\oplus} (\{\underline{B}\}_{ij_k} \otimes \beta_k) \qquad (i=1,\ldots, n)$$

But by hypothesis, $\{\underline{B}\}_{ij} = \eta_i^{-1} \otimes \{\underline{A}\}_{ij} \otimes \eta_j$ $\quad (i,j=1,\ldots, n)$ for certain finite scalars $\eta_i (i=1,\ldots, n)$. Hence:

$$\eta_i^{-1} \otimes \{\underline{A}\}_{ij_{s+1}} \otimes \eta_{j_{s+1}} = \sum_{k=1}^{s}{}_{\oplus}(\eta_i^{-1} \otimes \{\underline{A}\}_{ij_k} \otimes \eta_{j_k} \otimes \beta_k)$$

$$\eta_i^{-1} \otimes (\sum_{k=1}^{s}{}_{\oplus} (\{\underline{A}\}_{ij_k} \otimes \alpha_k)) \otimes \eta_{j_{s+1}}$$

where $\alpha_k = \eta_{j_k} \otimes \beta_k \otimes \eta_{j_{s+1}}^{-1}$. Pre- and post- multiplying, we obtain the desired result. The converse holds by the symmetry of the relation $\simeq$. The argument for the rows is similar. ●

24-3 . Structure of the Eigenspace

We are now in a position to prove our main result on the eigenproblem, which follows after the next lemma.

Lemma 24-8. Let E_1 be a linear blog and let $\underline{A} \in \mathcal{M}_{nn}$ for given integer $n\geqslant 1$. If $\alpha, \beta \in E_1$ are such that $\alpha \otimes \underline{A}$ and $\beta \otimes \underline{A}$ are both definite, then $\alpha = \beta$ (and are finite).

Proof. If σ is a circuit in $\Delta(\alpha \otimes \underline{A})$ with circuit product $p(\sigma)=\emptyset$, let τ be the circuit in $\Delta(\beta \otimes \underline{A})$ determined by the same nodes as σ. If $\alpha \neq \beta$, suppose without loss of generality that $\alpha<\beta$. Since the product $p(\tau)$ is obtained by replacing α by β in the expression for $p(\sigma)$, it follows from Corollary 4-12 that $p(\tau) > p(\sigma) = \emptyset$. But this contradicts the definiteness of $\beta \otimes \underline{A}$. Hence $\alpha=\beta$. And evidently α is finite, since $p(\sigma) = \emptyset$ contains α as a factor.

Theorem 24-9. Let E_1 be a linear blog and let $\underline{A} \in \mathcal{M}_{nn}$ for given integer $n \geq 1$. If the eigenproblem for $\underline{A}$ is finitely soluble then every finite eigenvector has the same unique corresponding finite eigenvalue λ. The matrix $\lambda^{-1} \otimes \underline{A}$ is definite, and all finite eigenvectors of $\underline{A}$ lie in the eigenspace of $\lambda^{-1} \otimes \underline{A}$. The non-equivalent fundamental eigenvectors which generate this space have the property that no one of them is linearly dependent on (any subset of) the others.

Proof. Suppose we can find $\lambda, \mu \in E_1$ and $\underline{\xi}$, $\underline{\eta} \in E_n$, all finite, such that $\underline{A} \otimes \underline{\xi} = \lambda \otimes \underline{\xi}$ and $\underline{A} \otimes \underline{\eta} = \mu \otimes \underline{\eta}$.

Now $$\sum_{j=1}^{n}{}_{\oplus}(\{\underline{A}\}_{ij} \otimes \{\underline{\xi}\}_j) = \underline{\lambda} \otimes \{\underline{\xi}\}_i \qquad (i=1,\ldots, n)$$

Hence, by the linearity of E_1 there is for each index $i=1,\ldots, n$ a (least) index $c(i)$ $(1 \leq c(i) \leq n)$ such that:

$$\{\underline{A}\}_{ic(i)} \otimes \{\underline{\xi}\}_{c(i)} = \lambda \otimes \{\underline{\xi}\}_i \qquad (i=1,\ldots, n)$$

But also $$\{\underline{A}\}_{ij} \otimes \{\underline{\xi}\}_j \leq \lambda \otimes \{\underline{\xi}\}_i \qquad (i,j=1,\ldots, n)$$

Hence: $$(\{\underline{\xi}\}_i)^{-1} \otimes \{\lambda^{-1} \otimes \underline{A}\}_{ic(i)} \otimes \{\underline{\xi}\}_{c(i)} = \emptyset \qquad (i=1,\ldots, n) \qquad (24\text{-}13)$$

and $$(\{\underline{\xi}\}_i)^{-1} \otimes \{\lambda^{-1} \otimes \underline{A}\}_{ij} \otimes \{\underline{\xi}\}_j \leq \emptyset \qquad (i,j=1,\ldots, n) \qquad (24\text{-}14)$$

Now if $\sigma=(i_o,i_1,\ldots, i_{t-1}, i_o)$ is any circuit in $\Delta(\lambda^{-1} \otimes \underline{A})$ (the graph associated with $\lambda^{-1} \otimes \underline{A}$) and $p(\sigma)$ is the circuit product for σ, we have:

$$\begin{aligned} &(\{\underline{\xi}\}_{i_o})^{-1} \otimes p(\sigma) \otimes \{\underline{\xi}\}_{i_o} \\ = \ &(\{\underline{\xi}\}_{i_o})^{-1} \otimes \{\lambda^{-1} \otimes \underline{A}\}_{i_o i_1} \otimes \{\underline{\xi}\}_{i_1} \\ \otimes \ &(\{\underline{\xi}\}_{i_1})^{-1} \otimes \{\lambda^{-1} \otimes \underline{A}\}_{i_1 i_2} \otimes \{\underline{\xi}\}_{i_2} \\ &\ldots \\ \otimes \ &(\{\underline{\xi}\}_{i_{t-1}})^{-1} \otimes \{\lambda^{-1} \otimes\}\underline{A}_{i_{t-1} i_o} \otimes \{\underline{\xi}\}_{i_o} \leq \emptyset \qquad \text{(by (24-14))}. \end{aligned}$$

Premultiplying by $\{\underline{\xi}\}_{i_o}$ and postmultiplying by $(\{\underline{\xi}\}_{i_o})^{-1}$, we have:

$$p(\sigma) \leq \emptyset \text{ for all circuits } \sigma \text{ in } \Delta(\lambda^{-1} \otimes \underline{A}) \qquad (24\text{-}15)$$

However, consider now the path $(1,c(1),c^2(1),\ldots, c^n(1))$ where $c^2(1)=c(c(1))$ etc. This must contain a circuit σ and by using (24-13) in place of (24-14) in the foregoing argument we have

$$p(\sigma) = \emptyset \quad \text{for at least one circuit in } \Delta(\lambda^{-1} \otimes \underline{A}) \qquad (24\text{-}16)$$

Evidently, (24-15) and (24-16) show that $\lambda^{-1} \otimes \underline{A}$ is definite.

Similarly, $\mu^{-1} \otimes \underline{A}$ is definite. Hence $\lambda=\mu$ by Lemma 24-8 so $\underline{A}$ can have only one value for its finite eigenvalues corresponding to finite eigenvectors.

Now from (24-13) and (24-14) we see that the matrix $\underline{B}$ is row-$\emptyset$-astic (and so $\emptyset$-astic definite) where:

$$\{\underline{B}\}_{ij} = (\{\underline{\xi}\}_i)^{-1} \otimes \{\lambda^{-1} \otimes \underline{A}\}_{ij} \otimes \{\underline{\xi}\}_j \qquad (i,j=1,\ldots, n)$$

Evidently also, $\underline{B} \sim \lambda^{-1} \otimes \underline{A}$.

If $\underline{u} \in E_n$ is the n-tuple having all components equal to $\emptyset$, then by the proof of Lemma 23-2 and Theorem 24-4, $\underline{u}$ is a linear combination of certain columns ($j_1,\ldots, j_s$, say) of $\underline{\Gamma}(\underline{B})$, these columns constituting non-equivalent fundamental eigenvectors of $\underline{B}$. Thus:

$$\emptyset = \{\underline{u}\}_i = \sum_{k=1}^{s}{}_{\oplus}(\{\underline{\Gamma}(\underline{B})\}_{ij_k} \otimes \alpha_k) \qquad (i=1,\ldots, n) \qquad (24\text{-}17)$$

for suitable $\alpha_k \in E_1$ $(k=1,\ldots, s)$

From Lemma 24-5, we may rewrite (24-17):

$$\emptyset = \sum_{k=1}^{s}{}_{\oplus}((\{\underline{\xi}\}_i)^{-1} \otimes \{\underline{\Gamma}(\lambda^{-1} \otimes \underline{A})\}_{ij_k} \otimes \{\underline{\xi}\}_{j_k} \otimes \alpha_k) \qquad (i=1,\ldots, n)$$

$$= (\{\underline{\xi}\}_i)^{-1} \otimes \sum_{k=1}^{s}{}_{\oplus}(\{\underline{\Gamma}(\lambda^{-1} \otimes \underline{A})\}_{ij_k} \otimes \beta_k) \qquad (i=1,\ldots, n)$$

where $\beta_k = \{\underline{\xi}\}_{j_k} \otimes \alpha_k$ $\quad (k=1,\ldots, n)$

$$\text{Hence } \{\underline{\xi}\}_i = \sum_{k=1}^{s}{}_{\oplus} (\{\underline{\Gamma}(\lambda^{-1} \otimes \underline{A})\}_{ij_k} \otimes \beta_k) \qquad (i=1,\ldots, n) \qquad (24\text{-}18)$$

But (24-18) says that $\underline{\xi}$ is a linear combination of columns in $\underline{\Gamma}(\lambda^{-1} \otimes \underline{A})$ having the same indices as those used in (24-17).

Since $\underline{B} \sim \lambda^{-1} \otimes \underline{A}$ and both $\underline{B}$ and $\lambda^{-1} \otimes \underline{A}$ are definite, Lemma 24-6 then implies that $\underline{\xi}$ is a linear combination of non-equivalent fundamental eigenvectors of $\lambda^{-1} \otimes \underline{A}$, i.e. that $\underline{\xi}$ lies in the eigenspace of $\lambda^{-1} \otimes \underline{A}$. Moreover Theorem 23-10 and Lemma 24-7 imply the required independence of the non-equivalent fundamental eigenvectors of $\lambda^{-1} \otimes \underline{A}$. ●

We shall call the unique scalar λ (when it exists) in Theorem 24-9 the <u>principal eigenvalue of $\underline{A}$</u>.

25. SOLVING THE EIGENPROBLEM

25-1. The Parameter $\lambda(\underline{A})$

Evidently, if we know the unique eigenvalue λ referred to in Theorem 24-7, then we can compute $\underline{\Gamma}(\lambda^{-1}\otimes\underline{A})$ and so have all finite solutions to the eigenproblem at our disposal. We investigate now conditions sufficient to guarantee the existence and computability of λ for a reasonably broad class of matrices.

We shall have to put extra conditions on E_1. Besides often assuming that E_1 is a commutative linear blog, we shall assume that E_1 is radicable, that is to say:

For each $a \in E_1$ and integer $t \geq 1$, there is a unique $x \in E_1$ such that:

$$x^t = a \tag{25-1}$$

The following lemma embodies what we might call the isotonicity of root extraction.

Lemma 25-1. Let E_1 be a linear radicable blog and for given integer $n \geq 1$ let $x,y,a,b \in E_1$ satisfy $x^n = a \geq b = y^n$. Then $x \geq y$. And $x \sim \phi$ if and only if $a \sim \phi$ where $\sim$ is $\geq$, $=$ or $\leq$.

Proof. Since E_1 is linear, either $x>y$ or $x \leq y$. In the latter case, $x^2 \leq x\otimes y \leq y^2$ and by iteration, $x^n \leq y^n$. So $a \leq b$, whence $a=b$. But then $x=y$ by the uniqueness of x and y entailed in (25-1). So $x \geq y$ in all cases.

Now suppose $a \geq \phi$. On putting $b=y=\phi$, we deduce $x \geq \phi$. And if $b \leq \phi$ we deduce $y \leq \phi$ on putting $x=a=\phi$. So on interchanging the roles of x and y, a and b we infer that $x \leq \phi$ if $a \leq \phi$, and hence that $x=\phi$ if $a=\phi$.

Conversely if $x \sim \phi$ where $\sim$ is $\geq$, $=$ or $\leq$ 0 then $x^n \sim \phi$ also, by isotonicity. ●

Now if E_1 is a radicable belt and $\underline{A} \in \mathcal{M}_{nn}$ for given integer $n \geq 1$, then for each circuit σ in $\Delta(\underline{A})$, of length $t(\sigma)$ and circuit product $p(\sigma)$ (say), we can compute a unique circuit mean $\mu(\sigma) \in E_1$ from:

$$(\mu(\sigma))^{t(\sigma)} = p(\sigma) \tag{25-2}$$

We define: $$\lambda(\underline{A}) = \Sigma_{\oplus}\mu(\sigma) \tag{25-3}$$

where the summation is taken over all elementary cycles in $\Delta(\underline{A})$. This notation will be standard in the sequel.

For example, if $\underline{A}$ is the matrix associated with the graph of Fig 22-1, we may compute the following data

Elem circuit	Circuit Product	Circuit length	Circuit mean
(1,1)	5	1	5
(2,2)	6	1	6
(3,3)	7	1	7
(1,2,1)	-5	2	-5/2
(2,3,2)	-2	2	-1
(3,1,3)	1	2	1/2
(1,2,3)	-6	3	-2
(3,2,1)	0	3	0

In the above table, we work in the principal interpretation. Thus for circuit (1,2,3) the circuit product is:

$$(-2) \otimes (-4) \otimes (0) = (-2) + (-4) + 0 = -6$$

And to get the corresponding circuit mean we must solve:

$$\mu^t = \mu \otimes \mu \otimes \mu = -6$$

i.e. (in conventional arithmetic): $3\mu = -6$ or $\mu = -2$.

Since we are working with a commutative belt, we have not included in the table any cyclic permutations of the circuits already recorded.

From the table, we see that $\lambda(\underline{A}) = 7$. It could happen in general, of course, that $\lambda(\underline{A})$ occurred in association with more than one circuit.

25-2. Properties of $\lambda(\underline{A})$

Lemma 25-2. Let E_1 be a radicable blog and let $\underline{A} \in \mathcal{M}_{nn}$, for given integer $n \geq 1$. If $\underline{A}$ is either row- or column-G-astic then $\lambda(\underline{A})$ is finite.

Proof. Suppose $\underline{A}$ is row-G-astic. Evidently then no circuit-product, and so no circuit mean, will equal $+\infty$. Moreover, for each index $i (i=1,\ldots, n)$ there is a (least) index $c(i)$ $(1 \leq c(i) \leq n)$ such that $\{\underline{A}\}_{ic(i)}$ is finite, so by picking out a circuit from the path $(1,c(1),c^2(1),\ldots, c^n(1))$ where $c^2(1) = c(c(1))$ etc., we can find a circuit with finite mean. Hence $\lambda(\underline{A})$ is finite. The proof is similar if $\underline{A}$ is column-G-astic. ●

Lemma 25-3. Let E_1 be a commutative linear radicable blog, and let $\underline{A} \in \mathcal{M}_{nn}$ for given integer $n \geq 1$. Then for each $\alpha \in E_1$ there holds:

$$\lambda(\alpha \otimes \underline{A}) = \alpha \otimes \lambda(\underline{A})$$

Proof. From the linearity of E_1, there is an elementary circuit σ in $\Delta(\alpha \otimes \underline{A})$ for which the maximum is attained in (25-3), i.e.:

$$\mu(\sigma) = \lambda(\alpha \otimes \underline{A}) \tag{25-4}$$

If σ is $(i_o, i_1, \ldots, i_{t-1}, i_o)$ (of length t) then the circuit product for σ is:

$$\begin{aligned} p(\sigma) &= \{\alpha \otimes \underline{A}\}_{i_o i_1} \otimes \ldots \otimes \{\alpha \otimes \underline{A}\}_{i_{t-1} i_o} \\ &= \alpha^t \otimes \{\underline{A}\}_{i_o i_1} \otimes \ldots \otimes \{\underline{A}\}_{i_{t-1} i_o} \quad \text{(by commutativity)} \\ &= \alpha^t \otimes p(\tau) \end{aligned} \tag{25-5}$$

where τ is the cycle in $\Delta(\underline{A})$ determined by the same nodes as determine σ, having circuit product $p(\tau)$, (and length t).

From (25-3), $\lambda(\underline{A}) \geq$ circuit mean for τ

whence $(\lambda(\underline{A}))^t \geq (\text{circuit mean for } \tau)^t$

$$= p(\tau) \tag{25-6}$$

Hence $(\mu(\sigma))^t = p(\sigma)$ (by definition)

$= \alpha^t \otimes p(\tau)$ (by (25-5))

$\leqslant \alpha^t \otimes (\lambda(\underline{A}))^t$ (by (25-6))

$= (\alpha \otimes \lambda(\underline{A}))^t$ (by commutativity)

So using Lemma 25-1:

$$\mu(\sigma) \leqslant \alpha \otimes \lambda(\underline{A})$$

i.e. $\lambda(\alpha \otimes \underline{A}) \leqslant \alpha \otimes \lambda(\underline{A})$ (by (24-22)) (25-7)

Now if we change the notation slightly so that τ becomes an elementary circuit in $\Delta(\underline{A})$ for which the maximum circuit mean is attained, then :

$(\lambda(\underline{A}))^t = (\text{circuit mean for } \tau)^t = p(\tau)$

so $(\alpha \otimes \lambda(\underline{A}))^t = \alpha^t \otimes (\lambda(\underline{A}))^t$ (by commutativity)

$= \alpha^t \otimes p(\tau)$

$= p(\sigma)$ (by (25-5))

where σ is the (now not necessarily maximal) circuit in $\Delta(\alpha \otimes \underline{A})$ determined by the same nodes as determine τ. Hence :

$$\mu(\sigma) = \alpha \otimes \lambda(\underline{A})$$

whence from (25-3): $\lambda(\alpha \otimes \underline{A}) \geqslant \alpha \otimes \lambda(\underline{A})$

Combining this with (25-7) yields the required result. ●

Corollary 25-4. Let E_1 be a commutative linear radicable blog and let $\underline{A} \in \mathcal{M}_{nn}$ for given integer $n \geqslant 1$. If $\lambda(\underline{A})$ is finite then the matrix $(\lambda(\underline{A}))^{-1} \otimes \underline{A}$ is definite.

Proof. By Lemma 25-3:

$$\lambda((\lambda(\underline{A}))^{-1} \otimes \underline{A}) = \emptyset$$

From (25-3) and the linearity of E_1 it follows that the circuit mean $\mu(\sigma) \leqslant \emptyset$ for every elementary circuit σ in $\Delta((\lambda(\underline{A}))^{-1} \otimes \underline{A})$ and $\mu(\sigma) = \emptyset$ for at least one such circuit. So by Lemma 25-1, the circuit product $p(\sigma) \leqslant \emptyset$ for every elementary circuit σ in $\Delta((\lambda(\underline{A}))^{-1} \otimes \underline{A})$, and $p(\sigma) = \emptyset$ for at least one such circuit. Thus $(\lambda(\underline{A}))^{-1} \otimes \underline{A}$ is definite, by Lemma 22-2. ●

25-3. Necessary and Sufficient Conditions

The following result connects the parameter $\lambda(\underline{A})$ with the eigenproblem for $\underline{A}$.

Theorem 25-5. Let E_1 be a commutative linear radicable blog and let $A \in \mathcal{M}_{nn}$ for given integer $n \geqslant 1$. If the eigenproblem for $\underline{A}$ is finitely soluble then $\lambda(\underline{A})$ is finite; and $\lambda(\underline{A})$ is then the only possible value for the eigenvalue in any finite solution to the eigenproblem for $\underline{A}$. (i.e. $\lambda(\underline{A})$ is the principal eigenvalue of $\underline{A}$).

Proof. Suppose the eigenproblem for $\underline{A}$ is finitely soluble. Then by Theorem 24-8, there is only one possible finite value, λ(say), for the eigenvalue corresponding to finite eigenvectors, and the matrix $\lambda^{-1} \otimes \underline{A}$ is definite. Hence by Lemma 22-2, for every elementary circuit σ in $\Delta(\lambda^{-1} \otimes \underline{A})$, the circuit product $p(\sigma) \leqslant \emptyset$, and $p(\sigma) = \emptyset$ for at least one cycle. Thus by Lemma 25-1 the circuit mean $\mu(\sigma) \leq \emptyset$ for every elementary circuit σ in $\Delta(\lambda^{-1} \otimes \underline{A})$ and $\mu(\sigma) = \emptyset$ for at least one elementary circuit. Hence from (25-3):

$$\lambda(\lambda^{-1} \otimes \underline{A}) = \emptyset$$

Hence $\lambda(\underline{A}) = \lambda$ (by Lemma 25-3).

And this shows in particular that $\lambda(\underline{A})$ is finite.

Theorem 25-5 shows that the finiteness of $\lambda(\underline{A})$ is a necessary condition for finite solubility of the eigenproblem for $\underline{A}$. It is not in general a sufficient condition. For example if:

$$\underline{A} = \begin{bmatrix} \emptyset & -\infty \\ -\infty & -1 \end{bmatrix}$$

Then $\lambda(\underline{A}) = \emptyset$, but $\underline{A}$ has the single fundamental eigenvector

$$\begin{bmatrix} \emptyset \\ -\infty \end{bmatrix}.$$

Hence every 2-tuple in the eigenspace has its second coordinate equal to $-\infty$. The following result establishes necessary and sufficient conditions for finite solubility.

Theorem 25-6. Let E_1 be a commutative linear radicable blog, and let $\underline{A} \in \mathcal{M}_{nn}$ for given integer $n \geq 1$. Then the eigenproblem for $\underline{A}$ is finitely soluble if and only if: $\lambda(\underline{A})$ is finite and $\Phi((\lambda(\underline{A}))^{-1} \otimes \underline{A})$ is doubly G-astic, where $\Phi((\lambda(\underline{A}))^{-1} \otimes \underline{A})$ is any matrix whose columns form a maximal set of non-equivalent fundamental eigenvectors for the definite matrix $(\lambda(\underline{A}))^{-1} \otimes \underline{A}$.

Proof. If the eigenproblem for $\underline{A}$ is finitely soluble, then by Theorem 25-5, $\lambda(\underline{A})$ is finite, and from the proof of Theorem 24-7, the matrix $(\lambda(\underline{A}))^{-1} \otimes \underline{A}$ is directly similar to a certain row-$\emptyset$-astic matrix $\underline{B}$, and hence $\Gamma((\lambda(\underline{A}))^{-1} \otimes \underline{A})$ is directly similar to $\Gamma(\underline{B})$ (Lemma 24-5). Now the matrix $\Phi(\underline{B})$, which takes the same columns from $\Gamma(\underline{B})$ as $\Phi((\lambda(\underline{A}))^{-1} \otimes \underline{A})$ takes from $\Gamma(\underline{A})$ is, by Lemma 24-6 and Theorem 23-11, doubly $\emptyset$-astic, and so certainly doubly G-astic. Since we can derive $\Phi((\lambda(\underline{A}))^{-1} \otimes \underline{A})$ from $\Phi(\underline{B})$ by multiplying matrix elements by finite scalars (by virtue of the direct similarity), it follows that $\Phi((\lambda(\underline{A})^{-1} \otimes \underline{A})$ is also doubly G-astic.

Conversely, if $\lambda(\underline{A})$ is finite, then $(\lambda(\underline{A}))^{-1} \otimes \underline{A}$ is definite by Corollary 25-4, so the definition of $\Phi((\lambda(\underline{A}))^{-1} \otimes \underline{A})$ given in the theorem statement is meaningful; and if this matrix has s columns, and $\underline{x} \in E_s$ is finite then so is $\underline{u} \in E_n$, where

$$\underline{u} = \Phi((\lambda(\underline{A}))^{-1} \otimes \underline{A}) \otimes \underline{x}$$

(by Corollary 12-7). So $\underline{u}$ is a finite element of the eigenspace of $(\lambda(\underline{A}))^{-1} \otimes \underline{A}$.

Thus, $(\lambda(\underline{A}))^{-1} \otimes \underline{A}) \otimes \underline{u} = \underline{u}$ (by Lemma 24-1)

Hence $\underline{A} \otimes \underline{u} = \lambda(\underline{A}) \otimes \underline{u}$

showing that the eigenproblem for $\underline{A}$ is finitely soluble.

We now draw some further corollaries from Theorem 25-6.

Corollary 25-7. Let E_1 be a commutative linear radicable blog and let $\underline{A} \in \mathcal{M}_{nn}$, for given integer $n \geq 1$, be finite. Then the eigenproblem for $\underline{A}$ is finitely soluble

Proof. Evidently $\lambda(\underline{A})$ is finite and so are $(\lambda(\underline{A}))^{-1} \otimes \underline{A}$ and $\Gamma((\lambda(\underline{A}))^{-1} \otimes \underline{A})$. Hence

$\underline{\Phi}((\lambda(\underline{A}))^{-1}\otimes\underline{A})$ is finite, so row-G-astic. ●

The foregoing corollary shows that the eigenproblem is finitely soluble for a substantial class of matrices of practical importance, namely finite (square) matrices under the principal interpretation for E_1: for the extended real numbers form a commutative linear radicable blog, as may easily be verified.

The next corollary identifies another class of matrices, which we shall discuss again later, for which the eigenproblem is always finitely soluble.

Corollary 25-8. Let E_1 be a commutative linear radicable blog and let $\underline{A} \in \mathcal{M}_{nn}$, for given integer $n \geq 1$, be row-G-astic. If $\Delta(\lambda(\underline{A}))^{-1}\otimes\underline{A})$ has n eigen-nodes (not necessarily non-equivalent) then $\underline{A}$ is doubly G-astic and the eigenproblem for $\underline{A}$ is finitely soluble.

Proof. Since $\underline{A}$ is row-G-astic, evidently then $\lambda(\underline{A})$ is finite by Lemma 25-2 and so $(\lambda(\underline{A}))^{-1}\otimes\underline{A}$ is definite by Corollary 25-4. Hence the term 'eigen-nodes' is permissible in connection with $\Delta((\lambda(\underline{A}))^{-1}\otimes\underline{A})$. If in fact $\Delta((\lambda(\underline{A}))^{-1}\otimes\underline{A})$ has n eigen-nodes then for each index $i(i=1,.., n)$ there is an index h $(1 \leq h \leq n)$ such that $\{(\lambda(\underline{A}))^{-1}\otimes\underline{A}\}_{hi}$ contributes to a circuit product equal to $\emptyset$. Hence $(\lambda(\underline{A}))^{-1}\otimes\underline{A}$ has a finite element on each column and so is doubly G-astic, and then so is any sum of its powers, by Theorem 12-3. Hence from (22-14), $\underline{\Gamma}((\lambda(\underline{A}))^{-1}\otimes\underline{A})$ is doubly G-astic and by hypothesis all its columns are (not necessarily non-equivalent) fundamental eigenvectors. Let these columns be $\underline{\xi}(1),.., \underline{\xi}(n)$, and define $\underline{u} = \sum_{j=1}^{n}{}_{\oplus}\, \underline{\xi}(j)$. Since $\underline{\Gamma}((\lambda(\underline{A}))^{-1}\otimes\underline{A})$ is row-G-astic, $\underline{u}$ is finite. We have:

$$((\lambda(\underline{A}))^{-1}\otimes\underline{A}) \otimes \underline{u} = \sum_{j=1}^{n}{}_{\oplus} \{(\lambda(\underline{A}))^{-1}\otimes\underline{A})\otimes\underline{\xi}(j)\}$$

$$= \sum_{j=1}^{n}{}_{\oplus}\, \xi(j) \qquad \text{(by Theorem 23-5)}$$

$$= \underline{u}$$

Hence $\underline{A}\otimes\underline{u} = \lambda(\underline{A}) \otimes \underline{u}$ with $\underline{u}$ finite. ●

25-4. The Computational Task. We come now to considering the actual mechanics of computing solutions to the eigenproblem. The following theorem characterises the set of solutions in a convenient way.

Theorem 25-9. Let E_1 be a commutative linear radicable blog and let $\underline{A} \in \mathcal{M}_{nn}$, for given integer $n \geq 1$, have $\lambda(\underline{A})$ finite. Let $\underline{\Phi}((\lambda(\underline{A}))^{-1}\otimes\underline{A}) \in \mathcal{M}_{ns}$ $(1 \leq s \leq n)$ have as its columns a maximal set of non-equivalent fundamental eigenvectors for $(\lambda(\underline{A})^{-1}\otimes\underline{A})$. The following two sets (if non-empty) are spaces over G, the group of E_1:

(i) The set of finite eigenvectors of $\underline{A}$ having $\lambda(\underline{A})$ as corresponding eigenvalue.

(ii) The set $\{\underline{\Phi}((\lambda(\underline{A}))^{-1} \otimes \underline{A}) \otimes \underline{x} \mid \underline{x} \in E_s$ is finite$\}$.

If the first of these spaces is non-empty, then the two spaces are identical.

Proof. The verification that the sets are spaces is routine. Evidently, set (i) may equivalently be characterised as set (iii): The set of finite eigenvectors of $(\lambda(\underline{A}))^{-1} \otimes \underline{A}$ having $\emptyset$ as corresponding eigenvalue. Now, if set (i) is non-empty then $\underline{A}$ has finitely soluble eigenproblem, so $\underline{\Phi}((\lambda(\underline{A}))^{-1} \otimes \underline{A})$ is doubly G-astic, by Theorem 25-6. So set (ii) consists of finite n-tuples by Corollary 12-7, and by Lemma 24-1 is a subset of set (iii) = set (i). Conversely, if $\underline{b}$ is in set (i) (and so finite) then the equation $\underline{\Phi}((\lambda(\underline{A}))^{-1} \otimes \underline{A}) \otimes \underline{x} = \underline{b}$ is soluble because by Theorem 24-9, $\underline{b}$ lies in the eigenspace of $(\lambda(\underline{A}))^{-1} \otimes \underline{A}$, which is generated by the columns of $\underline{\Phi}((\lambda(\underline{A}))^{-1} \otimes \underline{A})$.

But since $\underline{b}$ is finite and $\underline{\Phi}((\lambda(\underline{A}))^{-1} \otimes \underline{A})$ doubly G-astic, the equation $\underline{\Phi}((\lambda(\underline{A})^{-1} \otimes \underline{A}) \otimes \underline{x} = \underline{b}$ has a principal solution which is finite, by Theorem 14-4, so $\underline{b}$ lies in set (ii). ●

We shall refer to set (i) in the above theorem as the finite eigenspace of $\underline{A}$; it yields all finite solutions to the eigenproblem for $\underline{A}$. (Notice that the concept finite eigenspace is applicable to matrices in general, whereas the concept eigenspace applies only to definite matrices).

With the aid of the foregoing theorems, together with the results of previous chapters, we can now lay down a programme for the complete (finite) solution of the eigenproblem for a given matrix $\underline{A}$ (for the principal interpretation, say). First we calculate $\lambda(\underline{A})$, which is a well-defined function of the elements $\{\underline{A}\}_{ij}$. Then we evaluate $\underline{\Gamma}(\lambda(\underline{A})^{-1} \otimes \underline{A})$ and select columns j having $\{\underline{\Gamma}\}_{jj} = \emptyset$, which are the fundamental eigenvectors. By pairwise comparisons we can discover and eliminate equivalent fundamental eigenvectors. If the remaining columns do not form a doubly-G-astic matrix, then the finite eigenproblem for $\underline{A}$ is insoluble. If they do, then we have the entire finite eigenspace at our disposal by taking finite linear combinations of these columns.

All but one of these computational tasks are very straightforward to organise, are easily carried out by hand for matrices of low order, and take only modest amounts of computer time to execute for larger matrices. The exception is the evaluation of $\lambda(\underline{A})$ if one proceeds by direction examination of all elementary cycles. The following theorem, however, makes the task much more manageable.

Theorem 25-10. Taking the principal interpretation for E_1, let $\underline{A} \in \mathcal{M}_{nn}$ for given integer $n \geq 1$. If the eigenproblem for $\underline{A}$ is finitely soluble, then $\lambda(\underline{A})$ is the optimal value of λ in the following linear programming problem in the (n+1) real variables $\lambda, x_1, \ldots, x_n$:

Minimise λ

$$\text{Subject to } \lambda + x_i - x_j \geq \{\underline{A}\}_{ij} \tag{25-8}$$

where the inequality constraint is taken over all pairs i,j for which $\{\underline{A}\}_{ij}$ is finite.

<u>Proof</u>. Since the eigenproblem for $\underline{A}$ is finitely soluble, $\underline{A}$ must be row-G-astic (Lemma 23-2) so (25-8) is taken for all pairs i, j except where $\{\underline{A}\}_{ij} = -\infty$. (Since λ, x_i, x_j are real variables, a choice of $\{\underline{A}\}_{ij} = -\infty$ in (25-8) would give an empty constraint). If $\underline{x}$ is a finite eigenvector of $\underline{A}$ then for i, j=1,..., n we have $\lambda(\underline{A}) \otimes \{\underline{x}\}_i \geqslant \{\underline{A}\}_{ij} \otimes \{\underline{x}\}_j$, so in the notation of the principal interpretation the values:

$$\lambda = \lambda(\underline{A}), \qquad x_i = \{\underline{x}\}_i \qquad (i=1,\ldots, n) \tag{25-9}$$

are clearly feasibly for the above linear programming problem, and give the object function λ the value $\lambda(\underline{A})$.

Now the <u>dual</u> of the above linear programming problem [33] can be written in terms of real variables ω_{ij}, one for each finite $\{\underline{A}\}_{ij}$, as:

$$\begin{aligned} &\text{Maximise} \quad \sum_{i,j} \{\underline{A}\}_{ij}\omega_{ij} \\ &\text{Subject to} \quad \sum_{i,j} \omega_{ij} = 1 \\ &\qquad \sum_j \omega_{ij} - \sum_j \omega_{ji} = 0 \qquad (i=1,\ldots, n) \\ &\qquad \text{All } \omega_{ij} \geqslant 0 \end{aligned} \tag{25-10}$$

where summations are over all pairs i,j for which $\{\underline{A}\}_{ij}$ is finite.

Let $(i_o, i_1,\ldots, i_{t-1}, i_t)$ (with $i_t = i_o$) be a circuit in $\Delta(\underline{A})$ with circuit-mean equal to $\lambda(\underline{A})$ and length t. Choose values for the ω_{ij} as follows:

$$\left.\begin{aligned} &\omega_{i_{k-1}i_k} = \frac{1}{t} \qquad (k=1,\ldots, t) \\ &\omega_{ij} = 0 \qquad \text{otherwise} \end{aligned}\right\} \tag{25-11}$$

(Obviously $\{\underline{A}\}_{i_{k-1}i_k}$ is finite (k=1,..., t) since it contributes to $\lambda(\underline{A})$, so each $\omega_{i_{k-1}i_k}$ is actually a defined variable of the dual problem).

It is easily confirmed that (25-11) gives a set of values feasible for the dual problem (25-10), giving <u>its</u> object function the value:

$$\left(\frac{1}{t}\right) \sum_{k=1}^{t} \{\underline{A}\}_{i_{k-1}i_k} = \lambda(\underline{A})$$

Hence we have found feasible solutions for a linear program and for its dual, giving equal values to the two object functions. So by the theory of linear programming, this common object function value is in fact the optimal value for both problems. ●

Even if the eigenproblem for $\underline{A}$ is not finitely soluble, the linear program in Theorem 25-10 will always yield a finite solution if $\underline{A}$ is row-G-astic, for the following reason. If $\underline{A}$ is row-G-astic, then by Lemma 25-2, $\lambda(\underline{A})$ is finite. Now the dual object function (25-10) can take at least the value $\lambda(\underline{A})$ as the proof of Theorem 25-10 shows, but is not unbounded because the primal problem is always feasible (we take

$x_i = 0$ (i=1,..., n) and $\lambda \geq \max_{i,j}\{A\}_{ij}$). Hence the linear program in Theorem 25-10 has a solution, with optimal value of λ given by $\hat{\lambda} \geq \lambda(\underline{A})$.

Now, if the linear program yields $\hat{\lambda}>\lambda(\underline{A})$, we can always detect this fact, because (in max algebra notation) the matrix $\lambda^{-1}\otimes\underline{A}$ then has all circuit means $< \emptyset$ which is easily seen to entail that $\underline{\Gamma}(\lambda^{-1}\otimes\underline{A})$ has all diagonal elements $<\emptyset$. Hence we may draw the flow-diagram of Fig 25-1 so as to cover all contingencies.

25-5. An Extended Example

Let us suppose that an industrial process is given, in which four machines interact as described in Section 1-2.1 Let the following matrix $\underline{A}$ define the relevant data:

$$\underline{A} = \begin{bmatrix} 1 & 2 & -\infty & 6 \\ 1 & 3 & -\infty & -\infty \\ 1 & -\infty & 1 & -\infty \\ -\infty & -\infty & 2 & 2 \end{bmatrix} \qquad (25\text{-}12)$$

To simplify planning, the management wishes to schedule the activity so that it proceeds in regular steps at maximum speed, the same constant time λ elapsing between the initiation of consecutive cycles on every machine, with λ as small as possible. However, the marketing department has given delivery promises which entail that the next following cycles must be completed not later than times $\underline{b} = \begin{bmatrix} 7 \\ 5 \\ 8 \\ 4 \end{bmatrix}$.

In order not to have perishable finished goods ready too early, the maximum earliness relative to $\underline{b}$ must be minimised for these next following cycles, and after that the regular pattern of activity must ensue.

We calculate as follows:

By linear programming $\lambda(\underline{A}) = 3$

$$\text{Hence } (\lambda(\underline{A}))^{-1}\otimes\underline{A} = \begin{bmatrix} -2 & -1 & -\infty & 3 \\ -2 & 0 & -\infty & -\infty \\ -2 & -\infty & -2 & -\infty \\ -\infty & -\infty & -1 & -1 \end{bmatrix} = \underline{B} \text{ (say)} \qquad (25\text{-}13)$$

$$\underline{B}^2 = \begin{bmatrix} -3 & -1 & 2 & 2 \\ -2 & 0 & -\infty & 1 \\ -4 & -3 & -4 & 1 \\ -3 & -\infty & -2 & -2 \end{bmatrix}; \quad \underline{B}^3 = \begin{bmatrix} 0 & -1 & 1 & 1 \\ -2 & 0 & 0 & 1 \\ -5 & -3 & 0 & 0 \\ -4 & -4 & -3 & 0 \end{bmatrix}$$

$$\underline{B}^4 = \begin{bmatrix} -1 & -1 & 0 & 3 \\ -2 & 0 & 0 & 1 \\ -2 & -3 & -1 & -1 \\ -5 & -4 & -1 & -1 \end{bmatrix}; \quad \underline{\Gamma}(\underline{B}) = \begin{bmatrix} 0 & -1 & 2 & 3 \\ -2 & 0 & 0 & 1 \\ -2 & -3 & 0 & 1 \\ -3 & -4 & -1 & 0 \end{bmatrix} \qquad (25\text{-}14)$$

All columns of $\underline{\Gamma}(\underline{B})$ have diagonal element equal to $\emptyset$, so all are fundamental eigenvectors. We accept the first column of $\underline{\Gamma}(\underline{B})$ and by comparison we observe that columns three and four are scalar multiples of column one and so are equivalent. (This assertion is the converse of Theorem 23-6 and is easy to prove. For if $\{\underline{\Gamma}(\underline{B})\}_{hh} = \{\underline{\Gamma}(\underline{B})\}_{kk} = \emptyset$, and column k is a scalar multiple α of column h then $\{\underline{\Gamma}(\underline{B})\}_{kh} \otimes \alpha = \emptyset$ and $\emptyset \otimes \alpha = \{\underline{\Gamma}(\underline{B})\}_{hk}$. So $\{\underline{\Gamma}(\underline{B})\}_{kh} \otimes \{\underline{\Gamma}(\underline{B})\}_{hk} = \emptyset$ showing, by linearity of E_1, the existence of a circuit passing through both nodes h and k having circuit product $\emptyset$).

Column two remains. Hence the eigenspace is generated by:

$$\underline{\Phi}(\underline{B}) = \begin{bmatrix} 0 & -1 \\ -2 & 0 \\ -2 & -3 \\ -3 & -4 \end{bmatrix} \qquad (25\text{-}15)$$

This is row-G-astic so the finite eigenspace is non-empty.
To find a set of completion times prior to the given times $\underline{b}$ but so as to minimise the maximum earliness relative to $\underline{b}$, such that the system will move forward in regular steps, we must compute the projection of $\underline{b}$ on the column-space of $\underline{\Phi}(\underline{B})$:

$$\underline{\Phi}(\underline{B}) \otimes (\underline{\Phi}(\underline{B})^* \otimes' \underline{b}) = \begin{bmatrix} 7 \\ 5 \\ 5 \\ 4 \end{bmatrix} = \underline{c} \text{ (say)}$$

In order to calculate the times at which these next-following cycles must be initiated, we must subtract $\lambda(\underline{A})$ from each component of $\underline{c}$, to give

$$\begin{bmatrix} 4 \\ 2 \\ 2 \\ 1 \end{bmatrix} = \underline{d} \text{ (say)}$$

Since $\underline{A} \otimes \underline{d} = \lambda(\underline{A}) \otimes \underline{d} = \underline{c}$, the 4-tuple $\underline{d}$ gives the required start-times for the next-following cycle.

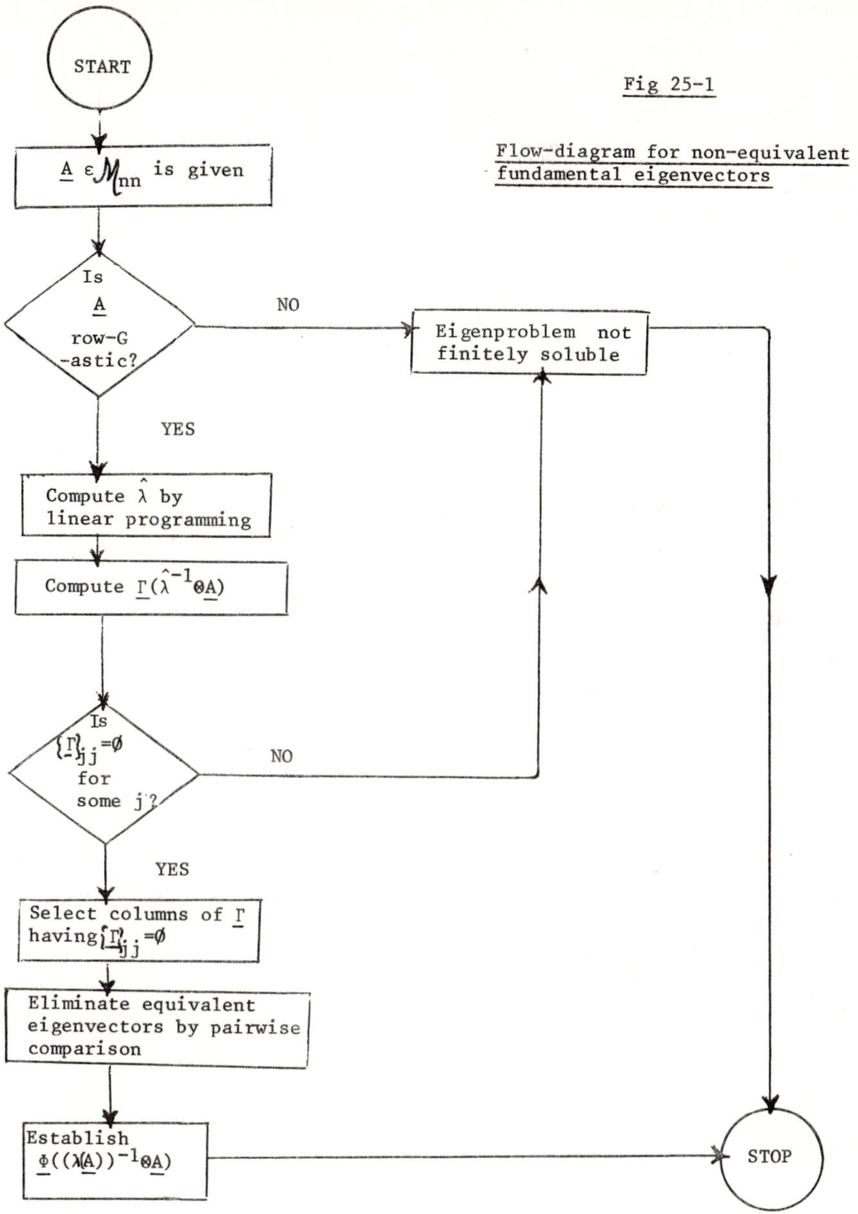

Fig 25-1

Flow-diagram for non-equivalent fundamental eigenvectors

26. SPECTRAL INEQUALITIES

26-1. Preliminary Inequalities One of the more famous results of conventional linear algebra is the spectral theorem which relates a self-conjugate linear operator $\underline{A}$ to a sum of projection operators associated with the eigenvectors of $\underline{A}$. In minimax algebra we can, somewhat analogously, prove certain spectral inequalities, for which in fact an assumption of self-conjugacy is not necessary.

We shall develop these spectral inequalities from a set of more general inequalities which we prove after the following introductory remarks.

There are eight different formal products which we can derive by inserting one $\otimes$ symbol and one $\otimes'$ symbol in the triple product $\underline{A}\underline{A}^*\underline{b}$ and then bracketing to give a well-formed formula. The associative law reduces the number of algebraically distinct products to six, of which two particular ones are the projections:

$\underline{A} \otimes (\underline{A}^* \otimes' \underline{b})$ and $\underline{A} \otimes' (\underline{A}^* \otimes \underline{b})$.

Our previous results enable us now to place these six products in order of magnitude and to relate them to the projections of $\underline{b}$ induced by individual elements of the column-space and dual column-space of $\underline{A}$.

Theorem 26-1. Let E_1 be a pre-residuated belt satisfying axiom X_{12}, let $\underline{A} \in \mathcal{M}_{mn}$, $\underline{u},\underline{v} \in E_n$ and $\underline{b} \in E_m$ for given integers $m,n \geq 1$. Let $\underline{\xi}=\underline{A}\otimes\underline{u}$ and $\underline{\eta}=\underline{A}\otimes'\underline{v}$. Consider the following sequence of relations:

$$\underline{A}\otimes\underline{A}^*\otimes\underline{b} \geq \underline{P}^*(\underline{\eta})\otimes\underline{b} = \underline{\eta}\otimes'(\underline{\eta}^*\otimes\underline{b}) \geq \underline{A}\otimes'(\underline{A}^*\otimes\underline{b}) \geq (\underline{A}\otimes'\underline{A}^*)\otimes\underline{b} \geq \underline{b}$$

$$\underline{A}\otimes'\underline{A}^*\otimes'\underline{b} \leq \underline{P}(\underline{\xi})\otimes'\underline{b} = \underline{\xi}\otimes(\underline{\xi}^*\otimes'\underline{b}) \leq \underline{A}\otimes(\underline{A}^*\otimes'\underline{b}) \leq (\underline{A}\otimes\underline{A}^*)\otimes'\underline{b} \leq \underline{b}$$

Then all inequalities to the right of the dotted line are valid. Moreover, if E_1 is a blog and $\underline{\xi}$, $\underline{\eta}$ are finite then the relations to the left of the dotted line are also valid.

Proof. We consider the second row of these inequalities, starting from the right. The first inequality follows from Lemma 8-3 and the second follows from Theorem 6-3 (axiom X_{12} for matrices). The next follows from Lemma 20-1, noting that $\underline{\xi}\otimes(\underline{\xi}^*\otimes'\underline{b})\leq\underline{b}$ from Theorem 8-8 with $\underline{\xi}$ in the role of $\underline{A}$.

Moreover, if E_1 is a blog and $\underline{\xi}$ is finite then $\underline{\xi}\otimes(\underline{\xi}^*\otimes'\underline{b})=\underline{P}(\underline{\xi})\otimes'\underline{b}$ by Theorem 21-7. But $\underline{P}(\underline{\xi})\geq\underline{A}\otimes'\underline{A}^*$ by Lemma 21-9, so the remaining inequality follows by isotonicity of matrix (dual) multiplication. The relations in the top row follow similarly. ●

Using the operators g and π as defined in (21-1), (21-2) and the identity operator i, we can re-express the relations of Theorem 26-1 in operator form, as follows.

Corollary 26-2. Let E_1 be a pre-residuated belt satisfying axiom X_{12}, let $\underline{A} \in \mathcal{M}_{mn}$, $\underline{u},\underline{v} \in E_n$ and $\underline{b} \in E_m$ for given integers $m,n\geq 1$. Let the columns of $\underline{A}$ be $\underline{a}(1),\ldots,\underline{a}(n) \in E_m$ in that order. Let $\underline{\xi}=\underline{A}\otimes\underline{u}$ and $\underline{\eta}=\underline{A}\otimes'\underline{v}$. Consider the following sequence of relations:

$$g_{\underline{A}\otimes\underline{A}^*} \geq \Big|\; \pi^*_{\underline{\eta}^*} \geq \pi^*_{\underline{A}^*} = \sum_{j\,\oplus'} \pi^*_{\underline{a}(j)^*} \geq g_{\underline{A}\otimes'\underline{A}^*} \geq i_{E_m}$$

$$g^*_{\underline{A}\otimes\underline{A}^*} \leq \Big|\; \pi_{\underline{\xi}} \leq \pi_{\underline{A}} = \sum_{j\,\oplus} \pi_{\underline{a}(j)} \leq g^*_{\underline{A}\otimes'\underline{A}^*} \leq i_{E_m}$$

Then all relations to the right of the dotted line are valid. Moreover, if E_1 is a blog and $\underline{\xi}$, $\underline{\eta}$ are finite then the inequalities to the left of the dotted line are also valid.

Proof. In the second row of given relations, everything follows by notational change in Theorem 26-1 except for the equality $\pi_{\underline{A}} = \sum_{j\oplus} \pi_{\underline{a}(j)}$, which is proved as follows:

For arbitrary $\underline{b} \in E_m$ and for i=1,..., m we have:

$$\{\pi_{\underline{A}}(\underline{b})\}_i = \{\underline{A}\otimes(\underline{A}^*\otimes'\underline{b})\}_i$$

$$= \sum_{j=1\,\oplus}^{n}(\{\underline{A}\}_{ij} \otimes \sum_{k=1\,\oplus'}^{m}(\{\underline{A}^*\}_{jk}\otimes'\{\underline{b}\}_k))$$

$$= \sum_{j=1\,\oplus}^{n}(\{\underline{a}(j)\}_i \otimes \sum_{k=1\,\oplus'}^{m}((\{\underline{a}(j)\}_k)^*\otimes'\{\underline{b}\}_k))$$

$$= \{\sum_{j=1\,\oplus}^{n}(\underline{a}(j) \otimes (\underline{a}(j)^*\otimes'\underline{b}))\}_i$$

$$=\{\sum_{j=1}^{n} \pi_{\underline{a}(j)}(\underline{b})\}_i$$

26-2 . Spectral Inequality

The following result presents a spectral inequality which applies fairly generally.

Theorem 26-3. Let E_1 be a pre-residuated belt satisfying axiom X_{12}, and let $\underline{A} \in \mathcal{M}_{nn}$ for given integer n≥1. If $\underline{\xi}(j_k) \in E_n$ (k=1,..., s) are eigenvectors of $\underline{A}$ with corresponding eigenvalues $\lambda_k \in E_1$ (k=1,..., s) respectively, then:

$$g_{\underline{A}} \geq \sum_{k=1\,\oplus}^{s}(\lambda_k\otimes\pi_{\underline{\xi}(j_k)})$$

Proof. If $\underline{\xi} \in E_n$ is any eigenvector of $\underline{A}$ with corresponding eigenvalue $\lambda \in E_1$ we have for arbitrary $\underline{b} \in E_n$:

$$\lambda \otimes \pi_{\underline{\xi}}(b) = \lambda \otimes \underline{\xi} \otimes (\underline{\xi}^*\otimes'\underline{b}) \qquad \text{(by definition)}$$

$$= \underline{A} \otimes \underline{\xi} \otimes (\underline{\xi}^*\otimes'\underline{b})$$

$$\leq A \otimes \underline{b} \qquad \text{(by Theorem 8-8)}$$

Hence $\quad g_{\underline{A}} \geq \lambda \otimes \pi_{\underline{\xi}}$ (26-1)

If we write (26-1) with $\lambda=\lambda_k$ and $\underline{\xi}=\underline{\xi}(j_k)$ and close with respect to k, we have the required result.

Theorem 8-5 furnishes an example of the foregoing result. For since $(\underline{A}\otimes'\underline{A}^*)\otimes\underline{A}=\underline{A}$, then each column $\underline{a}(j)$ (j=1,..., n) of $\underline{A}$ is an eigenvector of $\underline{A}\otimes'\underline{A}^*$ with corresponding eigenvector $\emptyset$. Hence by Theorem 26-3: $g_{\underline{A}\otimes'\underline{A}^*} \geq \sum_{j=1}^{n}{}_{\oplus}\pi_{\underline{a}(j)}$, a result already contained in Corollary 26-2.

Suppose now with the notation of Theorem 26-3 that we have another eigenvector $\underline{\xi}(j_{s+1}) \in E_n$ of $\underline{A}$ with corresponding eigenvalue λ_{s+1}. Evidently Theorem 26-3 implies:

$$g_{\underline{A}} \geq \sum_{k=1}^{s+1}{}_{\oplus}(\lambda_k \otimes \pi_{\underline{\xi}(j_k)})$$

$$=\left(\sum_{k=1}^{s}{}_{\oplus}(\lambda_k \otimes \pi_{\underline{\xi}(j_k)})\right) \oplus (\lambda_{s+1} \otimes \pi_{\underline{\xi}(j_{s+1})})$$

$$\geq \sum_{k=1}^{s}{}_{\oplus}(\lambda_k \otimes \pi_{\underline{\xi}(j_k)})$$

Hence on taking this extra eigenvector $\underline{\xi}(j_{s+1})$ we may get a "better" spectral inequality. However the following result shows that under particular circumstances a certain set of eigenvectors gives a "best" spectral inequality which cannot be improved by taking yet another eigenvector into consideration.

Theorem 26-4. Let E_1 be a pre-residuated belt satisfying X_{12} and let $\underline{A} \in \mathcal{M}_{nn}$ for given integer $n \geq 1$. If $\underline{\xi}(j_k) \in E_n$ (k=1,..., s) are eigenvectors of $\underline{A}$ with a common corresponding eigenvalue $\lambda \in E_1$, and $\underline{\xi} \in E_n$ lies in the space generated by $\underline{\xi}(j_1),\ldots,\underline{\xi}(j_s)$, then :

$$\sum_{k=1}^{s}{}_{\oplus}(\lambda \otimes \pi_{\underline{\xi}(j_k)}) \geq \lambda \otimes \pi_{\underline{\xi}}$$

In particular, this holds if E_1 is a linear blog, $\underline{\xi}$ and λ are finite, and $\underline{\xi}(j_1),\ldots,\underline{\xi}(j_s)$ are a maximal set of non-equivalent fundamental eigenvectors of $\lambda^{-1}\otimes\underline{A}$.

Proof. Evidently $\underline{\xi}$ is an eigenvector of $\underline{A}$ with corresponding eigenvalue λ.

From Corollary 26-2 with $\underline{\xi}(j_1),\ldots,\underline{\xi}(j_s)$ in the roles of $\underline{a}(1),\ldots,\underline{a}(n)$:

$$\pi_{\underline{\xi}} \leq \sum_{k=1}^{s}{}_{\oplus}\pi_{\underline{\xi}(j_k)}$$

On left-multiplying by λ we have the required result.

Evidently Theorem 26-4 implies, for a linear blog, that the spectral inequality which uses all the fundamental eigenvectors cannot be improved by using any element from the finite eigenspace of $\underline{A}$.

26-3. The Other Eigenproblems. Let us now agree that the title the eigenproblem for $\underline{A}$ (problem (23-1)) is an abbreviation of the right eigenproblem for $\underline{A}$. Then we may define the (right) dual eigenproblem and the left eigenproblem and left dual eigenproblem for $\underline{A} \in \mathcal{M}_{nn}$ as the following three problems respectively:

Find $\underline{\eta} \in E_n$ and $\mu \in E_1$ such that $\underline{A} \otimes' \underline{\eta} = \mu \otimes' \underline{\eta}$ (26-2)

Find $\underline{\zeta} \in \mathcal{M}_{1m}$ and $\nu \in E_1$ such that $\underline{\zeta} \otimes \underline{A} = \underline{\zeta} \otimes \nu$ (26-3)

Find $\underline{\chi} \in \mathcal{M}_{1m}$ and $\theta \in E_1$ such that $\underline{\chi} \otimes' \underline{A} = \underline{\chi} \otimes' \theta$ (26-4)

We shall refer to (23-1), (26-2), (26-3) and (26-4) for short as the four eigenproblems for $\underline{A}$.

We may also define left and right eigenproblems and dual eigenproblems for $\underline{A}^*$ but this leads to nothing new since e.g. the left dual eigenproblem for $\underline{A}^*$ is essentially just the right eigenproblem for $\underline{A}$. We shall call $\underline{\eta}$ in (26-2), $\underline{\zeta}$ in (26-3) and $\underline{\chi}$ in (26-4) a (right) dual eigenvector, left eigenvector, left dual eigenvector respectively of $\underline{A}$, with corresponding eigenvalue μ, ν, θ respectively. The epithet 'right' will be suppressed unless needed for emphasis. Finite solubility of an eigenproblem means solubility with both eigenvector and corresponding eigenvalue finite.

It is evident that appropriate left- and dual- variants of the material of Chapters 21 to 25 may be established, as in the following propositions.

Proposition 26-5. Let E_1 be a blog and let $\underline{B} \in \mathcal{M}_{nn}$, for given integer $n \geq 1$, be definite. If j is an eigen-node in $\Delta(\underline{B})$ then $\underline{\zeta}(j) \otimes \underline{B} = \underline{\zeta}(j)$ where $\underline{\zeta}(j)$ is the jth row of $\underline{\Gamma}(\underline{B})$. If k is an eigen-node in $\Delta(\underline{B})$ equivalent to j and $\underline{\zeta}(k)$ is the kth row of $\underline{\Gamma}(\underline{B})$ then $\underline{\zeta}(j)$, $\underline{\zeta}(k)$ are finite scalar multiples of one another. If E_1 is actually a linear blog and $\underline{B}$ is $\emptyset$-astic definite and $j_1, \ldots, j_s$ are pairwise non-equivalent eigen-nodes in $\Delta(\underline{B})$ and $\underline{\zeta}(j_1), \ldots, \underline{\zeta}(j_s)$ are the corresponding rows of $\underline{\Gamma}(\underline{B})$ then no one of $\underline{\zeta}(j_1), \ldots, \underline{\zeta}(j_s)$ is left linearly dependent on the others. Finally if E_1 is a linear blog and $\underline{B}$ is $\emptyset$-astic definite and $\underline{\zeta} \in \mathcal{M}_{1n}$ is a finite left eigenvector of $\underline{B}$ having corresponding eigenvalue $\emptyset$, then $\underline{\zeta}$ lies in the left row space of $\underline{\Psi}(\underline{B})$ where $\underline{\Psi}(\underline{B}) \in \mathcal{M}_{sn}$ is a matrix whose rows 1,..., s are rows $j_1, \ldots, j_s$ respectively of $\underline{\Gamma}(\underline{B})$, where $j_1, \ldots, j_s$ constitute a maximal set of non-equivalent eigen-nodes in $\Delta(\underline{B})$. ●

Proposition 26-6. Let E_1 be a linear blog and let $\underline{A} \in \mathcal{M}_{nn}$ for given integer $n \geq 1$. If the left eigenproblem for $\underline{A}$ is finitely soluble then every finite left eigenvector has the same unique corresponding finite eigenvalue λ. The matrix $\underline{A} \otimes \lambda^{-1}$ is definite and all finite left eigenvectors of $\underline{A}$ lie in the left row space of $\underline{\Psi}(\underline{A} \otimes \lambda^{-1})$ where $\underline{\Psi}(\underline{A} \otimes \lambda^{-1}) \in \mathcal{M}_{sn}$ is a matrix whose rows 1,..., s are rows $j_1, \ldots, j_s$ respectively of $\underline{\Gamma}(\underline{A} \otimes \lambda^{-1})$ where $j_1, \ldots, j_s$ constitute a maximal set of non-equivalent eigen-nodes in $\Delta(\underline{A} \otimes \lambda^{-1})$. The rows of $\underline{\Psi}(\underline{A} \otimes \lambda^{-1})$ have the property that no one of them is left linearly dependent on (any subset of) the others. ●

We shall call the unique scalar λ (when it exists) in Proposition 26-6 the <u>principal left eigenvalue of $\underline{A}$</u>. The notation $\underline{\Psi}(\underline{A}\otimes\lambda^{-1})$, with the significance explained Proposition 26-6, will be standard throughout the sequel.

We observe that the $\underline{\Gamma}(\underline{B})$ mentioned in Proposition 26-5 is the same as the $\underline{\Gamma}(\underline{B})$ in the corresponding results of Chapters 23 and 24, showing that the left and right eigenproblems for $\underline{A}$ are by no means unrelated.

In fact the notions of <u>circuit,</u> $\Delta(\underline{B})$ and $\underline{\Gamma}(\underline{B})$ are essentially row-column symmetric and so therefore are the notions <u>definite</u>, <u>$\emptyset$-astic definite</u>, $\lambda(\underline{A})$, <u>eigen-node</u> and so on.

This relationship between the left and right eigenproblems is brought out in the following result and its corollary.

<u>Lemma 26-7. Let E_1 be a linear blog and let $\underline{A} \in \mathcal{M}_{nn}$ for given integer $n\geq 1$. If $\alpha,\beta \in E_1$ are such that $\alpha\otimes\underline{A}$ and $\underline{A}\otimes\beta$ are both definite then $\alpha=\beta$ (are are finite).</u>

<u>Proof</u>. Since both α and β are contained as factors in some circuit product equal to $\emptyset$, both are finite. Now for $i,j=1,\ldots, n$:

$$\{\underline{A}\otimes\beta\}_{ij} = \beta^{-1}\otimes\{\beta\otimes\underline{A}\}_{ij} \otimes \beta$$

Hence $\underline{A}\otimes\beta \approx \beta\otimes\underline{A}$ and so by Lemma 24-6, $\beta\otimes\underline{A}$ is definite. But so is $\alpha\otimes\underline{A}$. Hence by Lemma 24-8, $\alpha=\beta$. ●

<u>Corollary 26-8. Let E_1 be a linear blog and let $\underline{A} \in \mathcal{M}_{nn}$ for given integer $n\geq 1$. Then the principal eigenvalue of $\underline{A}$ and the principal left eigenvalue of $\underline{A}$, if they both exist, are equal.</u>

<u>Proof</u>. If λ is the principal eigenvalue of $\underline{A}$ and μ the principal left eigenvalue of $\underline{A}$, then $\lambda^{-1}\otimes\underline{A}$ and $\underline{A}\otimes\mu^{-1}$ are both definite (Theorem 24-9 and Proposition 26-6). Hence $\lambda^{-1}=\mu^{-1}$ by Lemma 26-7, so $\lambda=\mu$. ●

If $\underline{A}$ is a matrix for which both the eigenproblem and the left eigenproblem are finitely soluble, then both $\lambda^{-1}\otimes\underline{A}$ and $\underline{A}\otimes\lambda^{-1}$ are definite and in fact $\lambda^{-1}\otimes\underline{A} \simeq \underline{A}\otimes\lambda^{-1}$, so $\underline{\Gamma}(\lambda^{-1}\otimes\underline{A}) \approx \underline{\Gamma}(\underline{A}\otimes\lambda^{-1})$ and any linear dependencies among the rows (or columns) of $\underline{\Gamma}(\lambda^{-1}\otimes\underline{A})$ are duplicated among the rows (or columns) of $\underline{\Gamma}(\underline{A}\otimes\lambda^{-1})$ and vice versa, by Lemma 24-7. Moreover, if we assume (in the terminology of Section 23-2) that $\lambda^{-1}\otimes\underline{A}$ is blocked then by Lemma 24-6, $\underline{A}\otimes\lambda^{-1}$ is also blocked in the same way.

An example will make this clear. Taking $\underline{A}$ as in (25-12) we obtain the definite matrix $\underline{B}=\lambda^{-1}\otimes\underline{A}$ as in (25-13) and then $\underline{\Gamma}(\underline{B})$ as in (25-14). We have already observed that columns three and four of $\underline{\Gamma}(\underline{B})$ are scalar multiples of column one but that columns two and one are not (right) linearly dependent. We can now confirm that exactly the same pattern of dependencies holds among the rows of $\underline{\Gamma}(\underline{B})$, namely that rows three and four are scalar multiples of row one but that rows two and one are not (left) linearly dependent.

The same would hold if we took $\underline{B}$ instead as $\underline{A}\otimes\lambda^{-1}$ but this would lead to nothing new in the present example since scalar multiplication is commutative in the principal interpretation.

In view of Corollary 26-8, we shall simply use the expression <u>principal eigenvalue</u> when discussing either the left or the right eigenproblem for a matrix over a linear blog.

<u>26-4. More Spectral Inequalities</u>. The spectral inequality of Theorem 26-3 related the operator $g_{\underline{A}}$ to the projection operators associated with the eigenvectors of $\underline{A}$, and followed from the identities of Chapter 8. Now if we assume that the eigenproblem for $\underline{A}$ is finitely soluble then we can obtain other spectral inequalities by quite different arguments. These inequalities relate the matrix $\underline{A}$ to the projection <u>matrices</u> associated with the eigenvectors of $\underline{A}$, although we can recast them in operator form.

It is convenient to introduce the following notation.

If $\underline{A}$, $\underline{B}$ are (m×n) matrices then we shall write $\underline{A} \ll \underline{B}$ or equivalently $\underline{B} \gg \underline{A}$ to mean:

$$\left.\begin{array}{l} \{\underline{A}\}_{ij} \leqslant \{\underline{B}\}_{ij} \ (i=1,\ldots, m;\ j=1,\ldots, n), \text{ but for each } i=1,\ldots, m \\ \text{there is at least one } j \ (1\leqslant j\leqslant n) \text{ such that } \{\underline{A}\}_{ij} = \{\underline{B}\}_{ij} \end{array}\right\} \qquad (26\text{-}5)$$

We can paraphrase (26-5) by "$\underline{A}\leqslant\underline{B}$ with equality at least once per row". We shall equivalently say that the inequality is <u>row-tight</u>. (26-6)

By interchanging the roles i,j in (26-5) we can in the obvious way define a relation $\ll'$ of <u>column-tight inequality</u> i.e. $\underline{A} \ll' \underline{B}$ if and only if $\underline{A}\leqslant\underline{B}$ with equality at least once per column. Extension of the notation to operators follows evidently.

We now prove some spectral inequalities, for which we recall the notation $\underline{P}(\underline{\xi})$, $\underline{P}^*(\underline{\xi})$ for projection matrices, introduced in Chapter 21.

<u>Theorem 26-9. Let E_1 be a linear blog and let $\underline{A} \in \mathcal{M}_{nn}$ for given integer $n\geqslant 1$. If the eigenproblem for $\underline{A}$ is finitely soluble, and $\underline{\xi}(j_k)$ (k=1,..., s) are any finite eigenvectors of $\underline{A}$ then:</u>

$$\underline{A} \quad \ll \quad \lambda \otimes' \sum_{k=1}^{s}{}_{\oplus'} \ \underline{P}(\underline{\xi}(j_k)) \qquad (26\text{-}7)$$

<u>where λ is the principal eigenvalue of $\underline{A}$.</u>

<u>Proof</u>. Let $\underline{\xi}$ be any finite eigenvector of $\underline{A}$.

From $\quad \sum_{j=1}^{n}{}_{\oplus} \ (\{\underline{A}\}_{ij} \otimes \{\underline{\xi}\}_j) = \lambda\otimes\{\underline{\xi}\}_i \qquad (i=1,\ldots, n)$

We can infer, an opening w.r.t.j:

$$\{\underline{A}\}_{ij} \otimes \{\underline{\xi}\}_j \ \leqslant \lambda\otimes\{\underline{\xi}\}_i \qquad (i,j=1,\ldots, n) \qquad (26\text{-}8)$$

i.e. $\{\underline{A}\}_{ij} \leqslant \lambda\otimes\{\underline{\xi}\}_i \otimes (\{\underline{\xi}\}_j)^{-1} \qquad (i,j=1,\ldots, n) \qquad (26\text{-}9)$

And by the linearity of E_1, equality holds in (26-8) (and so in (26-9)) for at least one index j per i=1,..., n (the attained maximum). In other words, (26-9) may be written

$$\underline{A} << \lambda \otimes \underline{P}(\underline{\xi}) \quad (26\text{-}10)$$

Since λ is finite we may rewrite:

$$\underline{A} << \lambda \otimes' \underline{P}(\underline{\xi}) \quad (26\text{-}11)$$

If we write (26-11) with $\underline{\xi}=\underline{\xi}(j_k)$ and close with respect to index k, we have the required result. ●

The operator version of Theorem 26-9 now follows.

Theorem 26-10. Let E_1 be a linear blog and let $\underline{A} \in \mathcal{M}_{nn}$ for given integer $n \geqslant 1$. If the eigenproblem for $\underline{A}$ is finitely soluble and $\underline{\xi}(j_k)$ $(k=1,\ldots, s)$ are any finite eigenvectors of $\underline{A}$ then:

$$g_{\underline{A}} \leqslant \lambda \otimes' \sum_{k=1}^{s} {}_{\oplus'} \pi^{*}(\underline{\xi}(j_k))^{*} \quad (26\text{-}12)$$

where λ is the principal eigenvalue of A.

Proof. From (26-10) we have for arbitrary $\underline{x} \in E_n$ and arbitrary finite eigenvector $\underline{\xi}$:

$$\underline{A} \otimes \underline{x} \leqslant \lambda \otimes \underline{P}(\underline{\xi}) \otimes \underline{x} \quad (26\text{-}13)$$

This follows by isotonicity, regarding (26-10) as a simple inequality. (Tight inequality is not necessarily preserved under multiplication). Since λ and $\underline{\xi}$ are finite we may (by Theorem 21-8) rewrite (26-13) as:

$$\underline{A} \otimes \underline{x} \leqslant \lambda \otimes' (\underline{P}^{*}(\underline{\xi}) \otimes \underline{x})$$

$$= \lambda \otimes' \pi^{*}_{\underline{\xi}^{*}}(\underline{x})$$

Since $\underline{x}$ is arbitrary we infer:

$$g_{\underline{A}} \leqslant \lambda \otimes' \pi^{*}_{\underline{\xi}^{*}} \quad (26\text{-}14)$$

If we write (26-14) with $\underline{\xi}=\underline{\xi}(j_k)$ and close with respect to index k, we have the required result. ●

It is evident that versions of the preceding inequalities hold for all four eigenproblems for $\underline{A}$. In particular, there will be a principal dual eigenvalue λ' for $\underline{A}$ by obvious analogy with λ.

26-5. The Principal Interpretation

If E_1 is a radicable belt with duality and $\underline{A} \in \mathcal{M}_{nn}$ for given integer $n \geqslant 1$, then for each circuit σ in $\Delta(\underline{A})$ of length $t(\sigma)$ and circuit dual product $p'(\sigma)$ (Section 22-1) we can compute a unique circuit dual mean $\mu'(\sigma) \in E_1$ from:

$$(\mu'(\sigma))^{t(\sigma)} = p'(\sigma) \tag{26-15}$$

We define $\lambda'(\underline{A}) = \Sigma_{\oplus'}\mu'(\sigma)$ (26-16)

where the summation is taken over all elementary cycles of $\Delta(\underline{A})$. This notation will be standard in the sequel.

Intuitively, $\lambda'(\underline{A})$ is the least circuit (dual) mean in $\Delta(\underline{A})$, just as $\lambda(\underline{A})$ is the greatest circuit mean. Thus, if $\underline{A}$ is the matrix associated with the graph of Fig 22-1 we infer from the data in the table in Section 25-1 that $\lambda'(\underline{A})=-5/2$.

Proposition 26-11. Let E_1 be a commutative linear radicable blog and let $\underline{A} \in \mathcal{M}_{nn}$ for given integer $n\geq 1$. If the left or right dual eigenproblem for $\underline{A}$ is finitely soluble, then the principal dual eigenvalue for $\underline{A}$ is $\lambda'(\underline{A})$. ●

In order to present our principal result, we require some notation. Accordingly, if $\underline{B} \in \mathcal{M}_{nn}$ is definite, and $j_1,\ldots, j_s$ are a maximal set of non-equivalent eigen-nodes in $\Delta(\underline{B})$, then $\square(\underline{B})$ will denote a certain set containing 2s n-tuples, namely columns $j_1,\ldots, j_s$ of $\underline{\Gamma}(\underline{B})$ and of $\underline{\Gamma}^*(\underline{B})=(\underline{\Gamma}(\underline{B}))^*$.

Theorem 26-12. Let E_1 be a commutative linear radicable blog and let $\underline{A} \in \mathcal{M}_{nn}$ for given integer $n\geq 1$, be finite. Then all four eigenproblems for $\underline{A}$ are finitely soluble and there holds $\lambda(\underline{A}^*)=(\lambda'(\underline{A}))^{-1}$; $\lambda'(\underline{A}^*)=(\lambda(\underline{A}))^{-1}$; and:

$$\lambda'(\underline{A})\otimes \sum_{\underline{\eta}}{}_{\oplus}\, \underline{P}(\underline{\eta}) \leq A \leq \lambda(\underline{A}) \otimes' \sum_{\underline{\xi}}{}_{\oplus'}\, \underline{P}(\underline{\xi}) \tag{26-17}$$

where the first summation is over $\square((\lambda(A^*))^{-1}\otimes\underline{A}^*)$ and the second summation is over $\square((\lambda(\underline{A}))^{-1}\otimes\underline{A})$ and both inequalities are both row-tight and column-tight.

Proof. The finite solubility of all four eigenproblems for $\underline{A}$ (and for $\underline{A}^*$) follows from Corollary 25-7 and its variants. And since $\underline{\zeta}\otimes\underline{A}=\underline{\zeta}\otimes\lambda(\underline{A})$ for some finite left eigenvector $\underline{\zeta}$, we have $\underline{A}^*\otimes'\underline{\zeta}^* = (\lambda(\underline{A}))^*\otimes'\zeta^*$ so $(\lambda(\underline{A}))^*$ (i.e. $(\lambda(\underline{A}))^{-1}$) must be $\lambda'(\underline{A}^*)$ by uniqueness of the principal dual eigenvalue for $\underline{A}^*$. Similarly $\lambda(\underline{A}^*) = (\lambda'(\underline{A}))^{-1}$.

Now if $\underline{\xi} \in \square((\lambda(\underline{A}))^{-1}\otimes\underline{A})$ then by definition there is an eigen-node j in $\Delta((\lambda(\underline{A}))^{-1}\otimes\underline{A})$ such that either $\underline{\xi}$ is column j of $\underline{\Gamma}((\lambda(\underline{A}))^{-1}\otimes\underline{A})$ or $\underline{\xi}^*$ is row j of $\underline{\Gamma}((\lambda(\underline{A}))^{-1}\otimes\underline{A})$. In the first case, $\underline{\xi}$ is an eigenvector of $\underline{A}$ which is finite because $\underline{\Gamma}((\lambda(\underline{A}))^{-1}\otimes\underline{A})$ is finite.

By (26-11): $\underline{A} << \lambda(\underline{A}) \otimes'\underline{P}(\underline{\xi})$ (26-18)

In the second case, $\underline{\xi}^*$ is a finite left eigenvector of $\underline{A}$ by Proposition 26-5, and the appropriate variant of the proof of Theorem 26-9 leads to:

$$\begin{aligned}\underline{A} &<<' \underline{P}(\underline{\xi})\otimes\lambda(\underline{A}) \\ &= \lambda(\underline{A})\otimes'\underline{P}(\underline{\xi})\end{aligned} \tag{26-19}$$

(since E_1 is commutative and $\lambda(\underline{A})$ is finite).

Combining (26-18) and (26-19) and closing with respect to the index j:

$$\underline{A} \leq \lambda(\underline{A}) \otimes' \sum_{\underline{\xi}}{}_{\oplus'}\underline{P}(\underline{\xi}) \tag{26-20}$$

And since (26-18) is row-tight and (26-19) is column-tight, evidently (26-20) is both row- and column-tight. Summation is over $\square((\lambda(\underline{A}))^{-1} \otimes \underline{A})$.

Now relation (26-20) for the matrix $\underline{A}^*$ is:

$$\underline{A}^* \leq \lambda(\underline{A}^*) \otimes' \sum_{\substack{\oplus' \\ \underline{n}}} \underline{P}(\underline{n}) \tag{26-21}$$

where summation is over $\square((\lambda(\underline{A}^*))^{-1} \otimes \underline{A}^*)$ and the inequality is both row-tight and column-tight. Dualising (26-21) gives

$$\sum_{\substack{\oplus \\ \underline{n}}} \underline{P}^*(\underline{n}) \otimes (\lambda(\underline{A}^*))^* \leq \underline{A} \tag{26-22}$$

Now $\underline{P}^*(\underline{n}) = \underline{P}(\underline{n})$ by Theorem 21-8 and $(\lambda(\underline{A}^*))^* = \lambda'(\underline{A})$. Hence, using the commutativity of E_1:

$$\lambda'(\underline{A}) \otimes \sum_{\substack{\oplus \\ \underline{n}}} \underline{P}(\underline{n}) \leq \underline{A} \tag{26-23}$$

where summation is over $\square((\lambda(A^*))^{-1} \otimes A^*)$ and the inequality is both row-tight and column-tight. Evidently (26-20) and (26-23) yield the required conclusion. ●

We remark that inequality (26-17) cannot be improved by using further eigenvectors belonging to finite solutions to any of the four eigenproblems for $\underline{A}$ because of Theorem 26-4.

Evidently Theorem 26-12 covers an important set of cases for the principal interpretation and we illustrate the theorem with the following example.

Suppose $\underline{A} = \begin{bmatrix} 2 & -1 & -3 \\ 6 & 3 & 4 \\ 1 & 1 & 3 \end{bmatrix}$. Then $\lambda(\underline{A}) = 3$ and $\lambda'(\underline{A}) = -1$.

We obtain $\underline{\Gamma}((\lambda(\underline{A}))^{-1} \otimes \underline{A}) = \begin{bmatrix} -1 & -4 & -3 \\ 3 & 0 & 1 \\ 1 & -2 & 0 \end{bmatrix}$

There are two non-equivalent eigen-nodes, namely 2 and 3. Hence we construct four projection matrices from:

Col 2: $\begin{bmatrix} 0 & -4 & -2 \\ 4 & 0 & 2 \\ 2 & -2 & 0 \end{bmatrix}$; Col 3: $\begin{bmatrix} 0 & -4 & -3 \\ 4 & 0 & 1 \\ 3 & -1 & 0 \end{bmatrix}$

Row 2: $\begin{bmatrix} 0 & -3 & -2 \\ 3 & 0 & 1 \\ 2 & -1 & 0 \end{bmatrix}$ Row 3: $\begin{bmatrix} 0 & -3 & -1 \\ 3 & 0 & 2 \\ 1 & -2 & 0 \end{bmatrix}$

The minimum of these four matrices is $\underline{P}$ where:

$$\lambda(\underline{A}) \otimes' \underline{P} = \begin{bmatrix} 3 & -1 & 0 \\ 6 & 3 & 4 \\ 4 & 1 & 3 \end{bmatrix}$$

So we confirm that $\underline{A} \leq \lambda(\underline{A}) \otimes ' \underline{P}$, the inequality being both row-tight and column-tight.

Now $\underline{A}^* = \begin{bmatrix} -2 & -6 & -1 \\ 1 & -3 & -1 \\ 3 & -4 & -3 \end{bmatrix}$ and $\lambda(\underline{A}^*) = 1$

We obtain $\underline{\Gamma}((\lambda(\underline{A}^*))^{-1} \otimes \underline{A}^*) = \begin{bmatrix} 0 & -7 & -2 \\ 0 & -4 & -2 \\ 2 & -5 & 0 \end{bmatrix}$

There are two eigen-nodes but we observe that column three is obtained by subtracting 2 from column one. Hence there is just one non-equivalent eigen-node, namely 1. We construct two projection matrices from:

Col 1: $\begin{bmatrix} 0 & 0 & -2 \\ 0 & 0 & -2 \\ 2 & 2 & 0 \end{bmatrix}$ Row 1: $\begin{bmatrix} 0 & -7 & -2 \\ 7 & 0 & 5 \\ 2 & -5 & 0 \end{bmatrix}$

The maximum of these two matrices is $\underline{Q}$ where:

$$\lambda'(A) \otimes Q = \begin{bmatrix} -1 & -1 & -3 \\ 6 & -1 & 4 \\ 1 & 1 & -1 \end{bmatrix}$$

So we confirm that $\lambda'(\underline{A}) \otimes \underline{Q} \leq \underline{A}$ the inequality being both row-tight and column-tight.

27. THE ORBIT

27-1. Increasing Matrices. In Section 1-2, we discussed sequences $\underline{A}$, $\underline{A}^2$,... of powers of a matrix $\underline{A}$. We shall now examine the properties of such sequences in the light of our theory.

We shall say that a square matrix $\underline{A} \in \mathcal{M}_{nn}$ is an increasing matrix if the operator $g_{\underline{A}}$ has the property:

$$g_{\underline{A}}(\underline{x}) \geq \underline{x} \quad \text{for all } \underline{x} \in E_n \tag{27-1}$$

Evidently this property is equivalent to:

$$\underline{A}\otimes\underline{B} \geq \underline{B} \text{ for all } \underline{B} \in \mathcal{M}_{np} \text{ for any integer } p\geq 1 \tag{27-2}$$

Lemma 27-1. Let E_1 be a blog and let $\underline{A} \in \mathcal{M}_{nn}$ for given integer $n\geq 1$. Then $\underline{A}$ is increasing if and only if $\{\underline{A}\}_{ii} \geq \emptyset$ $(i=1,\ldots, n)$.

Proof. If $\underline{A}$ is increasing then by (27-2):

$$\underline{A} = \underline{A} \otimes \underline{I}_n \geq \underline{I}_n$$

Hence $$\{\underline{A}\}_{ii} \geq \{\underline{I}_n\}_{ii} = \emptyset \quad (i=1,\ldots, n). \tag{27-3}$$

Conversely if (27-3) holds then for any $\underline{x} \in E_n$:

$$\begin{aligned} \{\underline{A}\otimes\underline{x}\}_i &= \sum_{j=1}^{n}{}_{\oplus} (\{\underline{A}\}_{ij}\otimes\{\underline{x}\}_j) && (i=1,\ldots, n) \\ &\geq \{\underline{A}\}_{ii} \otimes \{\underline{x}\}_i && (i=1,\ldots, n) \\ &= \{\underline{x}\}_i \end{aligned}$$

Hence $\underline{A}$ is increasing. ●

Lemma 27-2. Let E_1 be a blog and let $\underline{B} \in \mathcal{M}_{nn}$, for given integer $n\geq 1$, be increasing and definite. Then

$$\underline{B} \leq \underline{B}^2 \leq \ldots$$

and for each integer $r\geq(n-1)$:

$$\underline{B}^r = \Gamma(\underline{B})$$

Proof. In (27-2) we may put $\underline{A}=\underline{B}$. Hence $\underline{B}\leq\underline{B}^2$ and by iteration $\underline{B}\leq\underline{B}^2\leq \ldots \leq\underline{B}^n\leq\underline{B}^r (r>n)$ Thus $\Gamma(\underline{B}) = \underline{B}\oplus\underline{B}^2\oplus \ldots \oplus \underline{B}^n = \underline{B}^n$.

Moreover, $$\begin{aligned} \underline{B}^r &\leq \Gamma(\underline{B}) && \text{(by Theorem 22-6)} \\ &= \underline{B}^n \\ &\leq \underline{B}^r \text{ for } r>n \end{aligned}$$

We have thus proved that $\underline{B}^r=\Gamma(\underline{B})$ for $r>n$. But each diagonal element $\{\underline{B}^n\}_{ii}$ of $\underline{B}^n$ is a circuit product for the definite matrix $\underline{B}$ and so:

$$\{\underline{B}^n\}_{ii} \leq \emptyset = \{\underline{B}\}_{ii} \quad (i=1,\ldots, n)$$

And each element $\{\underline{B}^n\}_{ij}$ for $i\neq j$ is a sum of path products for paths of length n somewhere in $\Delta(\underline{B})$, which cannot be elementary. Hence each product consists of terms making up a product for a path of some length $r<n$ from i to j, together with term(s) making up a circuit product. Since $\underline{B}$ is definite we therefore have (by Proposition 22-1 and isotonicity) that:

$$\underline{B}^n \leqslant \underline{B} \oplus \ldots \oplus \underline{B}^{n-1}$$

$$\leqslant B^n \text{ and the result follows.}$$

●

<u>Corollary 27-3. Let E_1 be a blog and let $\underline{B} \in \mathcal{M}_{nn}$, for given integer $n\geqslant 1$, be increasing and definite. For arbitrary $\underline{\xi}(0) \in E_n$ define a sequence $\{\underline{\xi}(r)\}$ by $\underline{\xi}(r+1)=\underline{B}\otimes\underline{\xi}(r)$ $(r=0,1,\ldots)$. Then $\underline{\xi}(r)$ is an eigenvector of $\underline{B}$ for $r\geqslant(n-1)$, with corresponding eigenvalue $\emptyset$.</u>

<u>Proof</u>. By iteration $\underline{\xi}(n-1) = \underline{B}^{n-1}\otimes\underline{\xi}(0)$

$$= \underline{\Gamma}(B)\otimes\underline{\xi}(0) \qquad \text{(by Lemma 27-2)}$$

Hence $\underline{\xi}(n-1)$ belongs to the column-space of $\underline{\Gamma}(B)$. But since $\{\underline{B}\}_{ii} = \emptyset$ $(i=1,\ldots, n)$, every node i is an eigen-node, so all columns of $\underline{\Gamma}(\underline{B})$ are fundamental eigenvectors of $\underline{B}$. Hence $\underline{\xi}(n-1)$ lies in the eigenspace of $\underline{B}$, and Lemma 24-1 implies our result. ●

<u>Corollary 27-4. Let E_1 be a blog and let $\underline{B} \in \mathcal{M}_{nn}$, for given integer $n\geqslant 1$, have the following properties:</u>

(i) <u>$\{\underline{B}\}_{ij} < +\infty \qquad (i,j,\ldots, n)$</u>

(ii) <u>$\{\underline{B}\}_{ii} = \emptyset \qquad (i=1,\ldots, n)$</u>

(iii) <u>All elementary circuits in $\Delta(\underline{B})$ of length greater than unity have circuit product $-\infty$.</u>

(iv) <u>One column of $\underline{B}$ is finite</u>

<u>Then $\underline{B}$ has exactly one finite fundamental eigenvector $\underline{\eta}$. If $\underline{\xi}(0)\in E_n$ is an arbitrary finite n-tuple and we define the sequence $\{\underline{\xi}(r)\}$ by $\underline{\xi}(r+1) = \underline{B}\otimes\underline{\xi}(r)$ $(r=0,1,\ldots)$ then $\underline{\xi}(r)$ is a finite eigenvector of $\underline{B}$, for $r\geqslant(n-1)$.</u>

<u>Proof.</u> Evidently $\underline{B}$ is increasing and definite, so $\underline{\xi}(n-1)$ is an eigenvector of $\underline{B}$ by Corollary 27-3. But $\underline{B}$ is also row G-astic, so the $\underline{\xi}(r)$ are all finite by Corollary 12-7. Hence $\underline{\xi}(n-1)$ lies in the finite eigenspace of $\underline{B}$.

Suppose for simplicity that column one of $\underline{B}$ is finite. Then

$$\{\underline{B}\}_{1j} = -\infty \qquad (j=2,\ldots, n)$$

because $\{\underline{B}\}_{1j} \otimes \{\underline{B}\}_{j1} = -\infty$ $(j=2,\ldots, n)$ by hypothesis (iii). Hence each path from node 1 to node $j\neq 1$ in $\Delta(\underline{B})$ must have path product $-\infty$ since this product must involve a factor $\{\underline{B}\}_{1h}$ with $h\neq 1$. Hence $\{\underline{B}^r\}_{1j} = -\infty = \{\underline{\Gamma}(\underline{B})\}_{1j}$ for $r=1,2,\ldots$ and $j=2,\ldots, n$. Hence columns $2,\ldots, n$ of $\underline{\Gamma}(\underline{B})$ are not finite.

However, $\underline{\Gamma}(\underline{B}) \geq \underline{B}$ so column one of $\underline{\Gamma}(\underline{B})$ contains no $-\infty$. But $\underline{\Gamma}(\underline{B})$ is row-G-astic (because $\underline{B}$ and so $\underline{B}^{n-1}$ is row-G-astic). Hence column one of $\underline{\Gamma}(\underline{B})$ is finite.

Thus $\underline{\Gamma}(\underline{B})$ has a unique finite column $\underline{n}$ which is the only finite fundamental eigenvector of $\underline{B}$. ●

The foregoing corollary clears up a question raised in Section 1-2.3, for a matrix having properties (i) to (iv) would describe an activity network. Hypothesis (i) says that no infinitely great lead times exist; hypothesis (ii) says that an activity start-time coincides with itself; hypothesis (iii) says that there are no logically impossible temporal precedence relations; and hypothesis (iv) says that the activity network is connected in the sense that a certain activity (activity one, the "start" activity) must logically precede all others.

The iteration in Section 1-2.3 is of course dual to that of Corollary 27-4 since a different point was being illustrated in Section 1-2.3, namely backward planning from a last activity. But the conclusion is the same - the iteration terminates after (n-1) steps having generated a solution.

27-2. The Orbit. Given $\underline{\xi}(0) \in E_n$ and $\underline{A} \in \mathcal{M}_{nn}$, we may define the sequence $\{\underline{\xi}(r)\}$ $(r=0,1,\ldots)$ by the recurrence:

$$\underline{\xi}(r+1) = \underline{A} \otimes \underline{\xi}(r) \qquad (r=0,1,\ldots) \tag{27-4}$$

We call the sequence $\{\underline{\xi}(r)\}$ the orbit of $\underline{A}$ based on $\underline{\xi}(0)$ or simply the orbit (when $\underline{\xi}(0)$, $\underline{A}$ are fixed). We shall say that the orbit terminates finitely if, for some integer $r \geq 0$, $\underline{\xi}(r)$ is a finite eigenvector of $\underline{A}$, having a finite corresponding eigenvalue.
Now, Corollaries 27-3 and 27-4 evidently give particular answers to a question suggested by Section 1-2.1: what conditions must we lay on a real matrix $\underline{A}$ in order to guarantee that the orbit of $\underline{A}$ based on $\underline{\xi}(0)$ will terminate finitely for every finite $\underline{\xi}(0)$? We continue by analysing this question for matrices which are $\emptyset$-astic definite and finite.

Theorem 27-5. Let E_1 receive the principal interpretation and let $\underline{B} \in \mathcal{M}_{nn}$, for given integer $n \geq 1$, be $\emptyset$-astic definite and finite. Then a sufficient condition that the orbit of $\underline{B}$ based on $\underline{\xi}(0)$ shall terminate finitely for each finite $\underline{\xi}(0) \in E_n$ is that $\{\underline{B}\}_{jj} = \emptyset$ for each eigen-node j in $\Delta(\underline{B})$.

Proof. Suppose that $\{\underline{B}\}_{jj} = \emptyset$ for each eigen-node j. We consider $\{\underline{B}^N\}_{ij}$ for $1 \leq i \leq n$; $1 \leq j \leq n$, and some given integer $N \geq n$. Since E_1 is linear, we have from (22-14) that there exists an integer r_{ij} $(1 \leq r_{ij} \leq n)$ such that:

$$\{\underline{\Gamma}(\underline{B})\}_{ij} = \{\underline{B}^{r_{ij}}\}_{ij} \tag{27-5}$$

We must identify several cases.

<u>Case 1. Index j is an eigen-node</u>. Then (with an appropriate interpretation of the notation if $N=r_{ij}$).

$$\{\underline{B}^N\}_{ij} = \{\underline{B}^{r_{ij}} \otimes \underline{B}^{N-r_{ij}}\}_{ij}$$

$$= \sum_{\alpha,\beta,\gamma,\ldots\delta}{}_{\oplus} \left(\{\underline{B}^{r_{ij}}\}_{i\alpha} \otimes \{\underline{B}\}_{\alpha\beta} \otimes \{\underline{B}\}_{\beta\gamma} \otimes \ldots \otimes \{\underline{B}\}_{\delta j}\right)$$

$$\longleftarrow (N-r_{ij}) \text{ factors} \longrightarrow$$

$$\geq \{\underline{B}^{r_{ij}}\}_{ij} \otimes (\{\underline{B}\}_{jj})^{N-r_{ij}} \quad (\text{i.e. } \alpha = \ldots = \delta = j)$$

$$= \{\underline{\Gamma}(\underline{B})\}_{ij} \quad (\text{by (27-5) and hypothesis on } \{\underline{B}\}_{jj})$$

But $\{\underline{B}^N\}_{ij} \leq \{\underline{\Gamma}(\underline{B})\}_{ij}$ by Theorem 22-6, so in fact we conclude that if j is an eigen-node then $\{\underline{B}^N\}_{ij} = \{\underline{\Gamma}(\underline{B})\}_{ij}$ for all $i=1,\ldots, n$ and $N\geq n$.

<u>Case 2. Index i is an eigen-node</u>. Similarly we conclude that:

$$\{\underline{B}^N\}_{ij} = \{\underline{\Gamma}(\underline{B})\}_{ij} \quad \text{for all } j=1,\ldots, n \text{ and } N\geq n.$$

<u>Case 3. Neither i nor j is an eigen-node</u>. Let us define:

$$\gamma_{ij} = \sum_{\text{eigen-nodes } k}{}_{\oplus} \left(\{\underline{\Gamma}(\underline{B})\}_{ik} \otimes \{\underline{\Gamma}(\underline{B})\}_{kj}\right) \qquad (27\text{-}6)$$

(the index k runs over the eigen-nodes only, in $\Delta(\underline{B})$). We note that γ_{ij} is finite, since $\underline{B}$ is finite.

Suppose $N\geq 2n$, so that $(N-n)\geq n$. Then from the conclusions of Cases 1 and 2:

$$\gamma_{ij} = \sum_{\text{eigen-nodes } k}{}_{\oplus} \left(\{\underline{B}^n\}_{ik} \otimes \{\underline{B}^{N-n}\}_{jk}\right)$$

$$\leq \sum_{k=1}^{n}{}_{\oplus} \left(\{\underline{B}^n\}_{ik} \otimes \{\underline{B}^{N-n}\}_{kj}\right)$$

$$= \{\underline{B}^N\}_{ij} \qquad (27\text{-}7)$$

Now let δ be the greatest (E_1 is linear!) of the circuit products of all elementary circuits in $\Delta(\underline{B})$ not containing any eigen-node. Evidently $\delta<\emptyset$.

In conventional algebraic notation, δ is strictly negative and hence there must be some integer $q\geq 2$ such that $t\delta+\{\underline{\Gamma}(\underline{B})\}_{ij}<\gamma_{ij}$ when $t\geq q$. In max algebra notation:

$$\delta^t \otimes \{\underline{\Gamma}(\underline{B})\}_{ij} < \gamma_{ij} \text{ when } t\geq q \qquad (27\text{-}8)$$

So take $N\geq nq$. By the linearity of E_1, and Proposition 22-1, $\{\underline{B}^N\}_{ij}$ must be equal to a path product for some path τ of length N from i to j in $\Delta(\underline{B})$. We assert that τ must contain an eigen-node. For since $N \geq nq$, such a path must contain q elementary circuits, together with nodes making up a path σ from i to j of length $t(\sigma)$ and path product $p(\sigma)$ (say). If none of these elementary circuits contains an eigen-node we may therefore write:

$$\{\underline{B}^N\}_{ij} \leq \delta^q \otimes p(\sigma) \qquad (27\text{-}9)$$

However, from Proposition 22-1 and Theorem 22-6:

$$p(\sigma) \leq \{\underline{\Gamma}(\underline{B})\}_{ij} \qquad (27\text{-}10)$$

Evidently (27-7), (27-8), (27-9) and (27-10) establish a contradiction, and it follows that τ contains an eigen-node, say α $(1 \leq \alpha \leq n)$.

Hence $\{\underline{B}^N\}_{ij}$ is either of the form $\lambda_{i\alpha} \otimes \mu_{\alpha\alpha} \otimes \nu_{\alpha j}$ or of the form $\lambda_{i\alpha} \otimes \nu_{\alpha j}$ where $\lambda_{i\alpha}$ and $\nu_{\alpha j}$ (and $\mu_{\alpha\alpha}$) are products for paths (and a circuit) in $\Delta(\underline{B})$ from i to α and from α to j (and from α to α) respectively. Since α is an eigen-node, $\mu_{\alpha\alpha} = \emptyset$ and in all cases:

$$\begin{aligned} \{\underline{B}^N\}_{ij} &= \lambda_{i\alpha} \otimes \nu_{\alpha j} \\ &\leq \{\underline{\Gamma}(\underline{B})\}_{i\alpha} \otimes \{\underline{\Gamma}(\underline{B})\}_{\alpha j} \quad \text{(by Proposition 22-1 and Theorem 22-6)} \\ &\leq \gamma_{ij} \quad \text{(from 27-6)} \end{aligned} \qquad (27\text{-}11)$$

Evidently (27-7) and (27-11) imply:

$\{\underline{B}^N\}_{ij} = \gamma_{ij}$ when $N \geq qn$ $(\geq 2n)$ and i,j are not eigen-nodes. This concludes Case 3.

Now since $\{\underline{\Gamma}(\underline{B})\}_{ij}$ and γ_{ij} are constants depending only on i,j, it is evident from Cases 1,2 and 3 that $\underline{B}^N$ is constant for $N \geq qn$.

Then we have for finite $\underline{\xi}(0) \in E_n$ and $N \geq qn$:

$\underline{B} \otimes \underline{\xi}(N) = \underline{\xi}(N+1) = \underline{B}^{N+1} \otimes \underline{\xi}(0) = \underline{B}^N \otimes \underline{\xi}(0) = \underline{\xi}(N)$ (which is finite by Proposition 5-13)

Hence $\underline{\xi}(N)$ is a finite eigenvector of $\underline{B}$ and the orbit terminates finitely. ●

27-3. The Orbital Matrix. For the matrix $\underline{B}$ in Theorem 27-5, we proved that the matrix $\underline{B}^N$ is constant for N sufficiently large. The following theorem identifies this constant matrix in terms of the fundamental eigenvectors and left eigenvectors of $\underline{B}$.

Theorem 27-6. Let E_1 receive the principal interpretation and let $\underline{B} \in \mathcal{M}_{nn}$, for given integer $n \geq 1$, be $\emptyset$-astic definite and finite and have $\{\underline{B}\}_{jj} = \emptyset$ for each eigen-node j in $\Delta(\underline{B})$. Let $j_1, \ldots, j_s$ be a maximal set of non-equivalent eigen-nodes in $\Delta(\underline{B})$ and let $\underline{\xi}(j_1), \ldots, \underline{\xi}(j_s)$ and $\underline{\zeta}(j_1), \ldots, \underline{\zeta}(j_s)$ be respectively columns $j_1, \ldots, j_s$ and rows $j_1, \ldots, j_s$ of $\underline{\Gamma}(\underline{B})$. Then for each sufficiently large integer N there holds:

$$\underline{B}^N = \sum_{k=1}^{s}{}_{\oplus} (\underline{\xi}(j_k) \otimes \underline{\zeta}(j_k))$$

Proof. To simplify the notation, we may assume that $\underline{B}$ is blocked and that the eigen-nodes are 1,..., r where r≤n. With an obvious notation, we may write $\underline{\Gamma}(\underline{B})$ as a partitioned matrix:

$$\underline{\Gamma}(\underline{B}) = \begin{bmatrix} \underline{\Gamma}_1 & \underline{\Gamma}_2 \\ \underline{\Gamma}_3 & \underline{\Gamma}_4 \end{bmatrix} \quad \text{(where } \underline{\Gamma}_1 \text{ is } (r\times r)). \tag{27-12}$$

If we refer to the conclusions to Cases 1, 2 and 3 of the proof of Theorem 27-5, and to the definition (27-6), it is clear that the matrix $\underline{B}^N$ has the form:

$$\begin{bmatrix} \underline{\Gamma}_1 & \underline{\Gamma}_2 \\ \underline{\Gamma}_3 & \underline{\Gamma}_3 \otimes \underline{\Gamma}_2 \end{bmatrix} \tag{27-13}$$

Now $\underline{\Gamma}_1$ is an increasing (r×r) matrix, whence:

$$\underline{\Gamma}_1^2 \geqslant \underline{\Gamma}_1 \text{ and } \underline{\Gamma}_1 \otimes \underline{\Gamma}_2 \geqslant \underline{\Gamma}_2 \tag{27-14}$$

On the other hand, since by Proposition 26-5 all the rows of the (r×n) matrix $\left[\underline{\Gamma}_1, \underline{\Gamma}_2\right]$ are left eigenvectors of $\underline{B}$ and therefore of $\underline{\Gamma}(\underline{B}) = \underline{B} \oplus \ldots \oplus \underline{B}^n$, we have:

$$\left[\underline{\Gamma}_1, \underline{\Gamma}_2\right] = \left[\underline{\Gamma}_1, \underline{\Gamma}_2\right] \otimes \begin{bmatrix} \underline{\Gamma}_1 & \underline{\Gamma}_2 \\ \underline{\Gamma}_3 & \underline{\Gamma}_4 \end{bmatrix}$$

$$= \left[\underline{\Gamma}_1^2 \oplus (\underline{\Gamma}_2 \otimes \underline{\Gamma}_3),\ (\underline{\Gamma}_1 \otimes \underline{\Gamma}_2) \oplus (\underline{\Gamma}_2 \otimes \underline{\Gamma}_4)\right]$$

Hence $\underline{\Gamma}_1 \geqslant \underline{\Gamma}_1^2$ and $\underline{\Gamma}_2 \geqslant \underline{\Gamma}_1 \otimes \underline{\Gamma}_2$, which in view of (27-14) implies:

$$\underline{\Gamma}_1 = \underline{\Gamma}_1^2 \text{ and } \underline{\Gamma}_2 = \underline{\Gamma}_1 \otimes \underline{\Gamma}_2 \tag{27-15}$$

By an analogous argument using the columns of $\underline{\Gamma}(\underline{B})$, we have:

$$\underline{\Gamma}_3 = \underline{\Gamma}_3 \otimes \underline{\Gamma}_1 \tag{27-16}$$

Hence we may write the matrix (27-13) as:

$$\begin{bmatrix} \underline{\Gamma}_1^2 & \underline{\Gamma}_1 \otimes \underline{\Gamma}_2 \\ \underline{\Gamma}_3 \otimes \underline{\Gamma}_1 & \underline{\Gamma}_3 \otimes \underline{\Gamma}_2 \end{bmatrix} = \begin{bmatrix} \underline{\Gamma}_1 \\ \underline{\Gamma}_3 \end{bmatrix} \otimes \begin{bmatrix} \underline{\Gamma}_1 & \underline{\Gamma}_2 \end{bmatrix}$$

$$= \sum_{j=1}^{r} {}_{\oplus} (\underline{\xi}(j) \otimes \underline{\zeta}(j)) \tag{27-17}$$

where $\underline{\xi}(j)$, $\underline{\zeta}(j)$ are respectively the jth column and the jth row of $\underline{\Gamma}(\underline{B})$. However by Theorem 23-8, equivalent fundamental eigenvectors of $\underline{B}$ are equal, and an analogous situation holds for the rows $\underline{\zeta}(j)$. Hence we may confine the summation (27-17) to non-equivalent eigen-nodes and the desired result follows.

We give an illustration of the foregoing theorem as follows. Suppose:

$$\underline{B} = \begin{bmatrix} 0 & -1 & -5 \\ -2 & 0 & -3 \\ -4 & -3 & -1 \end{bmatrix}; \quad \underline{B}^2 = \begin{bmatrix} 0 & -1 & -4 \\ -2 & 0 & -3 \\ -4 & -3 & -2 \end{bmatrix}; \quad \underline{B}^3 = \begin{bmatrix} 0 & -1 & -4 \\ -2 & 0 & -3 \\ -4 & -3 & -3 \end{bmatrix}$$

$$\underline{B}^4 = \begin{bmatrix} 0 & -1 & -4 \\ -2 & 0 & -3 \\ -4 & -3 & -4 \end{bmatrix}; \quad \underline{B}^5 = \begin{bmatrix} 0 & -1 & -4 \\ -2 & 0 & -3 \\ -4 & -3 & -5 \end{bmatrix}; \quad \underline{B}^6 = \begin{bmatrix} 0 & -1 & -4 \\ -2 & 0 & -3 \\ -4 & -3 & -6 \end{bmatrix}$$

$\underline{B}^7 = \underline{B}^6$, so all higher powers equal $\underline{B}^6$.

$$\text{Now } \underline{\Gamma}(\underline{B}) = \begin{bmatrix} 0 & -1 & -4 \\ -2 & 0 & -3 \\ -4 & -3 & -1 \end{bmatrix}$$

We form

$$\left(\begin{bmatrix} 0 \\ -2 \\ -4 \end{bmatrix} \otimes \begin{bmatrix} 0 & -1 & -4 \end{bmatrix}\right) \oplus \left(\begin{bmatrix} -1 \\ 0 \\ -3 \end{bmatrix} \otimes \begin{bmatrix} -2 & 0 & -3 \end{bmatrix}\right)$$

$$= \begin{bmatrix} 0 & -1 & -4 \\ -2 & -3 & -6 \\ -4 & -5 & -8 \end{bmatrix} \oplus \begin{bmatrix} -3 & -1 & -4 \\ -2 & 0 & -3 \\ -5 & -3 & -6 \end{bmatrix}$$

$$= \begin{bmatrix} 0 & -1 & -4 \\ -2 & 0 & -3 \\ -4 & -3 & -6 \end{bmatrix} = \underline{B}^6 \text{ as required.}$$

For a given definite matrix $\underline{B}$, let us call the matrix $\sum_{k=1}^{s}{}_{\oplus}(\underline{\xi}(j_k) \otimes \underline{\zeta}(j_k))$, defined as in Theorem 27-6, the <u>orbital matrix</u> of $\underline{B}$. Thus the theorem asserts that if $\underline{B}$ is $\emptyset$-astic definite, then all sufficiently high powers of $\underline{B}$ equal the orbital matrix of $\underline{B}$.

<u>27-4 . A Practical Case</u>. Suppose now that we are given an arbitrary real finite matrix $\underline{A}$ and that we compute the orbit of $\underline{A}$. What will happen? Theorem 27-9 below presents the various cases which can arise - but first we prove a couple of useful lemmas.

<u>Lemma 27-7. Let E_1 be a blog and let $\underline{B} \in \mathcal{M}_{nn}$, for given integer $n \geq 1$, be definite. Then each index j ($1 \leq j \leq n$) which is an eigen-node in $\Delta(\underline{B})$ is also an eigen-node in $\Delta(\underline{B}^q)$ for each integer $q \geq 1$. Conversely if some index j ($1 \leq j \leq n$) is an eigen-node in $\Delta(\underline{B}^q)$ for some integer $q \geq 1$, and E_1 is linear, then j is also an eigen-node in $\Delta(\underline{B})$.</u>

<u>Proof</u>. If j is an eigen-node in $\Delta(\underline{B})$ then there is some elementary circuit of length $t(\tau)$ (say) ($1 \leq t(\tau) \leq n$) from j to j in $\Delta(\underline{B})$ whose circuit product $p(\tau) = \emptyset$. Let σ be a path consisting of q repetitions of τ, for given integer $q \geq 1$. Then σ is a circuit of length $qt(\tau)$ from j to j in $\Delta(\underline{B})$ whose circuit product $p(\sigma) = \emptyset$.

But σ may be regarded as a sequence of $t(\sigma)$ paths each of length q from j to i_1,

i_1 to $i_2, \ldots, i_{t(\sigma)-1}$ to j respectively in $\Delta(\underline{B})$ (for suitable $i_1, \ldots, i_{t(\sigma)-1}$). Hence by Proposition 22-1:

$$\{\underline{B}^q\}_{ji_1} \otimes \{\underline{B}^q\}_{i_1 i_2} \otimes \ldots \otimes \{\underline{B}^q\}_{i_{t(\sigma)-1}j} \geq (\emptyset)^{t(\sigma)} = \emptyset \qquad (27\text{-}18)$$

But (27-18) displays a circuit in $\Delta(\underline{B}^q)$ with circuit product at least equal to $\emptyset$, and therefore equal to $\emptyset$ since $\underline{B}^q$ is definite by Theorem 22-4. Hence (27-18) also shows that j is an eigen-node in $\Delta(\underline{B}^q)$.

Conversely if j is indeed an eigen-node in $\Delta(\underline{B}^q)$, some (27-18) holds (with equality). Each factor $\{\underline{B}^q\}_{\alpha\beta}$ in the left-hand side of (27-18) is a summation of path products for paths of length q in $\Delta(\underline{B})$. But if E_1 is linear, each such summation equals one of its summands. Hence using these summands instead in (27-18) we obtain a circuit product at least equal to $\emptyset$ for a circuit from j to j in $\Delta(\underline{B})$. But $\underline{B}$ is definite, so j must be an eigen-node in $\Delta(\underline{B})$. ●

Lemma 27-8. Let E_1 be a linear blog and let $\underline{B} \in \mathcal{M}_{nn}$, for given integer $n \geq 1$, be definite. Then there is an integer r such that $\{\underline{B}^r\}_{jj} = \emptyset$ for each eigen-node j in $\Delta(\underline{B}^r)$.

Proof. Repeat the proof of Lemma 27-7 as for as (27-18). Now (27-18) displays a circuit from j to j in $\Delta(\underline{B}^q)$ of length $t(\sigma)$ and circuit product $\emptyset$ so by Proposition 22-1, $\{(\underline{B}^q)^{t(\sigma)}\}_{jj} \geq \emptyset$. But $\underline{B}^q$ is definite (by Theorem 22-4). Hence:

$$\{\underline{B}^{qt(\sigma)}\}_{jj} = \emptyset \qquad (27\text{-}19)$$

Now let r be the least common multiple of all the numbers $t(\sigma)$ for all elementary circuits in $\Delta(\underline{B})$ having circuit product equal to $\emptyset$, and put $q=(r/t(\sigma))$ in (27-19) to get:

$$\{\underline{B}^r\}_{jj} = \emptyset \qquad (27\text{-}20)$$

Hence (27-20) holds for each eigen-node j in $\Delta(\underline{B})$. But since E_1 is linear, these are exactly the eigen-nodes in $\Delta(\underline{B}^r)$, by Lemma 27-7. ●

Theorem 27-9. Let E_1 receive the principal interpretation and let $\underline{A} \in \mathcal{M}_{nn}$, for given integer $n \geq 1$, be finite. Then there exist integers r and N such that for arbitrary $\underline{\xi} \in E_n$ and arbitrary $\underline{\eta} \in \mathcal{M}_{1n}$ we have that $\underline{A}^N \otimes \underline{\xi}$ and $\underline{\eta} \otimes \underline{A}^N$ are respectively an eigenvector and a left eigenvector of the matrix $\underline{A}^r$, with corresponding eigenvalue $(\lambda(A))^r$. In order that r=1, it is sufficient that we have $\{\underline{A}\}_{jj} = \lambda(\underline{A})$ for each eigen-node j in $\Delta((\lambda(\underline{A}))^{-1} \otimes \underline{A})$; and then for each integer $q \geq N$, the matrix $\underline{A}^q$ equals:

$$(\lambda(\underline{A}))^q \otimes (\text{Orbital matrix of } ((\lambda(\underline{A}))^{-1} \otimes \underline{A}))$$

Proof. Recall that E_1 is a commutative linear radicable blog and that the principal eigenvalue of $\underline{A}$ is $\lambda(\underline{A})$. Since $\underline{A}$ is finite, the four eigenproblems for $\underline{A}$ are finitely soluble by Corollary 25-7 and its dual and variants. As in the proof of Theorem 24-9, we therefore know that the definite matrix $(\lambda(\underline{A}))^{-1} \otimes \underline{A}$ is directly similar to a row-$\emptyset$-astic (and so $\emptyset$-astic definite) matrix $\underline{B}$. Thus there exist $\underline{P}, \underline{Q} \in \mathcal{M}_{nn}$ such that $\underline{P} \otimes \underline{Q} = \underline{Q} \otimes \underline{P} = \underline{I}_n$ and a $\emptyset$-astic definite matrix $\underline{B} \in \mathcal{M}_{nn}$ such that:

$$\underline{B} = \underline{P} \otimes ((\lambda(\underline{A}))^{-1} \otimes \underline{A}) \otimes \underline{Q} \tag{27-21}$$

Now if r is the integer referred to in Lemma 27-8, then $\underline{B}^r$ is $\emptyset$-astic definite by Theorem 23-8 and has $\{\underline{B}^r\}_{jj} = \emptyset$ for each eigen-node j in $\Delta(\underline{B}^r)$ by Lemma 27-8.

Hence by Theorem 27-6 (with $\underline{B}^r$ in the role of $\underline{B}$), $\underline{B}^{rm}$ equals the orbital matrix for $\underline{B}^r$, for each sufficiently large integer m. For such an integer m, therefore $\underline{B}^{r(m+1)} = \underline{B}^{rm}$ so for arbitrary $\underline{\xi} \in E_n$ we have:

$$\underline{B}^{r(m+1)} \otimes \underline{P} \otimes \underline{\xi} = \underline{B}^{rm} \otimes \underline{P} \otimes \underline{\xi}$$

Substituting from (27-21) we have (using Lemma 24-5):

$$\underline{P} \otimes ((\lambda(\underline{A}))^{-1} \otimes \underline{A})^{r(m+1)} \otimes \underline{Q} \otimes \underline{P} \otimes \underline{\xi}$$

$$= \underline{P} \otimes ((\lambda(\underline{A}))^{-1} \otimes \underline{A})^{rm} \otimes \underline{Q} \otimes \underline{P} \otimes \underline{\xi}$$

Left-multiplying by $\underline{Q}$ and using $\underline{Q} \otimes \underline{P} = \underline{I}_n$:

$$((\lambda(\underline{A}))^{-1} \otimes \underline{A})^{r(m+1)} \otimes \underline{\xi} = ((\lambda(\underline{A}))^{-1} \otimes \underline{A})^{rm} \otimes \underline{\xi}$$

Multiplying through by $(\lambda(\underline{A}))^{r(m+1)}$ and using the commutativity of scalar multiplication (since E_1 is commutative):

$$\underline{A}^{r(m+1)} \otimes \underline{\xi} = (\lambda(\underline{A}))^r \otimes \underline{A}^{rm} \otimes \underline{\xi} \tag{27-22}$$

Putting N = rm we thus have

$$\underline{A}^r \otimes (\underline{A}^N \otimes \underline{\xi}) = \underline{A}^{r(m+1)} \otimes \underline{\xi}$$

$$= (\lambda(\underline{A}))^r \otimes \underline{A}^N \otimes \underline{\xi}$$

which shows that $\underline{A}^N \otimes \underline{\xi}$ is an eigenvector of $\underline{A}^r$ with corresponding eigenvalue $(\lambda(\underline{A}))^r$. The argument for $\underline{\eta} \otimes \underline{A}^N$ is analogous.

In order that r=1 it is obviously sufficient that we have $\{\underline{B}\}_{jj} = \emptyset$ for each eigen-node j in $\Delta(\underline{B})$; but by the direct similarity, we have for suitable finite scalars η_j:

$$\eta_j^{-1} \otimes \{(\lambda(\underline{A}))^{-1} \otimes \underline{A}\}_{jj} \otimes \eta_j = \{\underline{B}\}_{jj}$$

On left-multiplying by η_j and right-multiplying by η_j^{-1} we obtain an equivalent sufficient condition:

$$(\lambda(\underline{A}))^{-1} \otimes \{\underline{A}\}_{jj} = \emptyset \tag{27-23}$$

for each eigen-node in $\Delta(\underline{B})$. Left multiplying by $\lambda(\underline{A})$ and using Lemma 24-6, we have as an equivalent sufficient condition:

$$\{\underline{A}\}_{jj} = \lambda(\underline{A}) \quad \text{for each eigen-node } j \text{ in } \Delta((\lambda(\underline{A}))^{-1}\otimes\underline{A})$$

If indeed r=1 then $\underline{B}^{m+1} = \underline{B}^m = \underline{B}^N$, so $\underline{B}^q$ is the orbital matrix of $\underline{B}$ for $q\geq N$, so by Theorem 27-6:

$$\underline{B}^q = \sum_{k=1}^{s}{}_{\oplus} (\underline{\xi}(j_k) \otimes \underline{\zeta}(j_k)) \qquad \text{for } q\geq N. \tag{27-24}$$

where $j_1,\ldots, j_s$ are a maximal set of non-equivalent eigen-nodes in $\Delta(\underline{B})$ and $\underline{\zeta}(j_k)$, $\underline{\xi}(j_k)$ are respectively the j_k^{th} row and the j_k^{th} column of $\underline{\Gamma}(\underline{B})$ (k=1,..., s).

Hence for i,j=1,..., we have:

$$\{\underline{B}^q\}_{ij} = \sum_{k=1}^{s}{}_{\oplus}(\{\underline{\Gamma}(\underline{B})\}_{ij_k} \otimes \{\underline{\Gamma}(\underline{B})\}_{j_kj}) \tag{27-25}$$

But since $\underline{\Gamma}(\underline{B}) = \underline{P}\otimes\underline{\Gamma}((\lambda(\underline{A}))^{-1}\otimes\underline{A})\otimes\underline{Q}$ (by Lemma 24-5),

we have for i,j=1,..., n and certain finite elements $\eta_1,\ldots,\eta_n$:

$$\{\underline{\Gamma}(\underline{B})\}_{ij_k} \otimes \{\underline{\Gamma}(\underline{B})\}_{j_kj}$$
$$= \eta_i^{-1}\otimes\{\underline{\Gamma}((\lambda(A))^{-1}\otimes\underline{A})\}_{ij_k} \otimes\{\underline{\Gamma}((\lambda(\underline{A}))^{-1}\otimes\underline{A})\}_{j_kj} \otimes \eta_j \tag{27-26}$$

From (27-25) and (27-26):

$$\{\underline{B}^q\}_{ij} = \eta_i^{-1} \otimes \{\sum_{k=1}^{s}{}_{\oplus}(\underline{\alpha}(j_k) \otimes \underline{\beta}(j_k))\}_{ij}\otimes\eta_j \tag{27-27}$$

where $\underline{\alpha}(j_k)$, $\underline{\beta}(j_k)$ are respectively the j_k^{th} row and the j_k^{th} column of $\underline{\Gamma}((\lambda(\underline{A}))^{-1}\otimes\underline{A})$ (k=1,..., s). But by Theorem 24-6, the indices $j_1,\ldots, j_s$ constitute a maximal set of non-equivalent eigen-nodes in $\Delta((\lambda(\underline{A}))^{-1}\otimes\underline{A})$ so (27-24) says:

$$\underline{B}^q = \underline{P} \otimes (\text{Orbital matrix of } ((\lambda(\underline{A}))^{-1}\otimes\underline{A})\otimes\underline{Q}$$

But $\underline{B}^q = \underline{P}\otimes((\lambda(\underline{A}))^{-1}\otimes\underline{A})^q\otimes\underline{Q}$. Hence left-multiplying by $\underline{Q}$, right-multiplying by $\underline{P}$ and using $\underline{P}\otimes\underline{Q} = \underline{Q}\otimes\underline{P}=\underline{I}_n$, we obtain:

$(\lambda(\underline{A}))^{-q} \otimes \underline{A}^q$ = Orbital matrix of $(\lambda(\underline{A}))^{-1} \otimes \underline{A}$, implying our final result. ●

<u>Corollary 27-10. Let E_1 receive the principal interpretation and let $\underline{A} \in \mathcal{M}_{nn}$, for given integer $n\geq1$, be finite. Then a sufficient condtion that the orbit of $\underline{A}$ based on $\underline{\xi}(0)$ shall terminate finitely for each finite $\underline{\xi}(0) \in E_n$ is that $\{\underline{A}\}_{jj} = \lambda(\underline{A})$ for each eigen-node j in $\Delta((\lambda(\underline{A}))^{-1}\otimes\underline{A})$.</u>

<u>Proof</u>. The given orbit terminates finitely if and only the orbit of $(\lambda(\underline{A}))^{-1}\otimes\underline{A}$ based on $\underline{\xi}(0)$ terminates finitely. If $\underline{B}$ is as in the proof of Theorem 27-9, then the orbit of $(\lambda(\underline{A}))^{-1}\otimes\underline{A}$ based on $\underline{\xi}(\otimes)$ is easily seen to terminate finitely if and only if

the orbit of $\underline{B}$ based on $\underline{\xi}(0)$ also terminates finitely since the mapping $\underline{\xi} \mapsto \underline{P} \otimes \underline{\xi}$ is a bijection (with inverse mapping $\underline{\eta} \mapsto \underline{Q} \otimes \underline{\eta}$). And by Theorem 27-5 this last happens if and only if $\{\underline{B}\}_{jj} = \emptyset$ for each eigen-node j in $\Delta(\underline{B})$, which happens if and only if $(\lambda(\underline{A}))^{-1} \otimes \underline{A} = \emptyset$ for each eigen-node j in $((\lambda(\underline{A}))^{-1} \otimes \underline{A})$ (by (27-20) and Lemma 24-6). ●

27-5. More general situations. Corollary 27-10 answers the question suggested by Section 1-2.1, for a useful class of matrices - real finite matrices. But what can be said when $\underline{A}$ is not finite? Or when E_1 is a more general kind of belt?

If we examine the proof of Theorem 27-5, we see that the argument for Cases 1 and 2 will go through for any commutative linear radicable blog E_1. Case 3, however, requires an Archimedean property of E_1, namely that a sufficiently high multiple of a negative number will be strictly less than a given finite number. We could therefore generalise E_1 to any commutative linear Archmedean radicable blog and use essentially the same proof for Theorem 27-5. (It is matter for conjecture as to how impressive a generalisation this is!)

The finiteness assumption on $\underline{A}$ enters the argument in two ways. First, it ensures that $\underline{A}$ has finitely soluble eigenproblem and so reduces the argument to a consideration of the matrix $\underline{B}$. Secondly, the finiteness of γ_{ij} is essential to the possibility of (27-8). Consider the case where $\underline{A} = \begin{bmatrix} \emptyset & -\infty \\ -\infty & -1 \end{bmatrix}$. It is easy to see that $\underline{A}^n = \begin{bmatrix} \emptyset & -\infty \\ -\infty & -n \end{bmatrix}$

and that the sequence $\underline{\xi}(0), \underline{\xi}(1), \ldots,$ does not terminate finitely when e.g.

$\underline{\xi}(0) = \begin{bmatrix} \emptyset \\ \emptyset \end{bmatrix}$. This failure results directly from the fact that $\gamma_{22} = -\infty$.

27-6. Permanents. If $\underline{A} \in \mathcal{M}_{nn}$ is a square matrix over a belt E_1, we define the permanent of $\underline{A}$ as the scalar Perm $\underline{A} \in E_1$ given by:

$$\text{Perm } \underline{A} = \sum_{\sigma}{}_{\oplus} \left(\prod_{i=1}^{n}{}_{\otimes} \{\underline{A}\}_{i\sigma(i)} \right)$$

where the summation is over all permutations σ in the symmetric group $\mathcal{G}_n$ of order n!

This definition follows exactly that of a permanent in conventional linear algebra, which enjoys many of the properties of a determinant. We shall not explore this analogy very far in the present memorandum since our present aim is to discuss properties of particular classes of matrices. There is, however, a connection between permanents and metric matrices, presented by e.g. Yoeli [6] and we exhibit this connection, in the terminology of the present work, in the following result.

First let us define the adjugate matrix of $\underline{A} \in \mathcal{M}_{nn}$ as the matrix Adj $\underline{A} \in \mathcal{M}_{nn}$ given by:

$$\{\text{Adj } \underline{A}\}_{ij} = \text{Cofactor } \{\underline{A}\}_{ji} \qquad (i,j=1,\ldots, n)$$

where Cofactor $\{\underline{A}\}_{ji}$ is the permanent of the $(n-1)\times(n-1)$ matrix obtained by deleting row j and column i from $\underline{A}$.

The definition fails if n=1, for which case we define $\underline{\text{Adj}}\ \underline{A} = \emptyset$ (if necessary adjoining the identity element $\emptyset$ to E_1).

The following connection between $\underline{\text{Adj}}\ \underline{A}$ and $\underline{\Gamma}(\underline{A})$ is essentially due to Yoeli.

Theorem 27-11. Let E_1 be a commutative blog and $\underline{A} \in \mathcal{M}_{nn}$, for given integer $n \geq 1$, be definite and increasing. Then $\underline{\text{Adj}}\ \underline{A} = \underline{\Gamma}(\underline{A})$.

Proof. If n=1 the theorem is trivial, so take n>1. For i=1,..., n it is evident that $\{\underline{\text{Adj}}\ \underline{A}\}_{ii}$ is a sum of products:

$$\{\underline{\text{Adj}}\ \underline{A}\}_{ii} = \sum_{\tau}{}_{\oplus} \Big(\prod_{\substack{r=1,\ldots,n \\ r\neq i}}{}_{\otimes} a_{r\tau(r)} \Big) \tag{27-28}$$

where the summation is over the symmetric group on the (n-1) letters in the set $\{x \mid 1 \leq x \leq n;\ x \neq i\}$.

Each such permutation is a cycle or a product of cycles and since E_1 is commutative, it follows that each summand in (27-28) may be rearranged as a circuit product, or product of circuit products, for suitable circuits in $\Delta(\underline{A})$.

For example if n=4 then $\{\underline{\text{Adj}}\ \underline{A}\}_{44}$ contains (with an obvious notation) a summand $a_{1\tau(1)} \otimes a_{2\tau(2)} \otimes a_{3\tau(3)}$. Suppose in fact that τ takes 1, 2, 3 into 3, 2, 1 respectively. The summand may be rearranged to give $a_{13} \otimes a_{31} \otimes a_{22}$ which is a product of circuit products.

Since $\underline{A}$ is definite we conclude from (27-28) and the foregoing that:

$$\{\underline{\text{Adj}}\ \underline{A}\}_{ii} \leq \emptyset \quad (i=1,\ldots, n) \tag{27-29}$$

On the other hand, if τ is in particular the idenitity permutation of the (n-1) letters then $a_{r\tau(r)} = a_{rr} = \emptyset$ because $\underline{A}$ is increasing (Lemma 27-1), whence:

$$\{\underline{\text{Adj}}\ \underline{A}\}_{ii} \geq \prod_{\substack{r=1,\ldots,n \\ r\neq i}}{}_{\otimes} a_{r\tau(r)} \quad \text{(from (27-28))}$$

$$= \emptyset \tag{27-30}$$

Now let $1\leq i\leq n$, $1\leq j\leq n$ with $i\neq j$. By definition, $\{\underline{\text{Adj}}\ \underline{A}\}_{ij}$ is the sum of products, each product containing all possible (n-1) terms of the form $\underline{A}_{r\sigma(r)}$ where $r\neq j$ and $\sigma(r)\neq i$, the summation being over all elements σ of the symmetric group on the letters 1,..., n. By the commutativity of E_1, such summand can be rearranged as a path product for a path from i to j in $\Delta(\underline{A})$, together perhaps with one or more circuit products. Since each circuit product is $\leq \emptyset$ and each path product from i to j is $\leq \{\underline{\Gamma}(\underline{A})\}_{ij}$ we conclude:

$$\{\underline{\text{Adj}}\ \underline{A}\}_{ij} \leqslant \{\underline{\Gamma}(\underline{A})\}_{ij} \qquad (i\neq j) \tag{27-31}$$

On the other hand if δ is any given elementary path of length $(n-1)$ from i to j in $\Delta(\underline{A})$, define a permutation σ of the letters $1,\ldots, n$ as follows:

$$\sigma(r) = s \quad \text{if } s \text{ immediately follows } r \text{ on } \delta.$$
$$\sigma(j) = i$$

(The mapping σ is a permutation because δ is elementary).

From the commutativity of E_1 and the fact that $\{\underline{A}\}_{rr} = \emptyset$ it is easy to see that the path length $p(\delta)$ of δ is given by

$$p(\delta) = \prod_{\substack{\otimes\\ r=1,\ldots,n\\ r\neq j\\ \sigma(r)\neq i}} \{\underline{A}\}_{r\sigma(r)}$$

Hence

$$p(\delta) \leqslant \sum_{\sigma}{}_{\oplus} \Big(\prod_{\substack{\otimes\\ r=1,\ldots,n\\ r\neq j\\ \sigma(r)\neq i}} \{\underline{A}\}_{r\sigma(r)}\Big)$$

$$= \{\underline{\text{Adj}}\ \underline{A}\}_{ij} \qquad \text{(when } \sigma \text{ ranges over the symmetric group on } 1,\ldots, n)$$

Since this holds for all such paths δ we have by Proposition 22-1:

$$\{\underline{A}^{n-1}\}_{ij} \leqslant \{\underline{\text{Adj}}\ \underline{A}\}_{ij} \qquad (i\neq j) \tag{27-32}$$

However, $\underline{\Gamma}(\underline{A}) = \underline{A}^{n-1}$ by Lemma 27-2. Hence from (27-31) and (27-32) we conclude:

$$\{\underline{\text{Adj}}\ \underline{A}\}_{ij} = \{\underline{\Gamma}(\underline{A})\}_{ij} \qquad (i\neq j) \tag{27-33}$$

Taken together with (27-29) and (27-30) this clearly implies that $\underline{\text{Adj}}\ \underline{A} = \underline{\Gamma}(\underline{A})$. ●

In conventional algebra, if $\underline{A}$ is a matrix whose eigenvalues all satisfy $|\lambda|<1$, then we have a valid expansion:

$$(\underline{I}_n - \underline{A})^{-1} = \underline{I}_n + \underline{A} + \underline{A}^2 + \ldots\ldots \tag{27-34}$$

Using the results of the present chapter, we can prove an analogue to (27-34) for max algebra. First we must define an analogue to the <u>inverse</u> of a matrix. For $\underline{B} \in \mathcal{M}_{nn}$, therefore, we define:

$$\underline{\text{Inv}}\ \underline{B} = (\text{Perm}\ \underline{B})^{-1} \otimes \underline{\text{Adj}}\ \underline{B}, \tag{27-35}$$

by direct analogy with $\underline{B}^{-1}$ in elementary linear algebra.

<u>Theorem 27-12. Let E_1 be a blog and let $\underline{A} \in \mathcal{M}_{nn}$, for given integer $n\geqslant 1$, have $\lambda(\underline{A}) \leqslant \emptyset$. Then:</u>

$$\text{Inv}\ (\underline{I}_n \oplus \underline{A}) = \underline{I}_n \oplus \underline{A} \oplus \underline{A}^2 \oplus \ldots \oplus \underline{A}^N$$

<u>for arbitrary large N.</u>

Proof. Since $\lambda(\underline{A}) \leq \emptyset$, we have $\{\underline{A}\}_{jj} \leq \emptyset$ $(j=1,\ldots, n)$, so:

$$\{\underline{I}_n \oplus \underline{A}\}_{jj} = \emptyset \qquad (j=1,\ldots, n) \tag{27-36}$$

Now for $i \neq j$, $\{\underline{I}_n \oplus \underline{A}\}_{ij} = -\infty \oplus \{\underline{A}\}_{ij} = \{\underline{A}\}_{ij}$ and from this and (27-36) it easily follows that the circuit product for any circuit in $\Delta(\underline{I}_n \oplus \underline{A})$ involving two or more distinct nodes equals the circuit product for the same circuit in $\Delta(\underline{A})$ and is therefore at most equal to $\emptyset$ since $\lambda(\underline{A}) \leq \emptyset$. This, together with (27-36) implies that $(\underline{I}_n \oplus \underline{A})$ is definite and increasing.

Hence by Theorem 27-11:

$$\underline{\text{Adj}}\ (\underline{I}_n \oplus \underline{A}) = \underline{\Gamma}(\underline{I}_n \oplus \underline{A})$$

So

$$\underline{\text{Inv}}\ (\underline{I}_n \oplus \underline{A}) = (\text{Perm}\ (\underline{I}_n \oplus \underline{A}))^{-1} \otimes \underline{\Gamma}(\underline{I}_n \oplus \underline{A}) \tag{27-37}$$

However, from its definition we see that Perm $(\underline{I}_n \oplus \underline{A})$ is a sum of (products of) circuit products for circuits in $\Delta(\underline{I}_n \oplus \underline{A})$ and is therefore at most equal to $\emptyset$ by our earlier reasoning. At the same time, Perm $(\underline{I}_n \oplus \underline{A})$ includes the summand:

$$\{\underline{I}_n \oplus \underline{A}\}_{11} \otimes \ldots \otimes \{\underline{I}_n \oplus \underline{A}\}_{nn}$$

which equals $\emptyset$. Hence:

$$\text{Perm}(\underline{I}_n \oplus \underline{A}) = \emptyset \tag{27-38}$$

Moreover by Lemma 27-2:

$$\underline{\Gamma}(\underline{I}_n \oplus \underline{A}) = (\underline{I}_n \oplus \underline{A})^N \quad \text{for} \quad N \geq (n-1)$$

$$= \underline{I}_n \oplus \underline{A} \oplus \ldots \oplus \underline{A}^N \tag{27-39}$$

Evidently (27-37), (27-38) and (27-39) imply the required result. ●

28. STANDARD MATRICES

28-1 . Direct Similarity

A classical topic in linear algebra concerns the establishment of certain special types of matrices, called canonical forms, to which more general classes of matrices may be reduced by particular classes of transformations. In the present chapter we shall derive a number of analogous results for matrices in minimax algebra, our "canonical forms" including various kinds of ϕ-astic matrices. Our first results use the concept of direct similarity defined in Section 24-2.

Theorem 28-1. Let E_1 be a linear blog and let $\underline{A}\varepsilon\mathcal{M}_{nn}$ for given integer $n\geq 1$. Then the eigenproblem for $\underline{A}$ is finitely soluble if and only if some finite scalar multiple of $\underline{A}$ is directly similar to a row-ϕ-astic matrix.

Proof. If for finite $\underline{\xi}\varepsilon E_n$ and finite $\lambda\varepsilon E_1$ there holds $\underline{A}\otimes\underline{\xi} = \lambda\otimes\underline{\xi}$ then from the proof of Theorem 24-9 it is clear that $\lambda^{-1}\otimes\underline{A}\sim\underline{B}$ where $\underline{B}$ is row-ϕ-astic. Conversely if $\lambda^{-1}\otimes\underline{A}\sim\underline{B}$ for given finite λ and row-ϕ-astic $\underline{B}$ then there exist by definition finite elements $\xi_1, \ldots, \xi_n\varepsilon E_1$ such that for $i=1, \ldots, n$:

$$\sum_{j=1}^{n}{}_{\oplus}\left(\xi_i^{-1}\otimes\{\lambda^{-1}\otimes\underline{A}\}_{ij}\otimes\xi_j\right) = \sum_{j=1}^{n}{}_{\oplus}\{\underline{B}\}_{ij}$$

$$= \phi \text{ (because } \underline{B} \text{ is row-}\phi\text{-astic)}$$

Hence
$$\sum_{j=1}^{n}{}_{\oplus}\left(\{\underline{A}\}_{ij}\otimes\xi_j\right) = \lambda\otimes\xi_i \quad (i=1, \ldots, n)$$

But this says that $\underline{\xi}$ is a (finite) eigenvector of $\underline{A}$ with corresponding eigenvalue λ, when $\underline{\xi}$ is defined by $\{\underline{\xi}\}_i = \xi_i \quad (i=1, \ldots, n)$

Let us say that a matrix $\underline{B}\varepsilon\mathcal{M}_{mn}$ over a blog E_1 is trapezoidal if it satisfies the following conditions:

$$m\geq n$$

$$\{\underline{B}\}_{ii} = \phi \qquad (i=1, \ldots, n)$$

If $m>1$ then $\{\underline{B}\}_{ij} \leq \phi \qquad (1\leq j<i\leq m)$

If $n>1$ then $\{\underline{B}\}_{ij} < \phi \qquad (1\leq i<j\leq n)$

Fig. 28-1 shows a trapezoidal matrix. Elements lying in the shaded trapezium may possibly equal ϕ. These in the unshaded triangle are less than ϕ.

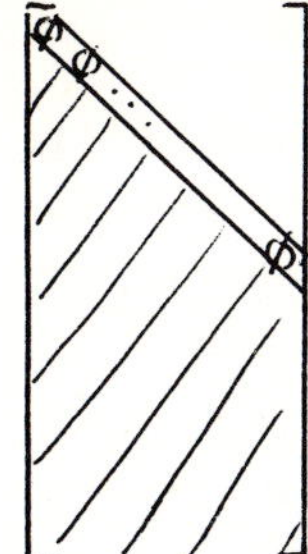

Fig. 28-1
A Trapezoidal Matrix

Evidently a trapezoidal matrix is column-ϕ-astic. Recall now the definition of strictly doubly ϕ-astic matrices from Section 15-3. The following theorem applies to square trapezoidal matrices, but will be generalised in Section 28-3.

Theorem 28-2. Let E_1 be a linear radicable blog other than ③ and let $\underline{B}\epsilon\mathcal{M}_{nn}$, for given integer $n\geq 1$, be trapezoidal. Then $\underline{B}$ is directly similar to a strictly doubly ϕ-astic matrix.

Proof. If n=1 the result is trivial, so assume n>1. Consider

$$\delta = \sum_{\substack{\oplus \\ 1\leq i<j\leq n}} \{\underline{B}\}_{ij}$$

Since E_1 is linear, δ is the "greatest of the above-diagonal elements" of $\underline{B}$, and certainly $\delta<\phi$. If $\delta=-\infty$ then redefine δ to be any finite element of E_1 satisfying $\delta<\phi$ (possible because $E_1\neq$ ③). Now use the fact that E_1 is radicable to define α from $\alpha^n = \delta$. Evidently α is finite and by Corollary 4-12 and Lemma 25-1:

$$\phi>\alpha>\alpha^2 \;\ldots.\; >\alpha^n \tag{28-1}$$

where

$$\alpha^n\geq\{\underline{B}\}_{ij} \qquad (1\leq i<j\leq n) \tag{28-2}$$

Now consider the matrix $\underline{C}\epsilon\mathcal{M}_{nn}$ defined by:

$$\{\underline{C}\}_{ij} = \alpha^i \otimes \{\underline{B}\}_{ij} \otimes \alpha^{-j} \qquad (i,j=1,\ldots, n) \tag{28-3}$$

Evidently $\underline{C}$ is directly similar to $\underline{B}$. From (28-3):

$$\{\underline{C}\}_{ii} = \phi \qquad (i=1, \ldots, n)$$

If $1 \leqslant j < i \leqslant n$:

$$\begin{aligned} \{\underline{C}\}_{ij} &= \alpha^{i} \otimes \{\underline{B}\}_{ij} \otimes \alpha^{-j} \\ &\leqslant \alpha^{(i-j)} \qquad (\text{since } \{\underline{B}\}_{ij} \leqslant \phi \\ &< \phi \qquad (\text{by } (28\text{-}1)) \end{aligned}$$

If $1 \leqslant i < j \leqslant n$:

$$\begin{aligned} \{\underline{C}\}_{ij} &= \alpha^{i} \otimes \{\underline{B}\}_{ij} \otimes \alpha^{-j} \\ &\leqslant \alpha^{i} \otimes \alpha^{n} \otimes \alpha^{-j} \qquad (\text{by } (28\text{-}2)) \\ &= \alpha^{n+i-j} \\ &< \phi \qquad (\text{by } (28\text{-}1)) \end{aligned}$$

Hence $\underline{C}$ has diagonal elements equal to ϕ and off-diagonal elements strictly less than ϕ and so is strictly doubly ϕ-astic. ●

28-2 · Invertible Matrices

In the sequel, we shall define the relation of equivalence, which is more general than that of direct similarity, and for this purpose we require to define invertible matrices, as follows.

If E_1 is a belt satisfying axioms X_8 and X_{10} then we define an invertible matrix over E_1 to be a square matrix $\underline{P} \varepsilon \mathcal{M}_{nn}$ for given integer $n \geqslant 1$, satisfying:

$$\text{There exists a matrix } \underline{Q} \varepsilon \mathcal{M}_{nn} \text{ such that } \underline{P} \otimes \underline{Q} = \underline{Q} \otimes \underline{P} = \underline{I}_n \tag{28-4}$$

We may characterise these invertible matrices quite closely. In fact, by analogy with the strictly doubly ϕ-astic matrices, let us define a strictly doubly G-astic matrix over a blog E_1 with group G as a square matrix $\underline{A} \varepsilon \mathcal{M}_{nn}$ for given integer $n \geqslant 1$, satisfying:

(i) $\{\underline{A}\}_{ij} < +\infty \, (i,j=1, \ldots, n)$

(ii) On each row and on each column of $\underline{A}$, we can find one and only one finite element. (28-5)

<u>Theorem 28-3. Let E_1 be a blog with group G and let $P \in \mathcal{M}_{nn}$ for given integer $n \geq 1$. Then $\underline{P}$ is invertible if and only if $\underline{P}$ is strictly doubly G-astic.</u>

<u>Proof</u>. If $\underline{P} \in \mathcal{M}_{nn}$ is strictly doubly G-astic then define $\underline{Q} \in \mathcal{M}_{nn}$ by:

$$\left.\begin{aligned}\{\underline{Q}\}_{ij} &= \left(\{\underline{P}\}_{ji}\right)^{-1} \quad \text{if } \{\underline{P}\}_{ji} \text{ is finite}\\ &= -\infty \qquad \text{otherwise}\end{aligned}\right\} \quad (i,j=1,\ldots,n) \qquad (28\text{-}6)$$

We may then confirm by direct computation that $\underline{P} \otimes \underline{Q} = \underline{Q} \otimes \underline{P} = \underline{I}_n$.

Conversely let $\underline{P} \in \mathcal{M}_{nn}$ be invertible, and let $\underline{Q}$ be as in (28-4). If n=1 then evidently $\underline{P} = [\eta]$ and $\underline{Q} = [\eta^{-1}]$ for some finite $\eta \in E_1$.

So let n>1. Now, for i,j=1,..., n we have:

$$\sum_{k=1}^{n}{}_{\oplus} \left(\{\underline{P}\}_{ik} \otimes \{\underline{Q}\}_{kj}\right) = \{\underline{I}_n\}_{ij} \qquad (28\text{-}7)$$

Using Proposition 4-2, it is clear from (28-7) that for each i=1,..., n there is a (least) index $c(i)$ $(1 \leq c(i) \leq n)$ such that $\{\underline{P}\}_{ic(i)}$ and $\{\underline{Q}\}_{c(i)i}$ are both finite, since $\{\underline{I}_n\}_{ii} = \emptyset$. Moreover, we cannot have $\{\underline{P}\}_{hc(i)}$ finite with $h \neq i$, since then:

$$\{\underline{P} \otimes \underline{Q}\}_{hi} \geq \{\underline{P}\}_{hc(i)} \otimes \{\underline{Q}\}_{c(i)i} > -\infty = \{\underline{I}_n\}_{hi}$$

It follows that the mapping $i \mapsto c(i)$ is a bijection and that each column of $\underline{P}$ is numbered c(i) for some i, contains exactly one finite element, and each row of $\underline{P}$ contains exactly one finite element. Moreover, we cannot have $\{\underline{P}\}_{ij} = +\infty$ for any i,j since then:

$$\{\underline{P}\}_{ij} \otimes \{\underline{Q}\}_{jk} = +\infty$$

if we choose k so that c(k) = j. But this implies $\{\underline{P} \otimes \underline{Q}\}_{ik} = +\infty$, contradicting (28-7). Hence $\underline{P}$ is strictly doubly G-astic. ●

If $\underline{P}$ is invertible then $\underline{Q}$ in (28-4) is unique (by the usual argument) and henceforth we shall write $\underline{P}^{-1}$ in place of $\underline{Q}$.

The intersection of the class of strictly doubly $\emptyset$-astic matrices and strictly doubly G-astic matrices we shall call the <u>permutation matrices</u>. Each matrix has an element equal to $\emptyset$ exactly once on each row and each column, and all other elements equal to

$-\infty$. It is clear that if $\underline{P}$ is a permutation matrix then the action of the operator $g_{\underline{P}}$ on any $\underline{\xi}\varepsilon E_n$ is to permute the elements of $\underline{\xi}$. Pre- or post-multiplication of a matrix $\underline{A}$ by $\underline{P}$ will permute the rows and columns of $\underline{A}$ respectively, and indeed these permutation matrices play a role exactly like that of their namesakes in conventional algebra.

Proposition 28-4. If E_1 is a blog then for given integer $n \geqslant 1$ the invertible matrices of order n form a group under the multiplication $\otimes$, containing $\underline{I}_n$ as identity element and having the permutation matrices as a subgroup isomorphic to the symmetric group on n letters.

28-3. Equivalence of Matrices

Let E_1 be a blog and let $\underline{A},\underline{B}\varepsilon \mathcal{M}_{mn}$, for given integers $m,n \geqslant 1$. We say that $\underline{A}$ and $\underline{B}$ are equivalent (written $\underline{A} \equiv \underline{B}$) if there exist invertible matrices $\underline{P}\varepsilon \mathcal{M}_{mm}$ and $\underline{Q}\varepsilon \mathcal{M}_{nn}$ such that $\underline{P}\otimes\underline{A}\otimes\underline{Q} = \underline{B}$. It is evident that direct similarity is a special case of equivalence, and that equivalence constitutes an equivalence relation (see Lemma 24-5).

Let us now say that $\underline{B}$ is obtained by an elementary operation on $\underline{A}$, if $\underline{B}$ is obtained from $\underline{A}$ by :

(i) Permuting the rows of $\underline{A}$, or

(ii) Permuting the columns of $\underline{A}$, or

(iii) Left-multiplying some row of $\underline{A}$ by a finite constant, or

(iv) Right-multiplying some column of $\underline{A}$ by a finite constant.

Lemma 28-5. Let E_1 be a blog and let $m,n \geqslant 1$ be given integers. Then the relation of equivalence is an equivalence relation on $\mathcal{M}_{mn}$. If $\underline{A},\underline{B}\varepsilon \mathcal{M}_{mn}$ then $\underline{A} \equiv \underline{B}$ if and only if there is a sequence $\underline{B}_0, \underline{B}_1, \ldots, \underline{B}_k$ such that $\underline{B}_0 = \underline{A}$ and $\underline{B}_k = \underline{B}$ and $\underline{B}_r$ is obtained by an elementary operation on $\underline{B}_{r-1}$ $(r=1, \ldots, k)$.

Proof. If $\underline{B}_r$ is obtained by permuting the rows of $\underline{B}_{r-1}$ then according to Proposition 28-4 there is a permutation matrix (so invertible) $\underline{P}$ such that $\underline{B}_r = \underline{P}\otimes\underline{B}_{r-1} = \underline{P}\otimes\underline{B}_{r-1}\otimes\underline{I}_n$, so $\underline{B}_r \equiv \underline{B}_{r-1}$. Similarly, permuting the columns of $\underline{B}_{r-1}$ can be achieved with a post-multiplying permutation matrix, giving $\underline{B}_r \equiv \underline{B}_{r-1}$. If $\underline{B}_r$ arises from $\underline{B}_{r-1}$ by left-multiplying some row of $\underline{B}_{r-1}$ by a finite constant λ then with a self-evident notation for a diagonal matrix:

$$\underline{B}_r = \underline{\mathrm{diag}}(\phi,\ldots,\phi,\lambda,\phi, \ldots,\phi)\otimes\underline{A}\otimes\underline{I}_n$$

so that $\underline{B}_r \equiv \underline{B}_{r-1}$. Similarly, right-multiplying a column of $\underline{B}_{r-1}$ can be achieved with a post-multiplying diagonal matrix, giving $\underline{B}_r \equiv \underline{B}_{r-1}$. Hence if $\underline{B}_0, \ldots, \underline{B}_k$ satisfy the conditions of the theorem then:

$$\underline{A} = \underline{B}_0 \equiv \ldots \equiv \underline{B}_k = \underline{B} \text{ , so } \underline{A} \equiv \underline{B}$$

Conversely if $\underline{A} \equiv \underline{B}$ then $\underline{B} = \underline{P} \otimes \underline{A} \otimes \underline{Q}$ with $\underline{P}, \underline{Q}$ invertible.
Since $\underline{P}$ is strictly doubly G-astic, we can permute the columns of $\underline{P}$ to obtain a diagonal matrix. Hence $\underline{P} = \underline{D} \otimes \underline{E}$ where $\underline{D}$ is diagonal and $\underline{E}$ is a permutation matrix. Similarly we may permute the rows of $\underline{Q}$ to obtain a diagonal matrix, so $\underline{Q} = \underline{H} \otimes \underline{K}$ where $\underline{H}$ is a permutation matrix and $\underline{K}$ a diagonal matrix. Hence:

$$\underline{B} = \underline{D} \otimes \underline{E} \otimes \underline{A} \otimes \underline{H} \otimes \underline{K}$$

Now, in the light of the known action of diagonal and permutation matrices: $\underline{A} \otimes \underline{H}$ can be obtained by permuting the columns of $\underline{A}$; $\underline{E} \otimes \underline{A} \otimes \underline{H}$ can be obtained by permuting the rows of $\underline{A} \otimes \underline{H}$; $\underline{D} \otimes \underline{E} \otimes \underline{A} \otimes \underline{H}$ can be obtained by left-multiplying certain rows of $\underline{E} \otimes \underline{A} \otimes \underline{H}$ by finite constants; and $\underline{D} \otimes \underline{E} \otimes \underline{A} \otimes \underline{H} \otimes \underline{K}$ can be obtained by right-multiplying certain columns of $\underline{D} \otimes \underline{E} \otimes \underline{A} \otimes \underline{H}$ by finite constants. Hence we can clearly define a sequence of matrices $\underline{B}_0, \ldots, \underline{B}_k$ having the required properties. ●

Lemma 28-6. Let E_1 be a blog with group G and let $\underline{A}, \underline{B} \in \mathcal{M}_{mn}$, for given integers $m,n \geq 1$, be equivalent. If one of $\underline{A}, \underline{B}$ is (row-, column-, or doubly) G-astic then so is the other.

Proof. Suppose $B = \underline{P} \otimes \underline{A} \otimes \underline{Q}$ with $\underline{P}, \underline{Q}$ invertible. Then by Theorem 28-3, $\underline{P}$ and $\underline{Q}$ are doubly G-astic, so the required results follow from Theorem 12-3. ●

Lemma 28-7. Let E_1 be a linear blog with group G and let $\underline{A} \in \mathcal{M}_{mn}$, for given integers $m,n \geq 1$. If a given row (or column) of $\underline{A}$ is G-astic then $\underline{A}$ is equivalent to a matrix in which that given row (or column) is ϕ-astic and all other rows (or columns) are identical with the corresponding rows (or columns) of $\underline{A}$. Hence if $\underline{A}$ is (row-, column- or doubly) G-astic then $\underline{A}$ is equivalent to a matrix which is (respectively row-, column-, or doubly) ϕ-astic.

Proof. Suppose row i $(1 \leq i \leq m)$ is G-astic. Define:

$$\beta = \sum_{\oplus\, j=1}^{n} \{\underline{A}\}_{ij}$$

Then β is finite.

Since E_1 is linear, we have that $\beta \geq \{\underline{A}\}_{ij}$ $(j=1, \ldots, n)$ but that there exists an index $k(1 \leq k \leq n)$ such that $\beta = \{\underline{A}\}_{ik}$. It follows that the set of elements $\beta^{-1} \otimes \{\underline{A}\}_{i1}, \ldots, \beta^{-1} \otimes \{\underline{A}\}_{in}$ is ϕ-astic. Hence if we left-multiply row i of $\underline{A}$ by β^{-1} we obtain a matrix which is equivalent to $\underline{A}$ (by Lemma 28-5), has row i ϕ-astic, and other rows unchanged. The other cases go through similarly. ●

The following result generalises Theorem 28-2

Lemma 28-8. Let E_1 be a linear radicable blog other than (3) with group G, and <u>let $\underline{B} \in \mathcal{M}_{mn}$, for given integers $m,n \geq 1$, be trapezoidal. The $\underline{B}$ is equivalent to a doubly ϕ-astic matrix which has an (nxn) strictly doubly ϕ-astic submatrix, if and only if $\underline{B}$ is row-G-astic.</u>

<u>Proof.</u> The "only if" follows from Lemma 28-6. So let $\underline{B}$ be row-G-astic.

If $n = 1$ or if $\sum_{\oplus}\{\underline{B}\}_{ij} = -\infty$ (summed over $1 \leq i < j \leq n$) then take δ arbitrarily to satisfy $-\infty < \delta < \phi$; else take $\delta = \sum_{\oplus}\{\underline{B}\}_{ij}$ (summed over $1 \leq i < j \leq n$). In both cases, $-\infty < \delta < \phi$ (by linearity of E_1).

Using the fact that E_1 is radicable, define α (finite) to satisfy $\alpha^m = \delta$ Arguing exactly as in Theorem 28-2, we readily confirm that the matrix $\underline{C} \in \mathcal{M}_{mn}$, where with an obvious notation:

$$\underline{C} = \underline{\mathrm{diag}}(\alpha, \ldots, \alpha^m) \otimes \underline{B} \otimes \underline{\mathrm{diag}}(\alpha^{-1}, \ldots, \alpha^{-n})$$

(which is equivalent to $\underline{B}$) has as its first n rows a strictly doubly ϕ-astic (nxn) submatrix. If $m > n$ we know that rows $n+1, \ldots, m$ are G-astic, by Lemma 28-6, so we may use Lemma 28-7 to find a matrix equivalent to $\underline{C}$, having rows $n+1, \ldots, m$ all ϕ-astic, and rows $1, \ldots, n$ as in $\underline{C}$. ●

<u>28-4. Equivalence and rank</u>. Let us recall from Chapter 17 the definitions of <u>row-</u>, <u>column-</u> and <u>ϕ-astic rank</u>. In the present section, we discuss the equality of rank of equivalent matrices

Lemma 28-9. <u>Let E_1 be a blog and let $\underline{A}, \underline{B} \in \mathcal{M}_{mn}$, for given integers $m,n \geq 1$, be equivalent. If the columns of $\underline{B}$ are SLI then so are the columns of $\underline{A}$.</u>

<u>Proof.</u> If $\underline{B} = \underline{P} \otimes \underline{A} \otimes \underline{Q}$ with $\underline{P}, \underline{Q}$ invertible, and $\underline{b} \in E_n$ is finite, then $\underline{P}^{-1} \otimes \underline{b}$ is finite, (using Theorem 28-3 and Corollary 12-7). Now suppose $\underline{x}\,\underline{y} \in E_m$ are such that $\underline{A} \otimes \underline{x} = \underline{A} \otimes \underline{y} = \underline{P}^{-1} \otimes \underline{b}$. Then

$$\underline{B} \otimes (\underline{Q}^{-1} \otimes \underline{x}) = \underline{P} \otimes \underline{A} \otimes \underline{Q} \otimes (\underline{Q}^{-1} \otimes \underline{x}) = \underline{P} \otimes \underline{A} \otimes \underline{x} = \underline{P} \otimes (\underline{P}^{-1} \otimes \underline{b}) = \underline{b}$$

and $$\underline{B}\otimes(\underline{Q}^{-1}\otimes\underline{y}) = \underline{P}\otimes\underline{A}\otimes\underline{Q}\otimes(\underline{Q}^{-1}\otimes\underline{y}) = \underline{P}\otimes\underline{A}\otimes\underline{y} = \underline{P}\otimes(\underline{P}^{-1}\otimes\underline{b}) = \underline{b}$$

Now, if the columns of $\underline{B}$ are SLI, then it is possible to choose $\underline{b}$ finite such that the equation $\underline{B}\otimes\underline{z} = \underline{b}$ has unique solution $\underline{z}$. Hence $\underline{Q}^{-1}\otimes\underline{x} = \underline{Q}^{-1}\otimes\underline{y}$ and so $\underline{x} = \underline{y}$. So there exists finite $\underline{c}\in E_n$ (namely $\underline{c} = \underline{P}^{-1}\otimes\underline{b}$) such that the equation $\underline{A}\otimes\underline{x} = \underline{c}$ has unique solution $\underline{x}$, showing that the columns of $\underline{A}$ are SLI. ●

Now, relation (17-1) may be re-written:

$$\underline{B} = \underline{\text{diag}}(\{\underline{y}\}_1, \ldots, \{\underline{y}\}_m)\otimes\underline{A}\otimes\underline{\text{diag}}(\{\underline{x}\}_1, \ldots, \{\underline{x}\}_n) \tag{28-8}$$

The following result is then an easy consequence of the definition of the term ϕ-astic rank, and of the arguments of Lemma 28-5.

Proposition 28-10. Let E_1 be a blog and let $\underline{A}\in\mathcal{M}_{mn}$ for given integers $m,n\geq 1$. Then $\underline{A}$ has ϕ-astic rank equal to r if and only if the following statement is true for k = r but not for k>r:

"$\underline{A}$ is equivalent to a doubly ϕ-astic matrix $\underline{C}$ which has a(k×k) strictly doubly ϕ-astic submatrix" ●

We can now draw some immediate inferences.

Corollary 28-11. Let E_1 be a linear radicable blog other than ③, with group G, and let $\underline{A}\in\mathcal{M}_{mn}$, for given integers $m,n\geq 1$, be a row-G-astic trapezoidal matrix. Then $\underline{A}$ has rank n.

Proof. By Lemma 28-8 and Proposition 28-10, $\underline{A}$ has ϕ-astic rank equal to n. But since $\underline{A}$ is column ϕ-astic (being trapezoidal) and so doubly G-astic, the ϕ-astic rank of $\underline{A}$ is also the rank of $\underline{A}$ (by Theorem 17-7). ●

Corollary 28-12. Let E_1 be a linear blog with group G and let $\underline{A},\underline{B}\in\mathcal{M}_{mn}$, for given integers $m,n\geq 1$, be equivalent. If either $\underline{A}$ or $\underline{B}$ has a rank then so does the other and the ranks are equal.

Proof. If $\underline{A}$ has a (ϕ-astic) rank then $\underline{A}$ is doubly G-astic (by Theorem 17-9) and so is $\underline{B}$ (by Lemma 28-6). Hence $\underline{B}$ has a (ϕ-astic) rank (by Theorem 17-9). Now by the transitivity of equivalence it is clear that $\underline{A}$ is equivalent to some matrix $\underline{C}$ as in Proposition 28-10, if and only if $\underline{B}$ is equivalent to the same $\underline{C}$. Hence the ϕ-astic ranks of $\underline{A}$ and $\underline{B}$ are equal, i.e. their ranks are equal (by Theorem 17-7). ●

In Section 17-1 we defined the notion of regularity of a square matrix. We can now collect together a number of conditions which are equivalent to regularity.

Theorem 28-13. Let E_1 be a linear blog with group G and let $\underline{A}\in\mathcal{M}_{nn}$, for given integer $n\geq 1$, be doubly G-astic. Then the following conditions are all equivalent:

(i) $\underline{A}$ is right column-regular.

(ii) $\underline{A}$ is left row-regular.

(iii) $\underline{A}$ is regular.

(iv) $\underline{A}$ has ϕ-astic rank equal to n.

(v) $\underline{A}$ is equivalent to a strictly doubly ϕ-astic matrix.

Moreover, if E_1 is radicable and different from ③, then conditions (i) to (v) are equivalent to:

(vi) $\underline{A}$ is equivalent to a trapezoidal matrix.

Proof. The equivalence of (i) to (iv) is established in Chapter 17. The equivalence of (iv) and (v) follows from Proposition 28-10. If E_1 is radicable and different from ③ then Theorem 28-2 shows that (vi) implies (v). And evidently (v) implies (vi) since a strictly doubly ϕ-astic matrix may be transformed into a trapezoidal matrix by a permutation of the columns which makes all diagonal elements equal to ϕ. ●

28-5. Rank of Γ

In Theorem 24-9, we have already noted the linear independence of the fundamental eigenvectors which generate an eigenspace. We are now in a position to prove a stronger result, applicable under the principal intepretation. First we need the following lemma.

Lemma 28-14. Let E_1 be a linear blog and let $\underline{B}\in\mathcal{M}_{nn}$, for given integer $n\geq 1$, be ϕ-astic definite. Let $j_1, \ldots, j_s (s>1)$ be a (not necessarily maximal) set of non-equivalent eigen-nodes in $\Delta(\underline{B})$, with corresponding fundamental eigenvectors $\underline{\xi}(j_1), \ldots, \underline{\xi}(j_s)$. Then for some h $(1\leq h\leq s)$ we have:

$$\{\underline{\xi}(j_k)\}_{j_h} < \phi \qquad (k=1, \ldots, s;\ k\neq h)$$

Proof. If the assertion is false then for each $k=1, \ldots, s$ there is a (least) index $c(k)\neq k$ $(1\leq c(k)\leq s)$ such that:

$$\{\underline{\xi}(j_{c(k)})\}_{j_k} = \phi$$

We may pick out a circuit from the path $(j_1, j_{c(1)}, j_{c^2(1)}, \ldots, j_{c^s(1)})$

where $c^2(1) = c(c(1))$, etc and then arguing as in the proof of Lemma 23-9, we shall find a circuit in $\Delta(\underline{B})$, one which at least two non-equivalent eigen-nodes occur, having circuit product equal to ϕ. Contradiction! ●

Theorem 28-15. Let E_1 be a commutative linear radicable blog other than ③ and let $\underline{A}\varepsilon\mathcal{M}_{nn}$, for given integer $n\geq 1$, have finitely soluble eigenproblem. Then the fundamental eigenvectors of $(\lambda(\underline{A}))^{-1}\otimes A$ corresponding to a maximal set of non-equivalent eigen-nodes in $\Delta((\lambda(\underline{A}))^{-1}\otimes\underline{A})$, are SLI.

Proof. By Theorem 25-5, $\lambda(\underline{A})$ is the principal eigenvalue of $\underline{A}$ and by the proof of Theorem 24-9, $(\lambda(\underline{A}))^{-1}\otimes\underline{A}$ is directly similar to $\underline{B}$ where $\underline{B}$ is row ϕ-astic and so ϕ-astic definite. Let $j_1, \ldots, j_s$ be a maximal set of non-equivalent eigen-nodes in $\Delta((\lambda(\underline{A}))^{-1}\otimes\underline{A})$, and hence (by Lemma 24-6) in $\Delta(\underline{B})$ also. Let the corresponding fundamental eigenvectors of $\underline{B}$ be $\underline{\xi}(j_1), \ldots, \underline{\xi}(j_s)$ and let $\underline{\Phi}\varepsilon\mathcal{M}_{ns}$ have $\underline{\xi}(j_k)$ as its kth column $(k = 1, \ldots, s)$. Then $\underline{\Phi}$ is doubly ϕ-astic by Lemma 23-11 and so doubly G-astic (where G is the group of E_1). We assert that it is possible to permute the rows and columns of $\underline{\Phi}$ to produce a trapezoidal matrix.

For if $s = 1$ then it suffices to permute row j_1 of $\underline{\Phi}$ into row 1. And if $s>1$ then according to Lemma 28-14 there is an index h_1 $(1\leq h_1\leq s)$ such that row j_{h_1} of $\underline{\Phi}$ has all its elements other than that in column h_1 strictly less than ϕ. We may then define indices $h_r (r = 2, \ldots, s)$ inductively as follows, using Lemma 28-14 to guarantee the existence of the defined indices:

> If $2\leq r<s$ then h_r is the least index h $(1\leq h\leq s;\ h\neq h_1, \ldots$ or $h_{r-1})$
> $$\text{with} \{\underline{\xi}(j_k)\}_{j_h} < \phi \quad (k = 1, \ldots, s;\ k\neq h_1, \ldots, h_{r-1} \text{ or } h) \tag{28-9}$$
> If $r = s$ then h_r is the index which remains when $h_1, \ldots h_{s-1}$ have been deleted from the list of indices $1, \ldots, s$.

Finally, we may then permute the rows of $\underline{\Phi}$ so that row j_{h_k} becomes row $k(k=1, \ldots, s)$ and permute the columns of $\underline{\Phi}$ so that column h_k becomes column $k(k=1, \ldots, s)$. A little thought will show that the matrix $\underline{\Psi}$ which so arises is trapezoidal (and still doubly G-astic). By Corollary 28-11, $\underline{\Psi}$ has rank s. But clearly $\underline{\Phi}$ is equivalent to $\underline{\Psi}$ and so $\underline{\Phi}$ also has rank s by Corollary 28-12. Hence its columns are SLI, ie. columns $j_1, \ldots, j_s$ of $\underline{\Gamma}(\underline{B})$ are SLI. But $\underline{\Gamma}((\lambda(\underline{A}))^{-1}\otimes\underline{A})$ is directly similar to $\underline{\Gamma}(\underline{B})$ by Lemma 24-5, ie. there exist finite elements $\eta_1, \ldots, \eta_n\varepsilon E_1$ such that:

$$\underline{\Gamma}(\underline{B}) = \underline{\text{diag}}(\eta_1^{-1}, \ldots, \eta_n^{-1}) \otimes \underline{\Gamma}((\lambda(\underline{A})^{-1} \otimes \underline{A}) \otimes \underline{\text{diag}}(\eta_1, \ldots, \eta_n) \qquad (28\text{-}10)$$

Now if $\underline{\Omega}$ is the matrix which has the $j_k{}^{th}$ column of $(\lambda(\underline{A}))^{-1} \otimes \underline{A}$ as its kth column, then from (28-10) and the definitions of $\underline{\Phi}$ and $\underline{\Omega}$:

$$\underline{\Phi} = \underline{\text{diag}}(\eta_1^{-1}, \ldots, \eta_n^{-1}) \otimes \underline{\Omega} \otimes \underline{\text{diag}}(\eta_{j_1}, \ldots, \eta_{j_s}) \qquad (28\text{-}11)$$

So Ω is equivalent to $\underline{\Phi}$ and hence $\underline{\Omega}$ also has rank s by Corollary 28-12. In other words the fundamental eigenvectors of $(\lambda(\underline{A}))^{-1} \otimes \underline{A}$ corresponding to $j_1, \ldots, j_s$ are SLI.

Corollary 28-16. Let E_1 be a commutative linear radicable blog other than ③, with group G, and let $\underline{A} \in \mathcal{M}_{nn}$, for given integer $n \geq 1$, be row G-astic. If $\Delta((\lambda(\underline{A}))^{-1} \otimes \underline{A})$ has n non-equivalent eigen-nodes, then $\underline{\Gamma}(\underline{A})$ is doubly G-astic and regular, and is directly similar to a strictly doubly ϕ-astic $(n \times n)$ matrix.

Proof. By Corollary 25-8, $\underline{A}$ is doubly G-astic, and then so is $\underline{\Gamma}(\underline{A})$. Moreover (by Corollary 25-8 again) the eigen-problem for $\underline{A}$ is finitely soluble; and $\lambda(\underline{A})$ is the principal eigenvalue (by Theorem 25-5). Clearly Theorem 28-15 indicates that $\underline{\Gamma}((\lambda(\underline{A}))^{-1} \otimes \underline{A})$ has n SLI columns, and so is regular.

Evidently from (iii) and (v) of Theorem 28-13 we may infer that $\underline{\Gamma}((\lambda(\underline{A}))^{-1} \otimes \underline{A})$ is equivalent to a strictly doubly ϕ-astic matrix, but we are asserting direct similarity, which is stronger. We may argue as follows.

Consider the proof of Theorem 28-15 when $s=n$. We have that:

$$\underline{\Phi} = \underline{\Gamma}(\underline{B}) \simeq \Gamma((\lambda(\underline{A}))^{-1} \otimes \underline{A}) \qquad (28\text{-}12)$$

Moreover, $j_1=1, \ldots, j_s=j_n=n$. Hence in transforming $\underline{\Phi}$ into $\underline{\Psi}$, we must carry out exactly the same permutations on the columns as on the rows.

From this it easily follows that $\underline{\Psi} = \underline{P} \otimes \underline{\Phi} \otimes \underline{P}^{-1}$ for a suitable permutation matrix $\underline{P}$. Hence $\underline{\Psi} \simeq \underline{\Phi}$, so $\underline{\Psi} \simeq \Gamma((\lambda(\underline{A}))^{-1} \otimes \underline{A})$ However, $\underline{\Psi}$ (being trapezoidal) is directly similar to a strictly doubly ϕ-astic matrix, by Theorem 28-2, and the result follows.

As an example, let $\underline{A} = \begin{bmatrix} 3 & 3 & -\infty \\ -\infty & 3 & -2 \\ -1 & 4 & 3 \end{bmatrix}$ Then $\lambda(\underline{A}) = 3$

Define $\underline{C} = (\lambda(\underline{A}))^{-1} \otimes \underline{A} = \begin{bmatrix} 0 & 0 & -\infty \\ -\infty & 0 & -5 \\ -4 & 1 & 0 \end{bmatrix}$

$$\underline{\Gamma}(\underline{C}) = \underline{C} \oplus \underline{C}^2 \oplus \underline{C}^3 = \begin{bmatrix} 0 & 0 & -5 \\ -9 & 0 & -5 \\ -4 & 1 & 0 \end{bmatrix}$$

Evidently $\underline{C}$ has 3 non-equivalent fundamental eigenvectors. Using e.g. the first of them (first column of $\underline{\Gamma}(\underline{C})$) we may transform $\underline{C}$ into a directly similar row-ϕ-astic matrix $\underline{B}$ as in Theorem 28-1. Thus:

$$\underline{B} = \underline{\text{diag}}(0,9,4) \otimes ((\lambda(\underline{A}))^{-1} \otimes \underline{A}) \otimes \underline{\text{diag}}(0,-9,-4)$$

$$= \begin{bmatrix} 0 & -9 & -\infty \\ -\infty & 0 & 0 \\ 0 & -4 & 0 \end{bmatrix}$$

Now $\underline{\Gamma}(\underline{B}) = \underline{\text{diag}}(0,9,4) \otimes \underline{\Gamma}(\underline{C}) \otimes \underline{\text{diag}}(0,-9,-4)$

$$= \begin{bmatrix} 0 & -9 & -9 \\ 0 & 0 & 0 \\ 0 & -4 & 0 \end{bmatrix} = \underline{\Phi} \text{ also}$$

Applying the procedure (28-9) we have: $h_1 = 1$; $h_2 = 3$; $h_3 = 2$. Hence we must interchange indices 2 and 3 to obtain $\underline{\Psi}$:

$$\underline{\Psi} = \begin{bmatrix} 0 & -\infty & -\infty \\ -\infty & -\infty & 0 \\ -\infty & 0 & -\infty \end{bmatrix} \otimes \underline{\Phi} \otimes \begin{bmatrix} 0 & -\infty & -\infty \\ -\infty & -\infty & 0 \\ -\infty & 0 & -\infty \end{bmatrix}$$

$$= \begin{bmatrix} 0 & -9 & -9 \\ 0 & 0 & -4 \\ 0 & 0 & 0 \end{bmatrix} \quad \text{which is trapezoidal.}$$

Now proceed as in the proof of Theorem 28-2. We have $\delta = -4$ so $\alpha = -4/3$. Then $\underline{\text{diag}}(-\frac{4}{3}, -\frac{8}{3}, -4) \otimes \underline{\Psi} \otimes \underline{\text{diag}}(\frac{4}{3}, \frac{8}{3}, 4)$

$$= \begin{bmatrix} 0 & -\frac{23}{3} & -\frac{19}{3} \\ -\frac{4}{3} & 0 & -\frac{8}{3} \\ -\frac{8}{3} & -\frac{4}{3} & 0 \end{bmatrix} \quad \text{which is strictly doubly } \phi\text{-astic.}$$

If we call this last matrix $\underline{D}$ then evidently $\underline{D} = \underline{P} \otimes \underline{\Gamma}((\lambda(\underline{A}))^{-1} \otimes \underline{A}) \otimes \underline{P}^{-1}$

where $\underline{P} = \underline{\text{diag}}(-\frac{4}{3}, -\frac{8}{3}, -4) \otimes \begin{bmatrix} 0 & -\infty & -\infty \\ -\infty & -\infty & 0 \\ -\infty & 0 & -\infty \end{bmatrix} \otimes \underline{\text{diag}}\,(0,9,4)$

$$= \begin{bmatrix} -\frac{4}{3} & -\infty & -\infty \\ -\infty & -\infty & \frac{4}{3} \\ -\infty & 5 & -\infty \end{bmatrix}$$

29. REFERENCES AND NOTATIONS

29-1. Previous Publications

In this section, we list in chronological order a number of earlier publications in which the authors develop or consider algebraic structures of the type which form the subject of the present work.

[1] A. Shimbel, "Structure in communication nets", *Proc. Symp. on Information Networks*, Polytechnic Institute of Brooklyn (1954) 119-203.

[2] F.E. Hohn, S. Seshu, and D.D. Aufenkamp, "The theory of nets", *Trans. I.R.E.* EC-6 (1957) 154-161.

[3] B. Roy, "Transitivité et connexité", *C.R. Acad. Sci. Paris 249* (1959) 216 - 218.

[4] R.A. Cuninghame-Green, "Process synchronisation in a steelworks - a problem of feasibility", in: Banbury and Maitland, ed., *Proc. 2nd Int. Conf. on Operational Research* (English University Press,1960) 323 - 328.

[5] B. Giffler, "Mathematical solution of production planning and scheduling problems", IBM ASDD Technical Report (1960).

[6] M. Yoeli, " A note on a generalization of boolean matrix theory", *Amer. Math. Monthly* 68 (1961) 552 - 557.

[7] R.A. Cuninghame-Green, "Describing industrial processes with interference and approximating their steady-state behaviour", *Operational Res. Quart.* 13 (1962) 95 - 100.

[8] S. Warshall, " A theorem on boolean matrices", *J. Assoc.Comput.Mach.* 9 (1962) 11 -18.

[9] T.S. Blyth, "Matrices over ordered algebraic structures", *J. London Math. Soc.* 39 (1964) 427-432.

[10] R.Cruon and Ph. Hervé, "Quelques resultats rélatifs à une structure algébrique et à son application au problème central de l'ordonnancement", *Rev. Francaise Recherche Opérationelle* 34 (1965) 3 - 19.

[11] V. Peteanu, "An algebra of the optimal path in networks", *Mathematica* 9 (1967) 335 - 342.

[12] C. Benzaken, "Structures algébriques des cheminements", in G. Biorci, ed. *Network and Switching Theory* (Academic Press, 1968) 40 - 57.

[13] B. Giffler, "Schedule algebra: a progress report", *Naval Res. Logist. Quart.* 15 (1968) 255 - 280.

[14] P. Robert and J. Ferland, "Généralisation de l'algorithme de Warshall, *Rev. Francaise Informat. Recherche Opérationelle* 2 (1968) 71 -85.

[15] I. Tomescu, "Sur l'algorithme matriciel de B. Roy", *Rev. Francaise Informat. Recherche Opérationelle* 2 (1968) 87 - 91.

[16] P. Brucker, "Verbände stetiger funktionen und kettenwertige homomorphismen", *Math. Ann.* 187 (1970) 77 - 84.

[17] B.A. Carré, "An algebra for network routing problems", *J. Inst. Math. Appl.* 7 (1971) 273 - 294.

[18] P. Brucker, "R-Netzwerke und matrixalgorithmen", *Computing* 10 (1972) 271 - 283.

[19] S. Rudeanu, "On boolean matrix equations", *Rev. Roumaine Math. Pures Appl.* 17 (1972) 1075 - 1090.

[20] D.R. Shier, "A decomposition algorithm for optimality problems in tree-structured networks", *Discrete Math.* 6 (1973) 175 - 189.

[21] E. Minieka and D.R. Shier, "A note on an algebra for the k best routes in a network", *J. Inst. Math. Appl.* 11 (1973) 145 - 149.

[22] M. Gondran, "Algèbre des chemins et algorithmes", NATO conference on combinatorial mathematics (Versailles, 1974).

[23] R.C. Backhouse and B.A. Carré, "Regular algebra applied to path-finding problems", *J. Inst. Math. Appl.* 15 (1975) 161 - 186.

[24] R.A. Cuninghame-Green, "Minimax algebra I: Bands and belts", Memorandum no. 70, Dept. of Applied Mathematics, T.H. Twente (Netherlands, 1975).

[25] R.A. Cuninghame-Green "Projections in minimax algebra", *Math.Programming* 10 (1976) 111 - 123.

29-2. Related References

Much relevant material is also contained in publications in the fields of game theory, graphs and networks, transportation, automata theory, theory of algorithms, discrete dynamic programming, machine scheduling, approximation theory, nonlinear optimisation and duality and convex programming.

The inspiration for some of the earlier results in the field can be directly traced to the work of Lunc and others in relation to switching theory and boolean matrices - for references see [6] and [33]. Related also is the work of Rockafeller [36] on bifunctions and convex processes. The following references give access to some of the wider issues.

[26] R. Bellman and W. Karush, "On a new functional transform in analysis: the maximum transform", *Bull. Amer. Math. Soc.* 67 (1961) 501 - 503.

[27] L. Fuchs, *Partially Ordered Algebraic Systems* (Pergamon, 1963).

[28] N.H. McCoy, *The Theory of Rings* (MacMillan, 1964).

[29] D.L. Bentley and K.L. Cooke, "Convergence of successive approximations in the shortest route problem", *J. Math. Anal. Appl.* 10 (1965) 269 - 274.

[30] R.G. Busacker and T.L. Saaty, *Finite Graphs and Networks* (McGraw Hill, 1965).

[31] G.B. Dantzig, "All shortest routes in a graph", *Theory of Graphs (International Symposium, Rome, 1966)* (Gordon & Breach,1966) 91 - 92.

[32] G. Birkhoff, *Lattice Theory* (American Mathematical Society, 1967)

[33] D.J. Wilde and C.S. Beightler, *Foundations of Optimization* (Prentice-Hall 1967).

[34] P.L. Hammer and S. Rudeanu, *Boolean Methods in Operations Research* (Springer Verlag, 1968).

[35] G. Birkhoff and T.C. Bartee, *Modern Applied Algebra* (McGraw-Hill 1969).

[36] K.L. Cooke and E. Halsey, "The shortest route through a network with time-dependent internodal transit", *J. Math. Anal. Appl.* 14 (1969) 493 - 498.

[37] J.J. Moreau,"Inf-convolution, sous-additivité, convexité des fonctions numériques", *J.Math.Pures Appl.* 49 (1970) 109 - 154.

[38] R.T. Rockafeller, *Convex Analysis* (Princeton University,1970)

[39] J.D. Murchland, "Shortest distances between all vertices in a network, and allied problems", *Veröffentlichungen des Instituts fur Stadtbauwesen, Heft 7* (Technical University of Braunschweig,1971) 1 - 24.

[40] P. Whittle, *Optimization under constraints* (Wiley-Interscience 1971).

[41] C.M. de Barros "Sur certaines catégories de couples d'applications croissantes", *Math. Nachr.* 54 (1972) 141 - 171.

[42] T.S. Blyth and M.F. Janowitz, *Residuation Theory* (Pergamon, 1972).

[43] A.H.G. Rinnooy Kan, "The machine scheduling problem", Mathematics Centre (Amsterdam, 1973).

[44] L. McLinden "An extension of Fenchel's duality theorem to saddle functions and dual minimax problems", *Pacific J. Math.* 50 (1974) 135 - 158.

[45] A. Brøndsted, "Convexification of conjugate functions", *Math. Scand.* 36 (1975) 131 - 136.

[46] F. Huisman, "Convergentieproblemen in minimax algebra" (Master's thesis, Twente University of Technology, Netherlands,1975).

[47] R.A. Cuninghame-Green, "Minimax algebra; Part V: Linear dependence" (Technical Report, Birmingham University,1976).

29-3. List of Notations

The most frequently used notations in the foregoing text are listed below.

$\{\alpha,\beta,\ldots,\gamma\}$	Intensional definition of a set: the set whose elements are $\alpha,\beta,\ldots,\gamma$.
$\{x \mid p(x)\}$	Extensional definition of a set: the set of all x such that proposition p(x) is true.
$x \in S$	Elementhood: x is an element of the set S.
$\cup, \cap$	Set-theoretical union and intersection.
$\supset, \subset$	Set inclusion (proper or improper).
$f: x \mapsto u$	Ostensive definition of a function: f maps the element x to the element u.
$f: U \to S$	Function f maps the set u into(possibly onto) the set S.
S^U	The set of all functions from the set U to the set S.
S^n	The set of all n-tuples of elements of the set S.
$\forall, \exists$	Universal and existential quantifiers.
$\oplus, \oplus'$	Addition and dual addition operations in a commutative band.
$\otimes, \otimes'$	Multiplication and dual multiplication operations in a belt.

$\sum_{\oplus}$, $\sum_{\oplus'}$	Iterated use of the operations $\oplus$ and $\oplus'$.
$\prod_{\otimes}$, $\prod_{\otimes'}$	Iterated use of the operations $\otimes$ and $\otimes'$.
$\hat{K}$	The composition algebra generated by the set K of functions.
R	The real numbers together with $-\infty$ and $+\infty$.
F	The (finite) real numbers.
+	Arithmetical addition operation.
$\max(x,y)$	The greater of x and y.
$\max_{k\varepsilon S}\{p(k)\}$	The greatest of the expressions p(k) for $k \varepsilon S$.
$\min(x,y)$	The smaller of x and y.
$\min_{k\varepsilon S}\{p(k)\}$	The least of the expressions p(k) for $k \varepsilon S$.
$x \geqslant y$	x is (componentwise where relevant) greater than or equal to y in a certain partial order.
$x \leqslant y$	Alternative notation for $y \geqslant x$.
$x > y$	$x \geqslant y$ but $x \neq y$.
$x < y$	Alternative notation for: $y > x$.
i_S	The identity mapping defined on the set S.
$g_a, g_{\underline{A}}$	The multiplication induced by the element a, or by the matrix $\underline{A}$.
$\emptyset$	The multiplicative identity element.
x^{-1}	The multiplicative inverse of the element x.
G	The group of a blog.
②	The 2-element Boolean algebra.
③	The 3-element blog.
E_n	A space of n-tuples.
$\langle \underline{a}(1),\ldots,\underline{a}(n)\rangle$	The space generated by the n-tuples $\underline{a}(1),\ldots\underline{a}(n)$.
$\mathcal{M}_{mn}$	The space of (m x n) matrices over E_1.
$\mathrm{Hom}_V(S,T)$	The set of all right-linear homomorphisms from S to T.
$\mathrm{Stab}_V(\dot{S},\dot{T})$	The subset of $\mathrm{Hom}_V(S,T)$ taking elements of $\dot{S}\subset S$ into $\dot{T}\subset T$.
$[a_{ij}]$	The matrix whose element in row i column j is a_{ij}.
$\{\underline{A}\}_{ij}$	The element in row i column j of the matrix $\underline{A}$.
$\{\underline{x}\}_j$	The jth element of the n-tuple $\underline{x}$.

$\underline{I}_n$	The (n x n) identity matrix.
$\dashv$	The foregoing statement applies only when the matrices involved are conformable for the relevant operations.
$x^*, \underline{A}^*, V^*$	The conjugates of the element x, the matrix $\underline{A}$ and the set V.
Dom f	The domain of the function f.
Ran f	The range of the function f.
fog	The composition of the functions f and g.
f\S	The restriction of the function f to the set S.
$\mathcal{W}_{mn}$	A given class of (m x n) matrices.
$\underline{A}^{n*}$	The nth power of $\underline{A}^*$ in min algebra.
M: S $\Longleftrightarrow$ T:N	M and N are conjugate sets of residuomorphisms between S and T.
$\mathcal{LM}_{mn}$	The set of left multiplications by (m x n) matrices.
$)\underline{x},\underline{y}($	Inner product of $\underline{x}$ and $\underline{y}$.
$(\sigma_{-\infty}, \sigma_\phi, \sigma_{+\infty})$	The subsets forming a trisection of a given belt.
$\mathcal{O}_{mn}$	The set of (m x n) matrices with elements in σ_ϕ, or with finite elements.
$\mathcal{N}_{mn}$	The set of (m x n) α-σ_ϕ-astic matrices.
$\mathcal{R}_{nr}(\underline{A})$	The set of all (n x r) matrices $\underline{X}$ for which $\underline{A} \otimes \underline{X}$ is /- defined.
$\mathcal{A}$	A special matrix for testing linear dependence.
$[\underline{a}(1),\ldots,\underline{a}(n)]$	The matrix whose columns are $\underline{a}(1),\ldots,\underline{a}(n)$.
$S(\tau,\delta)$	The set of elements $\underline{x}$ for which $\tau(\underline{x}) \leqslant \delta$.
ρ, ρ_R, ρ_C	The range, row-range and column-range seminorm respectively.
$\underline{I}_n\ (\delta^*)$	The (n x n) matrix with diagonal elements equal to δ^* and off-diagonal elements equal to $-\infty$.
$\bar{\rho}$	The co-range seminorm
$\pi_{\underline{A}}, \pi^*_{\underline{A}}$	The projection and the dual projection associated with the matrix $\underline{A}$.
$\underline{P}(\underline{\xi}), \underline{P}^*(\underline{\xi})$	The projection matrix and the dual projection matrix associated with the n-tuple $\underline{\xi}$.

$\Delta(\underline{A})$	The graph associated with a matrix $\underline{A}$.
$\underline{\Gamma}(\underline{A})$, $\underline{\Gamma}^*(\underline{A})$	The metric matrix and the dual metric matrix generated by a matrix $\underline{A}$.
$\tilde{\sim}$	The relation of direct similarity between matrices.
$\lambda(\underline{A})$	The principal eigenvalue of the matrix $\underline{A}$.
>> , <<	Tight inequality.
$\square(\underline{B})$	A maximal collection of non-equivalent principal eigenvectors from $\underline{\Gamma}(\underline{B})$ and $\underline{\Gamma}^*(\underline{B})$.
$\underline{\text{Adj}}\ \underline{A}$	The adjugate matrix of a matrix $\underline{A}$.
Perm $\underline{A}$	The permanent of a matrix $\underline{A}$.
$\underline{\text{Inv}}\ \underline{A}$	$(\text{Perm } \underline{A})^{-1} \otimes \underline{\text{Adj}}\ \underline{A}$.

29-4 . List of Definitions

We list below in alphabetical order, the more important technical terms introduced in the text, together with a reference to the page where each is defined, explained or first discussed. Terms which occur only occasionally, or which do not play a significant role, may not be listed.

Vol. 59: J. A. Hanson, Growth in Open Economies. V, 128 pages. 1971.

Vol. 60: H. Hauptmann, Schätz- und Kontrolltheorie in stetigen dynamischen Wirtschaftsmodellen. V, 104 Seiten. 1971.

Vol. 61: K. H. F. Meyer, Wartesysteme mit variabler Bearbeitungsrate. VII, 314 Seiten. 1971.

Vol. 62: W. Krelle u. G. Gabisch unter Mitarbeit von J. Burgermeister, Wachstumstheorie. VII, 223 Seiten. 1972.

Vol. 63: J. Kohlas, Monte Carlo Simulation im Operations Research. VI, 162 Seiten. 1972.

Vol. 64: P. Gessner u. K. Spremann, Optimierung in Funktionenräumen. IV, 120 Seiten. 1972.

Vol. 65: W. Everling, Exercises in Computer Systems Analysis. VIII, 184 pages. 1972.

Vol. 66: F. Bauer, P. Garabedian and D. Korn, Supercritical Wing Sections. V, 211 pages. 1972.

Vol. 67: I. V. Girsanov, Lectures on Mathematical Theory of Extremum Problems. V, 136 pages. 1972.

Vol. 68: J. Loeckx, Computability and Decidability. An Introduction for Students of Computer Science. VI, 76 pages. 1972.

Vol. 69: S. Ashour, Sequencing Theory. V, 133 pages. 1972.

Vol. 70: J. P. Brown, The Economic Effects of Floods. Investigations of a Stochastic Model of Rational Investment. Behavior in the Face of Floods. V, 87 pages. 1972.

Vol. 71: R. Henn und O. Opitz, Konsum- und Produktionstheorie II. V, 134 Seiten. 1972.

Vol. 72: T. P. Bagchi and J. G. C. Templeton, Numerical Methods in Markov Chains and Bulk Queues. XI, 89 pages. 1972.

Vol. 73: H. Kiendl, Suboptimale Regler mit abschnittweise linearer Struktur. VI, 146 Seiten. 1972.

Vol. 74: F. Pokropp, Aggregation von Produktionsfunktionen. VI, 107 Seiten. 1972.

Vol. 75: GI-Gesellschaft für Informatik e.V. Bericht Nr. 3. 1. Fachtagung über Programmiersprachen · München, 9.–11. März 1971. Herausgegeben im Auftrag der Gesellschaft für Informatik von H. Langmaack und M. Paul. VII, 280 Seiten. 1972.

Vol. 76: G. Fandel, Optimale Entscheidung bei mehrfacher Zielsetzung. II, 121 Seiten. 1972.

Vol. 77: A. Auslender, Problèmes de Minimax via l'Analyse Convexe et les Inégalités Variationelles: Théorie et Algorithmes. VII, 132 pages. 1972.

Vol. 78: GI-Gesellschaft für Informatik e.V. 2. Jahrestagung, Karlsruhe, 2.–4. Oktober 1972. Herausgegeben im Auftrag der Gesellschaft für Informatik von P. Deussen. XI, 576 Seiten. 1973.

Vol. 79: A. Berman, Cones, Matrices and Mathematical Programming. V, 96 pages. 1973.

Vol. 80: International Seminar on Trends in Mathematical Modelling, Venice, 13–18 December 1971. Edited by N. Hawkes. VI, 288 pages. 1973.

Vol. 81: Advanced Course on Software Engineering. Edited by F. L. Bauer. XII, 545 pages. 1973.

Vol. 82: R. Saeks, Resolution Space, Operators and Systems. X, 267 pages. 1973.

Vol. 83: NTG/GI-Gesellschaft für Informatik, Nachrichtentechnische Gesellschaft. Fachtagung „Cognitive Verfahren und Systeme", Hamburg, 11.–13. April 1973. Herausgegeben im Auftrag der NTG/GI von Th. Einsele, W. Giloi und H.-H. Nagel. VIII, 373 Seiten. 1973.

Vol. 84: A. V. Balakrishnan, Stochastic Differential Systems I. Filtering and Control. A Function Space Approach. V, 252 pages. 1973.

Vol. 85: T. Page, Economics of Involuntary Transfers: A Unified Approach to Pollution and Congestion Externalities. XI, 159 pages. 1973.

Vol. 86: Symposium on the Theory of Scheduling and its Applications. Edited by S. E. Elmaghraby. VIII, 437 pages. 1973.

Vol. 87: G. F. Newell, Approximate Stochastic Behavior of n-Server Service Systems with Large n. VII, 118 pages. 1973.

Vol. 88: H. Steckhan, Güterströme in Netzen. VII, 134 Seiten. 1973.

Vol. 89: J. P. Wallace and A. Sherret, Estimation of Product. Attributes and Their Importances. V, 94 pages. 1973.

Vol. 90: J.-F. Richard, Posterior and Predictive Densities for Simultaneous Equation Models. VI, 226 pages. 1973.

Vol. 91: Th. Marschak and R. Selten, General Equilibrium with Price-Making Firms. XI, 246 pages. 1974.

Vol. 92: E. Dierker, Topological Methods in Walrasian Economics. IV, 130 pages. 1974.

Vol. 93: 4th IFAC/IFIP International Conference on Digital Computer Applications to Process Control, Part I. Zürich/Switzerland, March 19–22, 1974. Edited by M. Mansour and W. Schaufelberger. XVIII, 544 pages. 1974.

Vol. 94: 4th IFAC/IFIP International Conference on Digital Computer Applications to Process Control, Part II. Zürich/Switzerland, March 19–22, 1974. Edited by M. Mansour and W. Schaufelberger. XVIII, 546 pages. 1974.

Vol. 95: M. Zeleny, Linear Multiobjective Programming. X, 220 pages. 1974.

Vol. 96: O. Moeschlin, Zur Theorie von Neumannscher Wachstumsmodelle. XI, 115 Seiten. 1974.

Vol. 97: G. Schmidt, Über die Stabilität des einfachen Bedienungskanals. VII, 147 Seiten. 1974.

Vol. 98: Mathematical Methods in Queueing Theory. Proceedings 1973. Edited by A. B. Clarke. VII, 374 pages. 1974.

Vol. 99: Production Theory. Edited by W. Eichhorn, R. Henn, O. Opitz, and R. W. Shephard. VIII, 386 pages. 1974.

Vol. 100: B. S. Duran and P. L. Odell, Cluster Analysis. A Survey. VI, 137 pages. 1974.

Vol. 101: W. M. Wonham, Linear Multivariable Control. A Geometric Approach. X, 344 pages. 1974.

Vol. 102: Analyse Convexe et Ses Applications. Comptes Rendus, Janvier 1974. Edited by J.-P. Aubin. IV, 244 pages. 1974.

Vol. 103: D. E. Boyce, A. Farhi, R. Weischedel, Optimal Subset Selection. Multiple Regression, Interdependence and Optimal Network Algorithms. XIII, 187 pages. 1974.

Vol. 104: S. Fujino, A Neo-Keynesian Theory of Inflation and Economic Growth. V, 96 pages. 1974.

Vol. 105: Optimal Control Theory and its Applications. Part I. Proceedings 1973. Edited by B. J. Kirby. VI, 425 pages. 1974.

Vol. 106: Optimal Control Theory and its Applications. Part II. Proceedings 1973. Edited by B. J. Kirby. VI, 403 pages. 1974.

Vol. 107: Control Theory, Numerical Methods and Computer Systems Modeling. International Symposium, Rocquencourt, June 17–21, 1974. Edited by A. Bensoussan and J. L. Lions. VIII, 757 pages. 1975.

Vol. 108: F. Bauer et al., Supercritical Wing Sections II. A Handbook. V, 296 pages. 1975.

Vol. 109: R. von Randow, Introduction to the Theory of Matroids. IX, 102 pages. 1975.

Vol. 110: C. Striebel, Optimal Control of Discrete Time Stochastic Systems. III. 208 pages. 1975.

Vol. 111: Variable Structure Systems with Application to Economics and Biology. Proceedings 1974. Edited by A. Ruberti and R. R. Mohler. VI, 321 pages. 1975.

Vol. 112: J. Wilhelm, Objectives and Multi-Objective Decision Making Under Uncertainty. IV, 111 pages. 1975.

Vol. 113: G. A. Aschinger, Stabilitätsaussagen über Klassen von Matrizen mit verschwindenden Zeilensummen. V, 102 Seiten. 1975.

Vol. 114: G. Uebe, Produktionstheorie. XVII, 301 Seiten. 1976.

Vol. 115: Anderson et al., Foundations of System Theory: Finitary and Infinitary Conditions. VII, 93 pages. 1976

Vol. 116: K. Miyazawa, Input-Output Analysis and the Structure of Income Distribution. IX, 135 pages. 1976.

Vol. 117: Optimization and Operations Research. Proceedings 1975. Edited by W. Oettli and K. Ritter. IV, 316 pages. 1976.

Vol. 118: Traffic Equilibrium Methods, Proceedings 1974. Edited by M. A. Florian. XXIII, 432 pages. 1976.

Vol. 119: Inflation in Small Countries. Proceedings 1974. Edited by H. Frisch. VI, 356 pages. 1976.

Vol. 120: G. Hasenkamp, Specification and Estimation of Multiple-Output Production Functions. VII, 151 pages. 1976.

Vol. 121: J. W. Cohen, On Regenerative Processes in Queueing Theory. IX, 93 pages. 1976.

Vol. 122: M. S. Bazaraa, and C. M. Shetty, Foundations of Optimization VI. 193 pages. 1976

Vol. 123: Multiple Criteria Decision Making. Kyoto 1975. Edited by M. Zeleny. XXVII, 345 pages. 1976.

Vol. 124: M. J. Todd. The Computation of Fixed Points and Applications. VII, 129 pages. 1976.

Vol. 125: Karl C. Mosler. Optimale Transportnetze. Zur Bestimmung ihres kostengünstigsten Standorts bei gegebener Nachfrage. VI, 142 Seiten. 1976.

Vol. 126: Energy, Regional Science and Public Policy. Energy and Environment I. Proceedings 1975. Edited by M. Chatterji and P. Van Rompuy. VIII, 316 pages. 1976.

Vol. 127: Environment, Regional Science and Interregional Modeling. Energy and Environment II. Proceedings 1975. Edited by M. Chatterji and P. Van Rompuy. IX, 211 pages. 1976.

Vol. 128: Integer Programming and Related Areas. A Classified Bibliography. Edited by C. Kastning. XII, 495 pages. 1976.

Vol. 129: H.-J. Lüthi, Komplementaritäts- und Fixpunktalgorithmen in der mathematischen Programmierung. Spieltheorie und Ökonomie. VII, 145 Seiten. 1976.

Vol. 130: Multiple Criteria Decision Making, Jouy-en-Josas, France. Proceedings 1975. Edited by H. Thiriez and S. Zionts. VI, 409 pages. 1976.

Vol. 131: Mathematical Systems Theory. Proceedings 1975. Edited by G. Marchesini and S. K. Mitter. X, 408 pages. 1976.

Vol. 132: U. H. Funke, Mathematical Models in Marketing. A Collection of Abstracts. XX, 514 pages. 1976.

Vol. 133: Warsaw Fall Seminars in Mathematical Economics 1975. Edited by M. W. Loś, J. Loś, and A. Wieczorek. V. 159 pages. 1976.

Vol. 134: Computing Methods in Applied Sciences and Engineering. Proceedings 1975. VIII, 390 pages. 1976.

Vol. 135: H. Haga, A Disequilibrium – Equilibrium Model with Money and Bonds. A Keynesian – Walrasian Synthesis. VI, 119 pages. 1976.

Vol. 136: E. Kofler und G. Menges, Entscheidungen bei unvollständiger Information. XII, 357 Seiten. 1976.

Vol. 137: R. Wets, Grundlagen Konvexer Optimierung. VI, 146 Seiten. 1976.

Vol. 138: K. Okuguchi, Expectations and Stability in Oligopoly Models. VI, 103 pages. 1976.

Vol. 139: Production Theory and Its Applications. Proceedings. Edited by H. Albach and G. Bergendahl. VIII, 193 pages. 1977.

Vol. 140: W. Eichhorn and J. Voeller, Theory of the Price Index. Fisher's Test Approach and Generalizations. VII, 95 pages. 1976.

Vol. 141: Mathematical Economics and Game Theory. Essays in Honor of Oskar Morgenstern. Edited by R. Henn and O. Moeschlin. XIV, 703 pages. 1977.

Vol. 142: J. S. Lane, On Optimal Population Paths. V, 123 pages. 1977.

Vol. 143: B. Näslund, An Analysis of Economic Size Distributions. XV, 100 pages. 1977.

Vol. 144: Convex Analysis and Its Applications. Proceedings 1976. Edited by A. Auslender. VI, 219 pages. 1977.

Vol. 145: J. Rosenmüller, Extreme Games and Their Solutions. IV, 126 pages. 1977.

Vol. 146: In Search of Economic Indicators. Edited by W. H. Strigel. XVI, 198 pages. 1977.

Vol. 147: Resource Allocation and Division of Space. Proceedings. Edited by T. Fujii and R. Sato. VIII, 184 pages. 1977.

Vol. 148: C. E. Mandl, Simulationstechnik und Simulationsmodelle in den Sozial- und Wirtschaftswissenschaften. IX, 173 Seiten. 1977.

Vol. 149: Stationäre und schrumpfende Bevölkerungen: Demographisches Null- und Negativwachstum in Österreich. Herausgegeben von G. Feichtinger. VI, 262 Seiten. 1977.

Vol. 150: Bauer et al., Supercritical Wing Sections III. VI, 179 pages. 1977.

Vol. 151: C. A. Schneeweiß, Inventory-Production Theory. VI, 116 pages. 1977.

Vol. 152: Kirsch et al., Notwendige Optimalitätsbedingungen und ihre Anwendung. VI, 157 Seiten. 1978.

Vol. 153: Kombinatorische Entscheidungsprobleme: Methoden und Anwendungen. Herausgegeben von T. M. Liebling und M. Rössler. VIII, 206 Seiten. 1978.

Vol. 154: Problems and Instruments of Business Cycle Analysis. Proceedings 1977. Edited by W. H. Strigel. VI, 442 pages. 1978.

Vol. 155: Multiple Criteria Problem Solving. Proceedings 1977. Edited by S. Zionts. VIII, 567 pages. 1978.

Vol. 156: B. Näslund and B. Sellstedt, Neo-Ricardian Theory. With Applications to Some Current Economic Problems. VI, 165 pages. 1978.

Vol. 157: Optimization and Operations Research. Proceedings 1977. Edited by R. Henn, B. Korte, and W. Oettli. VI, 270 pages. 1978.

Vol. 158: L. J. Cherene, Set Valued Dynamical Systems and Economic Flow. VIII, 83 pages. 1978.

Vol. 159: Some Aspects of the Foundations of General Equilibrium Theory: The Posthumous Papers of Peter J. Kalman. Edited by J. Green. VI, 167 pages. 1978.

Vol. 160: Integer Programming and Related Areas. A Classified Bibliography. Edited by D. Hausmann. XIV, 314 pages. 1978.

Vol. 161: M. J. Beckmann, Rank in Organizations. VIII, 164 pages. 1978.

Vol. 162: Recent Developments in Variable Structure Systems, Economics and Biology. Proceedings 1977. Edited by R. R. Mohler and A. Ruberti. VI, 326 pages. 1978.

Vol. 163: G. Fandel, Optimale Entscheidungen in Organisationen. VI, 143 Seiten. 1979.

Vol. 164: C. L. Hwang and A. S. M. Masud, Multiple Objective Decision Making – Methods and Applications. A State-of-the-Art Survey. XII, 351 pages. 1979.

Vol. 165: A. Maravall, Identification in Dynamic Shock-Error Models. VIII, 158 pages. 1979.

Vol. 166: R. Cuninghame-Green, Minimax Algebra. XI, 258 pages. 1979.